Organic-Inorganic Hybrid Materials III

Organic-Inorganic Hybrid Materials III

Jesús-María García-Martínez
Emilia P. Collar

Basel • Beijing • Wuhan • Barcelona • Belgrade • Novi Sad • Cluj • Manchester

Editors
Jesús-María García-Martínez
Institute of Polymer Science
and Technology (ICTP)
Spanish Council for Scientific
Research (CSIC)
Madrid
Spain

Emilia P. Collar
Institute of Polymer Science
and Technology (ICTP)
Spanish Council for Scientific
Research (CSIC)
Madrid
Spain

Editorial Office
MDPI AG
Grosspeteranlage 5
4052 Basel, Switzerland

This is a reprint of articles from the Special Issue published online in the open access journal *Polymers* (ISSN 2073-4360) (available at: www.mdpi.com/journal/polymers/special_issues/BRV433GFOM).

For citation purposes, cite each article independently as indicated on the article page online and as indicated below:

Lastname, A.A.; Lastname, B.B. Article Title. *Journal Name* **Year**, *Volume Number*, Page Range.

ISBN 978-3-7258-2608-7 (Hbk)
ISBN 978-3-7258-2607-0 (PDF)
doi.org/10.3390/books978-3-7258-2607-0

© 2024 by the authors. Articles in this book are Open Access and distributed under the Creative Commons Attribution (CC BY) license. The book as a whole is distributed by MDPI under the terms and conditions of the Creative Commons Attribution-NonCommercial-NoDerivs (CC BY-NC-ND) license.

Contents

About the Editors vii

Jesús-María García-Martínez and Emilia P. Collar
Current and Future Insights in Organic–Inorganic Hybrid Materials
Reprinted from: *Polymers* 2024, *16*, 3043, doi:10.3390/polym16213043 1

Lei Zhang, Haiqi Gao, Lixiang Guan, Yuchao Li and Qian Wang
Polyzwitterion–SiO$_2$ Double-Network Polymer Electrolyte with High Strength and High Ionic Conductivity
Reprinted from: *Polymers* 2023, *15*, 466, doi:10.3390/polym15020466 7

Solechan Solechan, Agus Suprihanto, Susilo Adi Widyanto, Joko Triyono, Deni Fajar Fitriyana and Januar Parlaungan Siregar et al.
Characterization of PLA/PCL/Nano-Hydroxyapatite (nHA) Biocomposites Prepared via Cold Isostatic Pressing
Reprinted from: *Polymers* 2023, *15*, 559, doi:10.3390/polym15030559 15

Vasily V. Spiridonov, Andrey V. Sybachin, Vladislava A. Pigareva, Mikhail I. Afanasov, Sharifjon A. Musoev and Alexander V. Knotko et al.
One-Step Low Temperature Synthesis of CeO$_2$ Nanoparticles Stabilized by Carboxymethylcellulose
Reprinted from: *Polymers* 2023, *15*, 1437, doi:10.3390/polym15061437 32

María Dolores Martín-Alonso, Valentina Salaris, Adrián Leonés, Víctor Hevilla, Alexandra Muñoz-Bonilla and Coro Echeverría et al.
Centrifugal Force-Spinning to Obtain Multifunctional Fibers of PLA Reinforced with Functionalized Silver Nanoparticles
Reprinted from: *Polymers* 2023, *15*, 1240, doi:10.3390/polym15051240 45

Biltu Mahato, Stepan V. Lomov, Aleksei Shiverskii, Mohammad Owais and Sergey G. Abaimov
A Review of Electrospun Nanofiber Interleaves for Interlaminar Toughening of Composite Laminates
Reprinted from: *Polymers* 2023, *15*, 1380, doi:10.3390/polym15061380 62

Adrián Leonés, Valentina Salaris, Ignacio Ramos Aranda, Marcela Lieblich, Daniel López and Laura Peponi
Thermal Properties and In Vitro Biodegradation of PLA-Mg Filaments for Fused Deposition Modeling
Reprinted from: *Polymers* 2023, *15*, 1907, doi:10.3390/polym15081907 82

Ayman M. Maqableh and Muhanad M. Hatamleh
Cohesive Zone Modeling of Pull-Out Test for Dental Fiber–Silicone Polymer
Reprinted from: *Polymers* 2023, *15*, 3668, doi:10.3390/polym15183668 97

Abdelrasoul Gadelmoula and Saleh Ahmed Aldahash
Dry Friction and Wear Behavior of Laser-Sintered Graphite/Carbon Fiber/Polyamide 12 Composite
Reprinted from: *Polymers* 2023, *15*, 3916, doi:10.3390/polym15193916 110

Rudá Aranha, Mario A. Albuquerque Filho, Cícero de Lima Santos, Viviane M. Fonseca, José L. V. Rivera and Antonio G. B. de Lima et al.
Water Sorption in Hybrid Polyester/Glass/Jute Composites Processed via Compression Molding and Vacuum-Assisted Resin Transfer Molding
Reprinted from: *Polymers* 2023, 15, 4438, doi:10.3390/polym15224438 126

Rudá Aranha, Mario A. Albuquerque Filho, Cícero de L. Santos, Tony Herbert F. de Andrade, Viviane M. Fonseca and Jose Luis Valin Rivera et al.
Effect of Water Absorption and Stacking Sequences on the Tensile Properties and Damage Mechanisms of Hybrid Polyester/Glass/Jute Composites
Reprinted from: *Polymers* 2024, 16, 925, doi:10.3390/polym16070925 143

Cristina Pina-Vidal, Víctor Berned-Samatán, Elena Piera, Miguel Ángel Caballero and Carlos Téllez
Mechanochemical Encapsulation of Caffeine in UiO-66 and UiO-66-NH_2 to Obtain Polymeric Composites by Extrusion with Recycled Polyamide 6 or Polylactic Acid Biopolymer
Reprinted from: *Polymers* 2024, 16, 637, doi:10.3390/polym16050637 162

Tamara M. Díez-Rodríguez, Enrique Blázquez-Blázquez, Ernesto Pérez and María L. Cerrada
Composites of Poly(3-hydroxybutyrate) and Mesoporous SBA-15 Silica: Crystalline Characteristics, Confinement and Final Properties
Reprinted from: *Polymers* 2024, 16, 1037, doi:10.3390/polym16081037 191

Adrián Leonés, Valentina Salaris, Laura Peponi, Marcela Lieblich, Alexandra Muñoz-Bonilla and Marta Fernández-García et al.
Bioactivity and Antibacterial Analysis of Plasticized PLA Electrospun Fibers Reinforced with MgO and $Mg(OH)_2$ Nanoparticles
Reprinted from: *Polymers* 2024, 16, 1727, doi:10.3390/polym16121727 207

About the Editors

Jesús-María García-Martínez

Jesús-María García-Martínez holds a Ph.D. in Chemistry (Chemical Engineering) from Universidad Complutense de Madrid (1995), two M.Sc.-level degrees (Industrial Chemistry and Polymer Science and Technology), and has completed more than 70 highly specialized courses. He is a Tenured Scientist at the Institute of Polymer Science and Technology (ICTP) of the Spanish National Research Council (CSIC). Since 1992, within the Polymer Engineering Group (GIP), he has co-authored more than 200 scientific and/or technical works on topics related to polymer engineering, chemical modification of polymers, heterogeneous materials based on polymers, interphases and interfaces, polymer composites, polyblends and alloys, organic–inorganic hybrid materials, polymer recycling, quality, and so on. Furthermore, he has participated in 32 research and industrial projects (national and international programs) and is co-author of one currently active industrial patent on polymer recycling. From 2000 to 2005, he also took on the position of Quality Director for the ICTP (CSIC) ISO 17025 Accreditation Project of ICTP Laboratories. Additionally, Dr. García-Martínez is actively reviewing tasks for WOS- and SCOPUS-indexed journals, with more than 450 reports in recent years, and has been awarded twice with the Publons Reviewer Award (2018 and 2019) and once with the POLYMERS Outstanding Reviewer Award (2019). He is an Editorial Member of polymer science-related journals and Editor of more than 100 scientific articles. He has been a Guest Editor for seven Special Issues of Q1 journals. Since 2016, he has served as the Head of the Department of Chemistry and Properties of Polymer Materials within the ICTP (CSIC).

Emilia P. Collar

Emilia P. Collar holds a Ph.D. in Industrial Chemistry (U. Complutense, 1986). Since 1990, she has been a permanent staff member (Tenured Scientist) at the Consejo Superior de Investigaciones Cientificas (CSIC). She was an CSIC Postdoc Fellow for two years (1986–88) and a Chemical Engineering Assistant Teacher at the Universidad Complutense de Madrid for one year (1989). In the Polymer Science and Technology Institute (ICTP/CSIC), she works within the Polymer Engineering Group (GIP), founded in 1982 by Prof. O. Laguna, and she has been the GIP's Head since 1999. Between 1990 and 2005, she supervised six doctoral theses and five post-graduate ones under fourteen publicly funded research projects and jointly supervised 18 research-funded contracts, issuing 53 technical reports for different companies. Furthermore, from 2001 to 2005, she performed the positions of Technical Director of the Physical and Mechanical Properties Laboratory and Deputy-Technical Director of the Thermal Properties Laboratory under the successful ISO 17025 Accreditation Project for the CSIC/ICTP's Laboratories, ACiTP, commanded by the ICTP's Head. She is the author and co-author of more than 20 book chapters, 2 currently active industrial patents on polymer recycling, and more than 150 papers, mainly in SCI journals. Her research interests include polymers and the environment under the general frame of heterogeneous materials based on polymers. From 2006 to date, she has participated in four public Spanish-funded research projects and one EU project under its seventh Framework Program.

Editorial

Current and Future Insights in Organic–Inorganic Hybrid Materials

Jesús-María García-Martínez * and Emilia P. Collar *

Polymer Engineering Group (GIP), Polymer Science and Technology Institute (ICTP), Spanish National Research Council (CSIC), C/Juan de la Cierva, 3, 28006 Madrid, Spain
* Correspondence: jesus.maria@ictp.csic.es (J.-M.G.-M.); ecollar@ictp.csic.es (E.P.C.)

Citation: García-Martínez, J.-M.; Collar, E.P. Current and Future Insights in Organic–Inorganic Hybrid Materials. *Polymers* **2024**, *16*, 3043. https://doi.org/10.3390/polym16213043

Received: 13 September 2024
Revised: 23 October 2024
Accepted: 25 October 2024
Published: 29 October 2024

Copyright: © 2024 by the authors. Licensee MDPI, Basel, Switzerland. This article is an open access article distributed under the terms and conditions of the Creative Commons Attribution (CC BY) license (https://creativecommons.org/licenses/by/4.0/).

The text below outlines some current and future possibilities for organic–inorganic hybrid materials. This third installment builds on the success of the previous Special Issues and presents new original contributions and approaches to the topic [1–4]. These investigations align with the definition of hybrid materials recommended by the IUPAC (International Union of Pure and Applied Chemistry). This definition signifies that a hybrid material comprises a close mixture of inorganic, organic, or both components, typically interpenetrating scales of less than one micrometer [5]. The potential for organic–inorganic hybrid materials to enhance properties in various fields is vast, offering substantial research and design opportunities for material scientists. However, it is necessary to note that functional hybrid materials are not just physical mixtures [6–10]. They are nanocomposites at the molecular scale, with at least one organic or inorganic component, having a characteristic length on the nanometer scale (a few Å to several tens of nanometers) [7]. The properties of hybrid materials are not just the sum of the individual contributions of their components but also arise from the strong synergy created by a hybrid interface [8–10].

This synergy is a key aspect of hybrid materials, highlighting the unique properties that can be achieved through careful design. These, such as enhanced electrical, optical, mechanical, separation capacity, catalysis, sensing capability, and chemical and thermal stability properties, make hybrid materials a valuable area of research. The nature of the inorganic–organic interface, including the types of interactions present, the surface energy, and the presence of labile bonds, plays a strong role in controlling these properties. Additionally, it must be noted that hybridization is a multifaceted strategy. In some cases, conducting organic polymers just act as a solid polymeric support for active species. In contrast, in other hybrid systems, the activity of organic and inorganic species combines to reinforce or modify each other. However, in every case, the work on these hybrid materials involves the underlying use and sometimes even the explicit search for synergy. Hybrid organic–inorganic materials, in general, represent the natural interface between two worlds of chemistry, each with very significant contributions to the field of materials science and each with characteristic properties that result in distinct advantages and limitations. Researching the topic of hybrid materials has challenges and opportunities. The main challenge is synthesizing hybrid materials that keep or enhance each component's best properties while reducing their limitations. Undertaking this challenge allows one to develop new materials with synergic behavior, improving performance and novel valuable properties. Indeed, hybrid materials frequently involve a combination of components thoroughly studied in their respective fields. Still, they provide an additional dimension to their properties when becoming part of the hybrid compound [11].

Various functional hybrid materials can be categorized into two main families depending on the nature of the interface, combining organic components and inorganic materials [7,8,10–12]. Class I deals with hybrid systems where the organic and inorganic parts interact by weak bonds, including Van der Waals, electrostatic, or hydrogen bonds. Class II indicates hybrid materials in which covalent or ionic-covalent chemical bonds link these components. Many hybrid materials combine jointly strong and weak interfaces [1–4,10,13,14]. Still, due to the significance of the presence of strong chemical bonds on the final properties, these hybrids are grouped into Class II. Hybrid Class I compounds have exciting features such as ease of material synthesis, easy removal of the organic phase to create functional architectures by self-assembly, et cetera. However, there is increasing development of Class II hybrid materials due to the advantages of covalent bonds between organic and mineral components, including the potential to synthesize entirely new materials, minimization of phase separation, and better definition of the organic–inorganic interface [7–10]. The effective grafting of organic functionality to the inorganic network also avoids a drawback of the hybrid compounds of Class I, which is the potential departure of organic components while the material is in use. Furthermore, categorizing organic–inorganic hybrid systems into Class-I and Class-II, based on the type of interactions between the phases, highlights these materials' diverse and promising nature. Combining weak and strong interactions in the same hybrid system opens new avenues for innovation and development. In summary, organic–inorganic materials are shown as multi-component compounds with at least one component in the nano-metric size domain. This yields significantly enhanced properties, positioning them as valuable components in various applications. To finalize, in essence, it can be said that the organic–inorganic materials are multi-component compounds with at least one of their organic (the polymer) or inorganic components in the nano-metric-size domain, which confers the material as a whole of greatly enhanced properties respecting the constitutive parts in isolation [6–10].

It is well worth mentioning here the so-called hybrid fiber reinforced polymer composites (HFRPCs) since they are organic–inorganic hybrid advanced materials combining two or more different types of fibers (organic and inorganic) within a polymer matrix, aiming to leverage the strengths of each type of fiber, addressing their individual limitations to achieve enhanced mechanical properties and performance, making them appropriate for automotive, aerospace, and construction applications due to their outstanding performance and cost-effectiveness derived both from the combination of materials and the processing operations choice [15–18].

Of the nineteen manuscripts submitted to this Special Issue, just 13 were published after the rigorous revision processes of *Polymers*. This Special Issue, therefore, includes these highly relevant and exciting works that are of utmost importance in this scientific field. Each article compiled in this volume fully matches the topic's fundamentals, underscoring their significance. Since this editorial aims not to elaborate on each text but to encourage the reader to browse them in depth, these contributions have been briefly described. Consequently, a brief resume of each one is reported to awaken interest in each of the contributions to this exciting Special Issue of *Polymers* rather than provide an exhaustive description of the articles themselves.

So, the contribution by Wang, Li, and coworkers [19] introduces a unique approach to designing and preparing high-performance polymer-based electrolytes for solid-state energy storage devices. The authors have developed a novel double-network PE based on the nonhydrolytic sol–gel reaction of tetraethyl orthosilicate and in situ polymerization of zwitterions. The resulting material, with its high strength and stretchability, represents a significant advancement in the field. This is largely due to the efficient dissipation of energy in the inorganic network. This unique property also allows it to act as a Lewis acid to adsorb trace impurities, resulting in an electrolyte with a high electrochemical window. The elastic characteristics of the polymer network further enhance its performance. Notably, the new PE demonstrates excellent interface compatibility with a Li metal electrode, an essential requirement for solid-state energy storage devices.

In their article, Solechan et al. [20] present a significant study on biocomposite scaffolds obtained from a blend of polylactic acid (PLA) and polycaprolactone (PCL) using cold isostatic pressing and incorporating hydroxyapatite (nHA) as an osteoconductive filler (0 up to 30%). This research, focusing on medical applications, aims to determine the effects of nHA on compound performance. Different characterization techniques were employed from both the micro and macro scales, such as FTIR, XRD, SEM, density, porosity, tensile, and flexural properties. The study concludes that incorporating nHA into the PLA/PCL blend induces an irregular structure, making the crystallization process more challenging. Higher amounts of nHA result in more porous materials. These findings have practical implications for the development of biomaterials in medical applications.

In their paper, Spiridonov et al. [21] presented synthesized nanocomposites of cerium-containing nanoparticles stabilized by carboxymethyl cellulose (CMC) macromolecules through a novel and elegant method. These products were further characterized, aiding in determining the type of crystal structure of inorganic nanoparticles and the mechanism of nanoparticle formation. They demonstrated and confirmed that neither the size nor the shape of the nanoparticles present in the nanocomposites depends on the ratio of the initial reagents, reassuring the method's robustness.

The article by Peponi, López, et al. [22] is a meticulous study that delves into the design and development of multifunctional fibers. This involves the incorporation of functionalized nanoparticles into matrices obtained by spinning techniques. The authors present a green protocol for obtaining functionalized silver nanoparticles, using it as a reducing agent. These nanoparticles, once obtained, were integrated into PLA solutions to explore the production of multifunctional polymeric fibers by centrifugal force-spinning. The study focused on nanoparticle concentration between 0 and 3.5 wt%, investigating the effect of nanoparticle incorporation and fiber preparation method on morphology, thermomechanical properties, biodisintegration, and antimicrobial behavior. The authors' conclusion that 1% of nanoparticles is the optimal amount for enhancing thermomechanical behavior while conferring high antibacterial activity to the PLA fibers and that 2% significantly enhances the material's shape memory underscores the meticulousness of this study.

In this article authored by Mahato, Abaimov, and colleagues [23], they have presented a comprehensive overview of the application of nanofiber polymeric veils as toughening interleaves in fiber-reinforced composite laminates. They note that the interest of these veils is to prevent delamination caused by the poor out-of-plane properties of composite laminates. They also identify and discuss the toughening mechanisms induced by polymeric veils. The results and conclusions claimed to be helpful in all the stages of the material designs, from the material selection to the modelization of the delamination process, provide a solid foundation for further research and application.

The article by Leonés, López, Peponi, and coworkers [24] is dedicated to the obtention of filaments based on polylactic acid (PLA) doped with varying amounts of magnesium microparticles. The authors employed a two-step extrusion process to investigate the impact of processing on the thermal degradation of the filaments. They also studied the in vitro degradation of the filaments and their subsequent 3D printing. The study culminated in the successful production of micro-composites via a double-extrusion process, with no degradation of the materials and a commendable dispersion of the microparticles into the PLA matrix, all without any chemical or physical modification of the microparticles.

Thus, the contribution by Maqableh et al. [25] has developed some analytical methods to evaluate the bond strength of fiber–matrix systems. For this purpose, they investigate the debonding mechanism of a fiber–silicone pull-out specimen and further validate the experimental data using 3D-FEM and a cohesive element approach. The comparison between the experimental values and the results from the finite element simulations shows that the proposed cohesive zone model accurately reproduces the experimental results, being considered almost identical to the experimental observations about the interface. This accuracy of the cohesive zone model reassures the reliability of the research.

The article authored by Godelmoula and colleagues [26] is a significant study on using carbon fiber-reinforced polymers (CFRPs) in fabricating complex geometries through selective laser sintering. It is crucial to note that while carbon fiber reinforcement enhances the mechanical properties of polymers, it also reduces tribological wear resistance, necessitating fillers as solid lubricants. The authors' exploration of graphite-filled carbon fiber-reinforced polyamide 12 (CFR-PA12) specimens, prepared using the selective laser sintering process, provides a deep understanding of the composite's dry sliding friction and wear characteristics, leading to engaging concluding remarks.

The aim of the investigation of the paper by Aranha, Carvalho et al. [27] is the determination of water sorption in hybrid polyester/glass fabric/jute fabric composites molded via compression and vacuum-assisted resin transfer molding, obtained at different stacking sequences and water sorption testing at different temperatures. The authors concluded that the manufacturing process does not affect water sorption at the saturation point, as the main factors influencing the absorbing moisture are the presence and content of jute fibers in the system, which are jointly affected by the immersion temperature. Additionally, in contribution by Aranha, Rivera, and colleagues [28], these authors have studied the mechanical behavior under tensile stress of neat and hydrothermally aged samples at different temperatures, evaluating fracture surfaces by SE, and it was concluded that exposure to the aqueous ambient led to a reduction in mechanical properties, both for the molding technique and the stacking sequence, observing a broad spectrum of defects such as delamination, fiber pull-out, fiber/matrix detachment, voids, and matrix removal correlated to the experimental conditions followed.

The article authored by Tellez et al. [29] has investigated encapsulated caffeine (CAF), which has anti-cellulite properties, in zirconium-based metal–organic frameworks (MOFs) by liquid-assisted milling, resulting in different textural properties and chemical functionalization. These capsules have been incorporated into recycled polyamide 6 (PA6) and a biopolymer based on polylactic acid (PLA) using extrusion. The resulting materials have been fully characterized, confirming the caffeine encapsulation, the preservation of caffeine during the extrusion process, and the good contact between the polymer and the MOFs. This research opens up exciting possibilities for using these materials in various applications, inspiring further exploration and development in the field.

The research performed by Cerrada et al. [30] meticulously investigated composites based on poly(3-hydroxybutyrate) (PHB) and mesoporous SBA-15 silica, a study conducted with utmost care and precision. The thermal stability, phase transitions, and crystalline details of these composites were studied in great detail, focusing on the confinement of the PHB polymeric chains in the silica's mesopores. The influence of nano-silica in the composites' thermal, morphologic, and dynamic mechanical performance not only confirms the observed confinement but also provides a comprehensive understanding of the influence of the filler in temperatures above the glass transition.

Finally, the last article, authored by Peponi, López, and coworkers [31], is devoted to a comprehensive study that thoroughly explores the bioactivity and antibacterial behavior of PLA-based electrospun fibers doped with both MgO and Mg(OH)$_2$ nanoparticles (NPs). The study meticulously tracks the evolution of these fibers in terms of morphology, infrared spectra, X-ray diffraction, and visual appearance. The authors discuss the bioactivity of hydroxyapatite growth after 28 days of immersion in simulated body fluid (SBF), noting an increase in the number of precipitated crystals with the amount of both NPs. The chemical composition of the precipitated crystals, characterized in terms of the Ca/P molar ratio after T28 of immersion in SBF, indicates the presence of hydroxyapatite on the surface of the reinforced fibers, along with a decrease in the average diameter of the PLA-based fibers. The study also reveals the promising antibacterial activity of the MgO and Mg(OH)$_2$ nanoparticles in the fibers against Escherichia coli and Staphylococcus aureus, providing a comprehensive understanding of their behavior.

To conclude, and as the guest editors of this captivating Special Issue, we are excited to announce that "Organic–Inorganic Hybrid Materials" has emerged as a crucial framework in the field of polymer science and technology, both currently and in the near future. This excitement fuels our anticipation for the fourth Special Issue on the topic, slated for 2025 in POLYMERS. We are now open for submissions and eagerly await contributions from those who share our passion for this vibrant scientific field.

Conflicts of Interest: The authors declare no conflict of interest.

References

1. García-Martínez, J.M.; Collar, E.P. Organic–Inorganic Hybrid Materials. *Polymers* **2021**, *13*, 86. [CrossRef] [PubMed]
2. García-Martínez, J.M.; Collar, E.P. Organic–Inorganic Hybrid Materials II: Some Additional Contributions to the Topic. *Polymers* **2021**, *13*, 2390. [CrossRef] [PubMed]
3. García-Martínez, J.M.; Collar, E.P. (Eds.) *Organic–Inorganic Hybrid Materials*; MDPI Books: Basel, Swizertland, 2021; ISBN 978-3-0365-1301-0 (Hbk), ISBN2 978-3-0365-1302-7 (PDF).
4. García-Martínez, J.M.; Collar, E.P. (Eds.) *Organic–Inorganic Hybrid Materials II*; MDPI Books: Basel, Swizertland, 2021; ISBN 978-3-0365-7079-2 (Hbk), ISBN2 978-3-7078-5 (PDF).
5. Alemán, J.; Chadwick, A.V.; He, J.; Hess, M.; Horie, K.; Jones, R.G.; Kratochvíl, P.; Meisel, I.; Mita, I.; Moad, G.; et al. Definitions of terms relating to the structure and processing of sols, gels, networks, and inorganic-organic hybrid materials (IUPAC recommendations 2007). *Pure Appl. Chem.* **2007**, *79*, 1801–1829. [CrossRef]
6. Pielichowski, K.; Majka, T.M. *Polymer Composites with Functionalized Nanoparticles: Synthesis, Properties, and Applications*; Elsevier Inc.: Amsterdam, The Netherlands, 2019; pp. 1–504.
7. Judeinstain, P.; Sanchez, C. Hybrid organic-inorganic materials: A land of multidisciplinarity. *J. Mater. Chem.* **1996**, *6*, 511. [CrossRef]
8. Sanchez, C.; Soler-Illia, G.J.A.A.; Ribot, F.; Grosso, D. Design of functional nano-structured materials through the use of controlled hybrid organic–Inorganic interfaces. *C. R. Chim.* **2003**, *6*, 1131. [CrossRef]
9. Sanchez, C.; Soler-Illia, G.J.A.A.; Ribot, F.; Lalot, T.; Mayer, C.R.; Cabuil, V. Designed Hybrid Organic−Inorganic Nanocomposites from Functional Nanobuilding Blocks. *Chem. Mater.* **2001**, *13*, 3061. [CrossRef]
10. Gomez-Romero, P.; Pokriyal, A.; Rueda-Garcia, D.; Bengoa, L.N.; González-Gil, R.M. Hybrid Materials: A Metareview. *Chem. Mater.* **2024**, *36*, 8–27. [CrossRef]
11. Mir, S.H.; Nagahara, L.A.; Thundat, T.; Mokarian-Tabari, P.; Furukawa, H.; Khosla, A. Review—Organic-Inorganic Hybrid Functional Materials: An Integrated Platform for Applied Technologies. *J. Electrochem. Soc.* **2015**, *165*, B3137–B3156. [CrossRef]
12. Faustini, M.; Nicole, L.; Ruiz-Hitzky, E.; Sanchez, C. History of Organic-Inorganic Hybrid Materials: Prehistory, Art, Science, and Advanced Applications. *Adv. Funct. Mater.* **2018**, *28*, 1704158. [CrossRef]
13. Pogrebnjak, A.D.; Beresnev, V.M. *Nanocoatings Nanosystems Nanotechnologies*; Bentham Books: Sharjah, United Arab Emirates, 2012.
14. Collar, E.P.; Areso, S.; Taranco, J.; García-Martínez, J.M. Heterogeneous Materials based on Polypropylene. In *Polyolefin Blends*, 1st ed.; Nwabunma, D., Kyu, T., Eds.; Wiley-Interscience: Hoboken, NJ, USA, 2008; pp. 379–410.
15. Sathishkumar, T.P.; Naveen, J.; Satheeshkumar, S. Hybrid fiber reinforced polymer composites—A review. *J. Reinf. Plast. Comp.* **2014**, *33*, 454–471. [CrossRef]
16. Seydibeyoğlu, M.Ö.; Dogru, A.; Wang, J.; Rencheck, M.; Han, Y.; Wang, L.; Seydibeyoğlu, E.A.; Zhao, X.; Ong, K.; Shatkin, J.A.; et al. Review on Hybrid Reinforced Polymer Matrix Composites with Nanocellulose, Nanomaterials, and Other Fibers. *Polymers* **2023**, *15*, 984. [CrossRef] [PubMed]
17. Maiti, S.; Islam, R.; Abbas-Uddin, M.; Afroj, S.; Eichhorn, S.J.; Karim, N. Sustainable Fiber-Reinforced Composites: A Review. *Adv. Sustainable Syst.* **2022**, *6*, 2200258. [CrossRef]
18. Islam, T.; Chaion, M.H.; Jalil, M.A.; Rafi, A.S.; Mushtari, F.; Dhar, A.K.; Hossain, S. Advancements and challenges in natural fiber-reinforced hybrid composites: A comprehensive review. *SPE Polym.* **2024**, *5*, 481–506. [CrossRef]
19. Zhang, L.; Gao, H.; Guan, L.; Li, Y.; Wang, Q. Polyzwitterion–SiO$_2$ Double-Network Polymer Electrolyte with High Strength and High Ionic Conductivity. *Polymers* **2023**, *15*, 466. [CrossRef]
20. Solechan, S.; Suprihanto, A.; Widyanto, S.A.; Triyono, J.; Fitriyana, D.F.; Siregar, J.P.; Cionita, T. Characterization of PLA/PCL/Nano-Hydroxyapatite (nHA) Biocomposites Prepared via Cold Isostatic Pressing. *Polymers* **2023**, *15*, 559. [CrossRef]
21. Spiridonov, V.V.; Sybachin, A.V.; Pigareva, V.A.; Afanasov, M.I.; Musoev, S.A.; Knotko, A.V.; Zezin, S.B. One-Step Low Temperature Synthesis of CeO2 Nanoparticles Stabilized by Carboxymethylcellulose. *Polymers* **2023**, *15*, 1437. [CrossRef]
22. Martín-Alonso, M.D.; Salaris, V.; Leonés, A.; Hevilla, V.; Muñoz-Bonilla, A.; Echeverría, C.; Fernández-García, M.; Peponi, L.; López, D. Centrifugal Force-Spinning to Obtain Multifunctional Fibers of PLA Reinforced with Functionalized Silver Nanoparticles. *Polymers* **2023**, *15*, 1240. [CrossRef]
23. Mahato, B.; Lomov, S.V.; Shiverskii, A.; Owais, M.; Abaimov, S.G. A Review of Electrospun Nanofiber Interleaves for Interlaminar Toughening of Composite Laminates. *Polymers* **2023**, *15*, 1380. [CrossRef]
24. Leonés, A.; Salaris, V.; Ramos Aranda, I.; Lieblich, M.; López, D.; Peponi, L. Thermal Properties and In Vitro Biodegradation of PLA-Mg Filaments for Fused Deposition Modeling. *Polymers* **2023**, *15*, 1907. [CrossRef]
25. Maqableh, A.M.; Hatamleh, M.M. Cohesive Zone Modeling of Pull-Out Test for Dental Fiber–Silicone Polymer. *Polymers* **2023**, *15*, 3668. [CrossRef]
26. Gadelmoula, A.; Aldahash, S.A. Dry Friction and Wear Behavior of Laser-Sintered Graphite/Carbon Fiber/Polyamide 12 Composite. *Polymers* **2023**, *15*, 3916. [CrossRef] [PubMed]
27. Aranha, R.; Filho, M.A.A.; de Lima Santos, C.; Fonseca, V.M.; Rivera, J.L.V.; de Lima, A.G.B.; de Amorim, W.F., Jr.; Carvalho, L.H. Water Sorption in Hybrid Polyester/Glass/Jute Composites Processed via Compression Molding and Vacuum-Assisted Resin Transfer Molding. *Polymers* **2023**, *15*, 4438. [CrossRef] [PubMed]
28. Aranha, R.; Filho, M.A.A.; Santos, C.d.L.; de Andrade, T.H.F.; Fonseca, V.M.; Rivera, J.L.V.; dos Santos, M.A.; de Lima, A.G.B.; de Amorim, W.F., Jr.; de Carvalho, L.H. Effect of Water Absorption and Stacking Sequences on the Tensile Properties and Damage Mechanisms of Hybrid Polyester/Glass/Jute Composites. *Polymers* **2024**, *16*, 925. [CrossRef] [PubMed]

29. Pina-Vidal, C.; Berned-Samatán, V.; Piera, E.; Caballero, M.Á.; Téllez, C. Mechanochemical Encapsulation of Caffeine in UiO-66 and UiO-66-NH2 to Obtain Polymeric Composites by Extrusion with Recycled Polyamide 6 or Polylactic Acid Biopolymer. *Polymers* **2024**, *16*, 637. [CrossRef]
30. Díez-Rodríguez, T.M.; Blázquez-Blázquez, E.; Pérez, E.; Cerrada, M.L. Composites of Poly(3-hydroxybutyrate) and Mesoporous SBA-15 Silica: Crystalline Characteristics, Confinement and Final Properties. *Polymers* **2024**, *16*, 1037. [CrossRef]
31. Leonés, A.; Salaris, V.; Peponi, L.; Lieblich, M.; Muñoz-Bonilla, A.; Fernández-García, M.; López, D. Bioactivity and Antibacterial Analysis of Plasticized PLA Electrospun Fibers Reinforced with MgO and Mg(OH)$_2$ Nanoparticles. *Polymers* **2024**, *16*, 1727. [CrossRef]

Disclaimer/Publisher's Note: The statements, opinions and data contained in all publications are solely those of the individual author(s) and contributor(s) and not of MDPI and/or the editor(s). MDPI and/or the editor(s) disclaim responsibility for any injury to people or property resulting from any ideas, methods, instructions or products referred to in the content.

Article

Polyzwitterion–SiO₂ Double-Network Polymer Electrolyte with High Strength and High Ionic Conductivity

Lei Zhang [1], Haiqi Gao [2,*], Lixiang Guan [3], Yuchao Li [4,*] and Qian Wang [3,*]

[1] School of Materials and Chemical Engineering, Chuzhou University, 1528 Fengle Avenue, Chuzhou 239099, China
[2] State Key Laboratory of Chemistry and Utilization of Carbon Based Energy Resources, College of Chemistry, Xinjiang University, Urumqi 830017, China
[3] Institute of Energy Innovation, College of Materials Science and Engineering, Taiyuan University of Technology, Taiyuan 030024, China
[4] School of Materials Science and Engineering, Liaocheng University, Liaocheng 252000, China
* Correspondence: iamhqgao@njupt.edu.cn (H.G.); liyuchao@lcu.edu.cn (Y.L.); qianwang19930825@163.com (Q.W.)

Abstract: The key to developing high-performance polymer electrolytes (PEs) is to achieve their high strength and high ionic conductivity, but this is still challenging. Herein, we designed a new double-network PE based on the nonhydrolytic sol–gel reaction of tetraethyl orthosilicate and in situ polymerization of zwitterions. The as-prepared PE possesses high strength (0.75 Mpa) and high stretchability (560%) due to the efficient dissipation energy of the inorganic network and elastic characteristics of the polymer network. In addition, the highest ionic conductivity of the PE reaches 0.44 mS cm^{-1} at 30 °C owning to the construction of dynamic ion channels between the polyzwitterion segments and between the polyzwitterion segments and ionic liquids. Furthermore, the inorganic network can act as Lewis acid to adsorb trace impurities, resulting in a prepared electrolyte with a high electrochemical window over 5 V. The excellent interface compatibility of the as-prepared PE with a Li metal electrode is also confirmed. Our work provides new insights into the design and preparation of high-performance polymer-based electrolytes for solid-state energy storage devices.

Keywords: inorganic–organic double-network; polyzwitterion; polymer electrolyte; ionic conductivity

1. Introduction

Polymer electrolytes (PEs) have attracted much attention recently due to their non-volatility, flame resistance, ease of processing, and low cost [1–6]. More importantly, PEs have the ability to inhibit the dendrite growth of solid metal (e.g., Na, Li) batteries, which are expected to fundamentally solve the safety problems of metal-based cells [7–11]. Whether it is a solid-state lithium metal battery or a solid-state sodium ion battery [12,13], polyethylene oxide (PEO) and its derivatives are currently the dominant PE matrices due to PEO's good solubility for metal salts and its ability to transport metal ions [14–16]. However, PEO is known to be highly crystalline at room temperature and usually has a limited ability to transport metal ions; for example, the room temperature ionic conductivity of PEO–lithium salt systems is typically in the range of 10^{-6}–10^{-8} S cm^{-1} [17], which is far from the desired ionic conductivity of 10^{-3}–10^{-4} S cm^{-1} expected for PEs. In order to improve the room temperature ionic conductivity of PEO-based electrolytes, the key is to inhibit the crystallization of the polymer and thus improve the motility of the polymer chain segments. Physical blending [18–20], grafting [21], copolymerization [22,23], cross-linking [24], and branching [12] have been used to inhibit the crystallization of polymers. Although these methods can improve the ionic conductivity of PEs, the extent of their improvement is usually very limited, e.g., Appetecchi et al. made the room temperature ionic conductivity of the corresponding PEs reach ~1.1×10^{-6} S cm^{-1} using the method of blending modification [25], Niitani et al. made the room temperature ionic conductivity of PEs increase to ~1×10^{-5} S cm^{-1} using the strategy of copolymerization [26], and a PE with room temperature ionic conductivity of 8×10^{-5} S cm^{-1} was obtained using the branching method by wang et al [27]. To further improve the ionic conductivity of PEs, attempts have been made to introduce plasticizers such as liquid electrolytes and ionic liquids into PEs [28,29], which has significantly improved the room temperature ionic conductivity of PEs, but usually at the expense of the mechanical properties such as strength and tensile properties.

Materials with a double-network structure have been pioneered in the study of hydrogels [30], and better mechanical properties are usually obtained for polymers with a double-network structure compared to those with a single-network structure. The first network of materials with a dual-network structure usually consists of a rigid network in which there are hydrogen bonds, ionic bonds, or physical cross-linking points as sacrificial bonds, etc. The sacrificial bonds are reversibly broken and generated during the stretching and rebounding process as a mechanism to dissipate energy and improve the mechanical strength of the material [31]; the second network is usually a flexible polymer network connected by covalent bonds, which provides elasticity and maintains the basic architecture of the dual network [32]. In the field of ionogels, Kamio et al. first demonstrated that an ionogel based on an organic–inorganic double network has much higher mechanical strength than the single-network ionogel [33]. Inspired by this, Yu et al. constructed an organic–inorganic double-network solvate ionogel with high toughness (80 MPa) and high ionic conductivity (0.12 mS cm^{-1}) for lithium-metal batteries [34]. However, there are few reports on organic–inorganic dual-network ionogel electrolytes, and the polymer network is limited to polyacrylamide, resulting in unsatisfactory mechanical properties, such as the strain of only ~170% that was reported by yu et al [34]. There is an urgent need to develop more types of new dual-network ionic gels.

Herein, we designed a new double-network polymer electrolyte (PE) based on the nonhydrolytic sol–gel reaction of tetraethyl orthosilicate and the in situ polymerization of zwitterions to synergistically achieve high strength, high tensile, and high ionic conductivity of PEs. Among them, the inorganic network is a physical crosslinking point composed of Si nanoparticles that can reversibly dissipate the energy generated by stretching, thus enhancing the tensile strength of the electrolyte; the elastic polyzwitterion network gives the electrolyte high tensile capacity. The ion–dipole forces between the polyzwitterion segments and between the polyzwitterion segments and ionic liquid in the organic network provide dynamic nano-conducting channels to facilitate ion transport, and the inorganic network can further enhance the electrochemical stability of the electrolyte, ultimately enabling the construction of high-performance electrolytes. Excellent interfacial compatibility of the PEs with lithium metal electrodes has also been demonstrated.

2. Experimental Section

2.1. Materials

2-methacryloyloxyethyl phosphorylcholine (Macklin, Shanghai, China), Tetraethyl orthosilicate (Aladdin, Shanghai, China), formic acid (FA, 88%, Aladdin) 1-ethyl-3-methylimidazoliumbis(trifluoromethylsulfonyl)imide (Aladdin, Shanghai, China), bis(trifluoro methane) sulfonimide lithium (Aladdin, Shanghai, China), 3-[Dimethyl-[2-(2-methylprop-2-enoyloxy) ethyl]azaniumyl]propane-1-sulfonate (Macklin, Shanghai, China), and 1-hydroxycyclohexyl phenyl ketone (Aladdin, Shanghai, China) were used directly without other treatment.

2.2. Synthesis of Organic–Inorganic Double-Network PEs

The double-network PEs were prepared using the sol–gel reaction of tetraethyl orthosilicate and in situ polymerization of zwitterions. Detailed information on the sample can be found in Table S1. Typically, 2-methacryloyloxyethyl phosphorylcholine (MPC, 0.572 g, 1.94 mmol) and 3-[Dimethyl-[2-(2-methylprop-2-enoyloxy)ethyl]azaniumyl]propane-1-sulfonate (DPS, 0.064 g, 0.23 mmol) were dissolved in 1-ethyl-3-methylimidazoliumbis(triflu oromethylsulfonyl)imide (IL)/lithium salt (0.8 g, 30 wt% LiTFSI), followed by adding TEOS (120 µL), FA (160 µL), and a photoinitiator (1-hydroxycyclohexyl phenyl ketone, 0.018 mg, 0.09 mmol). The mixture was put into a homemade mold and underwent photopolymerization for 30 min, and then it was transferred to a 50 °C oven for 48 h. Finally, the double-network PE was obtained.

2.3. Characterization and Testing

Structural information on the monomers and PEs was recorded using FTIR spectroscopy (L1600400 Spectrum TWO DTGS, MA, USA). Thermal properties of the PEs were acquired using a thermal gravimetric analyzer (NETZSCH STA 2500, Selb, Germany) and differential scanning calorimetry (DSC, Shimadzu DSC-60A equipment, Tokyo, Japan). A Field Emission Scanning Electron Microscope (FESEM, Hitachi SU8010, Tokyo, Japan) was used to observe the microstructures of the PEs. Stress–strain curves of the PEs were obtained using the Instron 3300 electronic universal material testing system. The ionic conductivity (σ) of the PEs was obtained from impedance measurements using Zennium Electrochemical workstation (ZahnerEnnium) and calculated using $\sigma = l/(RS)$, where l is the thickness of an electrolyte, R is bulk resistance, and S is the contact area between the stainless steel (SS) electrode and PE [35–37]. Linear sweep voltammetry was used to test the electrochemical stability of the PEs.

3. Results and Discussion

Figure 1a shows the preparation route of the novel organic–inorganic double-network polymer electrolyte based on the nonhydrolytic sol–gel reaction of tetraethyl orthosilicate and in situ polymerization of zwitterion (see the detailed preparation method in the experimental section). Figure 1b and 1c shows the IR spectra of Tetraethyl orthosilicate (TEOS), DPS, MPC, double-network PE (DPE) and single-network PE (SPE) form different wavelength ranges. According to the FTIR spectra, the obvious characteristic peaks of $-CH=CH_2$ from MPC (3029 cm^{-1} (Figure 1b) and 1635 cm^{-1} (Figure 1c)) and DPS (3038 cm^{-1} (Figure 1b) and 1635 cm^{-1} (Figure 1c)) completely disappeared in the DPE (M9D1T30)) and SPE (M9D1T0), without SiO_2 inorganic network), demonstrating the sufficient polymerization of zwitterions.

The thermal stability of related PEs was further acquired using a thermal gravimetric analyzer. Here, we denote the prepared PE by MxDyTz, where x, y, and z represent the weight ratios of MPC, DPS, and TEOS, respectively. Specifically, in this work, we prepared the samples M5D5T20, M5D0T20, M9D1T15, M9D1T20, M9D1T25, M9D1T30, and M9D1T35 to explore their differences in mechanical and electrochemical properties (see Table S1). The TGA curves in Figure 2a indicate that the prepared PEs have a high thermal decomposition temperature of over 200 °C. When the content of inorganic components reaches a certain level, their ability to improve thermal stability is demonstrated. Especially, M9D1T30 shows a thermal decomposition temperature of ~300 °C, which is the highest among all the samples. The high thermal decomposition temperature of PEs indicates that they can be used over a wide range of temperatures.

The mechanical properties of electrolyte membranes, such as mechanical strength and stretchability, are among the most important indicators of safe, high-performance solid-state batteries, and we further investigated the effect of the content of each component on the mechanical properties. We first investigated the effect of amphoteric monomer content on the mechanical properties of PEs. Comparing the stress–strain curves of M10D0T20 (stress: 0.45 MPa; strain: 367%), M9D1T20 (stress: 0.55 MPa: strain: 439%) and M5D5T20 (stress: 0.30 MPa: strain: 195%), it can be seen that the stress and strain of the materials increase and then decrease with the increase of DPS content. At a fixed MPC and DPS content, i.e., the samples M9D1Tz (z = 15, 20, 25, 30 and 35), the strain of the material increases as the TEOS content increases, and overall, the strength of the material increases, and when the TEOS content exceeds 30 (i.e., corresponding to M9D1T35), both the strain and stress of the material decrease. The best sample (M9D1T30) can reach a strain of 569%

and a stress of 0.75 MPa. It can be seen that the content of amphoteric monomers has an important influence on the mechanical properties, which is due to the large difference in the stiffness and flexibility of the corresponding two polymer segments, and only in a reasonable ratio can the common advantages of both be exploited. In addition, there is also a threshold value for the content of inorganic networks, which is due to the fact that too high a content of inorganic networks usually agglomerates and causes a decrease in the mechanical properties of the material.

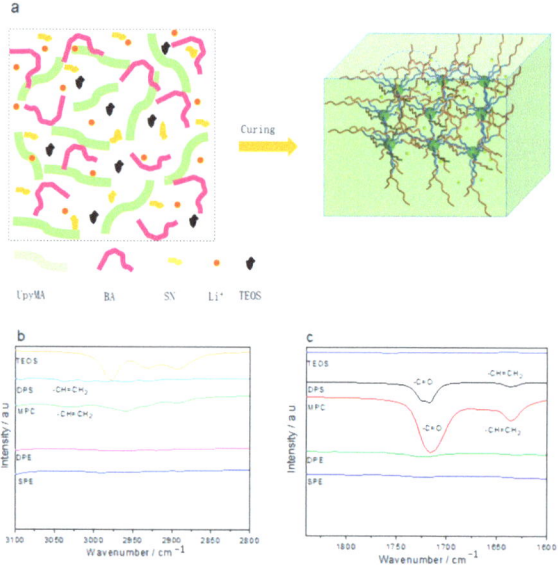

Figure 1. (**a**) Preparation route of a double-network PE. (**b**,**c**) FTIR characterization of related monomers and PEs.

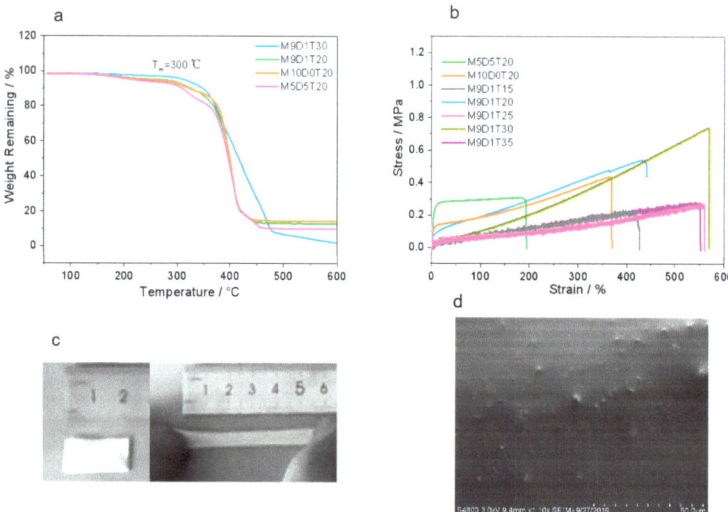

Figure 2. TGA curves (**a**) and Stress–strain curves (**b**) of double-network PEs with different ratios of TEOS and zwitterions. (**c**) Digital images of M9D1T20 before and after stretching. (**d**) Surface SEM image of M9D1T20.

The digital images of M9D1T20 before and after stretching are shown in Figure 2c. It is clearly seen that the film is highly stretchable. In addition, the SEM image in Figure 2d shows a uniform distribution of nanoparticles on the surface of the films, which can be attributed to the inorganic networks of SiO_2 formed through the non-hydrolytic-sol-gel of TEOS.

Based on the mechanical property analysis, M9D1T30, M9D1T20, M10D0T20, and M5D5T20 were preferably selected to further test their ionic conductivity between 30 and 80 °C. Overall, the ionic conductivity of the PEs decreases with increasing DPS content, as can be seen from the curves of ionic conductivity and temperature (Figure 3a). For instance, the ionic conductivity of M9D1T30, M9D1T20, and M10D0T20 at 30 °C is 0.30, 0.31, and 0.44 mS cm^{-1}, respectively. Specifically, M5D5T20 with the highest content of DPS shows the lowest conductivity in the whole temperature range (0.14 mS cm^{-1} at 30 °C). This can most likely be attributed to the fact that polyMPC and lithium ions have a more reasonable coordination interaction, which facilitates the migration of lithium ions. According to the fitted results, M9D1T30, M9D1T20, M10D0T20, and M5D5T20 show activation energies of 8.00, 6.68, 5.65, and 3.57 kJ mol^{-1}, respectively, and all the corresponding curves follow the VTF model [23], which suggests that the polyzwitterion segments have an important role in regulating ion transport. We also compared the ionic conductivity of M9D1T20 and M9D1T30 throughout the temperature interval (Figure 3a) and found that the overall effect of the SiO_2 content on the ionic conductivity is not significant. However, we found that the content of SiO_2 strongly affects the electrochemical stability of the PE. As shown in Figure 3b, M9D1 without SiO_2 has a very unstable electrochemical window (below 3 V). In contrast, M9DT20 exhibits a high electrochemical window of close to 5 V. This is due to the ability of SiO_2 to act as Lewis acid and adsorb trace impurities [38], which facilitates the electrochemical stability. Figure 3c shows that the electrochemical windows of the electrolytes after the addition of the SiO_2 inorganic networks all exceeded 4 V, and in contrast to the relationship with conductivity, the increase in DPS content is overall favorable for the electrochemical stability. Meanwhile, M9D1T30 also shows a wide electrochemical window of 4.9 V (Figure S2). In conclusion, M9D1T30 is the best sample in terms of mechanical and electrochemical properties. The material has a strain of 569%, a tensile strength of 0.75 MPa, and a high ionic conductivity of 0.30 mS/cm at 30 °C, which is much higher than that reported for other PEs [5,34].

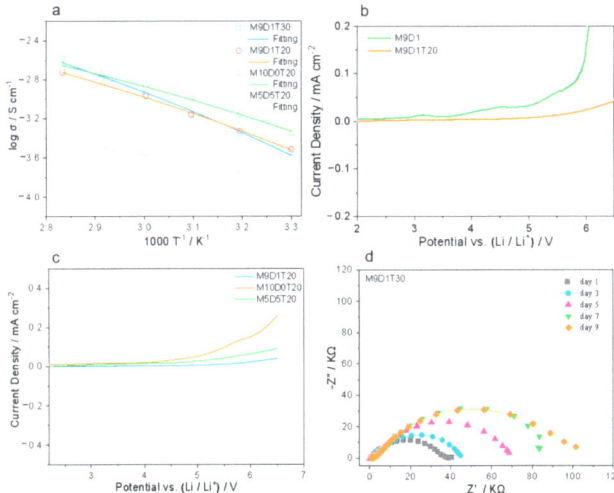

Figure 3. (**a**) Temperature dependencies of the ion conductivities of PEs. LSV curves (**b**,**c**) using double-network PEs with different ratios of TEOS and zwitterions. (**d**) EIS of Li/Li cell using M9D1T30.

The interfacial stability of Li/Li symmetric cells is one of the key indicators to assess whether the electrolyte can be applied to lithium metal batteries. As shown in Figure 3d, it can be seen that the interfacial resistance of the Li/M9D1T30/Li cell gradually increases in the initial 1–7 days, which is due to the formation of SEI film on the surface of lithium metal. Then, the interface reaches stability after 9 days of shelving, corresponding to an interfacial resistance of about 100 Ω, indicating good interfacial compatibility between the electrolyte and lithium metal. As also shown in Figure S1, Li/ Li/M9D1T20/Li also has a stable interface after 7 days, further confirming the advantages of using the novel double-network PEs.

To further understand the ionic transport mechanism of the as-prepared PEs, we further obtained the DSC curves of M9D1T30, M9D1T20, and M10D0T20. Regardless of the sample, there is only one melt peak belonging to the ionic liquid at about 17 °C [39], and no crystalline peak was observed in the sample, indicating that the sample is amorphous, and this amorphous feature facilitates the transport of lithium ions. In addition, there are ion–dipole interactions between the polyzwitterion segments and between the polyzwitterion segments and ionic liquid, which facilitate the formation of dynamic ion transport channels in the electrolyte. In addition, the interface between the inorganic network and the organic network also facilitates the formation of an interfacial transport layer, which promotes the transport of lithium ions. Therefore, we propose the ion transport model for this dual-network electrolyte (Figure 4b).

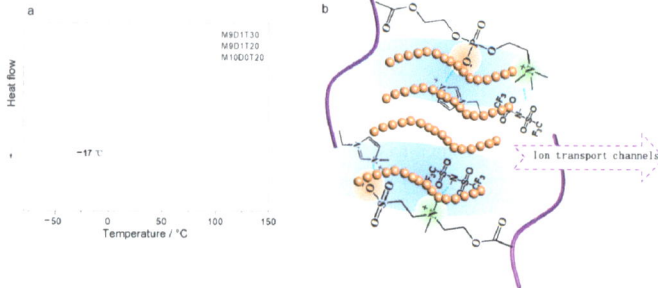

Figure 4. (**a**) DSC curves of double-network PEs with different ratios of TEOS and zwitterions. (**b**) Ion transport model of the prepared double-network PE.

4. Conclusions

We designed and prepared a novel organic–inorganic dual-network electrolyte using the non-hydrolytic sol–gel reaction and in situ photopolymerization. The effects of organic and inorganic components on the mechanical properties, ionic conductivity, and electrochemical stability of the electrolytes were systematically investigated by regulating the content of organic and inorganic networks. A polyzwitterion–SiO$_2$ double-network PE with high mechanical strength (tensile strength: 0.75 MPa), good tensile property (strain: 569%), high conductivity (0.30 mS/cm at 30 °C), and wide electrochemical window (close to 5 V) was prepared. The electrolyte also had good interfacial compatibility with lithium metal and was expected to be used in solid-state metal batteries. The possible ion transport mechanism of the electrolyte was further given using the phase transition analysis of the polymer electrolyte and the unique molecular conformation of the polyzwitterion segments. The present work provides new ideas for the design and synthesis of novel organic–inorganic dual-network PEs for high-performance solid-state energy storage devices.

Supplementary Materials: The following supporting information can be downloaded at: https://www.mdpi.com/article/10.3390/polym15020466/s1, Figure S1: EIS of Li/Li cell using M9D1T20; Figure S2: Electrochemical stability of M9D1T30; Table S1: Weight percentage of each component of MxDyTz, the strength, stress and ionic conductivity also are shown.

Author Contributions: Conceptualization, L.Z. and Q.W.; methodology, L.Z., Y.L. and L.G.; software, H.G.; validation, L.Z., Q.W., H.G. and Y.L.; formal analysis, L.Z. and L.G.; investigation, L.Z.; resources, L.Z.; data curation, L.Z.; writing—original draft preparation, L.Z.; writing—review and editing, L.Z., Y.L., L.G. and Q.W.; visualization, L.Z.; supervision, L.Z.; project administration, L.Z.; funding acquisition, L.Z., Y.L. and Q.W. All authors have read and agreed to the published version of the manuscript.

Funding: This research was funded by Research Project of Chuzhou University (2022XJYB08), Natural Science Foundation of China (52177020), Fundamental Research Program of Shanxi Province (202103021222006), Research Project Supported by Shanxi Scholarship Council of China (HGKY2019085), General Project of Natural Science Research in Universities of Anhui Province (KJ2021B15) and Chuzhou University Scientific Research Start Foundation Project (2020qd04).

Institutional Review Board Statement: Not applicable.

Data Availability Statement: Not applicable.

Conflicts of Interest: There are no conflict to declare.

References

1. Martinez, M.C.; Boaretto, N.; Naylor, A.; Alcaide, F.; Salian, G.; Palombardini, F.; Ayerbe, E.; Borras, M.; Casas-Cabanas, M. Are Polymer-Based Electrolytes Ready for High-Voltage Lithium Battery Applications? An Overview of Degradation Mechanisms and Battery Performance. *Adv. Energy Mater.* **2022**, *12*, 2201264. [CrossRef]
2. Choudhury, S.; Tu, Z.; Nijamudheen, A.; Zachman, M.; Stalin, S.; Deng, Y.; Zhao, Q.; Vu, D.; Kourkoutis, L.; Mendoza-Cortes, J.; et al. Stabilizing polymer electrolytes in high-voltage lithium batteries. *Nat. Commun.* **2019**, *10*, 3091. [CrossRef]
3. Fan, L.-Z.; He, H.; Nan, C.-W. Tailoring inorganic-polymer composites for the mass production of solid-state batteries. *Nat. Rev. Mater.* **2021**, *6*, 1003–1019. [CrossRef]
4. Bachman, J.; Muy, S.; Grimaud, A.; Chang, H.-H.; Pour, N.; Lux, S.; Paschos, O.; Maglia, F.; Lupart, S.; Lamp, P.; et al. Inorganic Solid-State Electrolytes for Lithium Batteries: Mechanisms and Properties Governing Ion Conduction. *Chem. Rev.* **2016**, *116*, 140–162. [CrossRef]
5. Wang, H.; Wang, Q.; Cao, X.; He, Y.; Wu, K.; Yang, J.; Zhou, H.; Liu, W.; Sun, X. Thiol-Branched Solid Polymer Electrolyte Featur-ing High Strength, Toughness, and Lithium Ionic Conductivity for Lithium-Metal Batteries. *Adv. Mater.* **2020**, *32*, 2001259. [CrossRef] [PubMed]
6. Wang, S.; Cheng, T.; Zhang, Y.; Wu, X.; Xiao, S.; Lai, W.-Y. Deformable lithium-ion batteries for wearable and implantable elec-tronics. *Appl. Phys. Rev.* **2022**, *9*, 041310. [CrossRef]
7. Alarco, P.; Abu-Lebdeh, Y.; Abouimrane, A.; Armand, M. The plastic-crystalline phase of succinonitrile as a universal matrix for solid-state ionic conductors. *Nat. Mater.* **2004**, *3*, 476–481. [CrossRef]
8. Zhang, L. Robust thiol-branched all-solid-state polymer electrolyte featuring high ionic conductivity for lithium-metal batter-ies. *Ionics* **2021**, *27*, 599–605. [CrossRef]
9. Wang, S.; Li, Q.; Bai, M.; He, J.; Liu, C.; Li, Z.; Liu, X.; Lai, W.-Y.; Zhang, L. A dendrite-suppressed flexible polymer-in-ceramic elec-trolyte membrane for advanced lithium batteries. *Electrochim. Acta* **2020**, *353*, 136604. [CrossRef]
10. Zhang, L.; Gao, H.; Xiao, S.; Li, J.; Ma, T.; Wang, Q.; Liu, W.; Wang, S. In-Situ Construction of Ceramic–Polymer All-Solid-State Electrolytes for High-Performance Room-Temperature Lithium Metal Batteries. *ACS Mater. Lett.* **2022**, *4*, 1297–1305. [CrossRef]
11. Wang, S.; Li, X.; Cheng, T.; Liu, Y.; Li, Q.; Bai, M.; Liu, X.; Geng, H.; Lai, W.-Y.; Huang, W. Highly conjugated three-dimensional co-valent organic frameworks with enhanced Li-ion conductivity as solid-state electrolytes for high-performance lithium metal batteries. *J. Mater. Chem. A* **2022**, *10*, 8761–8771. [CrossRef]
12. Su, Y.; Rong, X.; Gao, A.; Liu, Y.; Li, J.; Mao, M.; Qi, X.; Chai, G.; Zhang, Q.; Suo, L. Rational design of a topological polymeric solid electrolyte for high-performance all-solid-state alkali metal batteries. *Nat. Commun.* **2022**, *13*, 4181. [CrossRef] [PubMed]
13. Wang, X.; Zhang, C.; Sawczyk, M.; Sun, J.; Yuan, Q.; Chen, F.; Mendes, T.; Howlett, P.; Fu, C.; Wang, Y. Ultra-stable all-solid-state sodium metal batteries enabled by perfluoropolyether-based electrolytes. *Nat. Mater.* **2022**, *21*, 1057–1065. [CrossRef] [PubMed]
14. Atik, J.; Diddens, D.; Thienenkamp, J.; Brunklaus, G.; Winter, M.; Paillard, E. Cation-Assisted Lithium-Ion Transport for High-Performance PEO-based Ternary Solid Polymer Electrolytes. *Angew. Chem. Int. Ed.* **2021**, *60*, 11919–11927. [CrossRef]
15. Tan, J.; Ao, X.; Dai, A.; Yuan, Y.; Zhuo, H.; Lu, H.; Zhuang, L.; Ke, Y.; Su, C.; Peng, X.; et al. Polycation ionic liquid tailored PEO-based solid polymer electrolytes for high temperature lithium metal batteries. *Energy Storage Mater.* **2020**, *33*, 173–180. [CrossRef]
16. Lopez, J.; Mackanic, D.; Cui, Y.; Bao, Z. Designing polymers for advanced battery chemistries. *Nat. Revi. Mater.* **2019**, *4*, 312–330. [CrossRef]
17. Christie, A.; Lilley, S.; Staunton, E.; Andreev, Y.; Bruce, P. Increasing the conductivity of crystalline polymer electrolytes. *Nature* **2005**, *433*, 50. [CrossRef]
18. Croce, F.; Appetecchi, G.; Persi, L.; Scrosati, B. Nanocomposite polymer electrolytes for lithium batteries. *Nature* **1998**, *394*, 456. [CrossRef]

19. Isaac, J.; Devaux, D.; Bouchet, R. Dense inorganic electrolyte particles as a lever to promote composite electrolyte conductiv-ity. *Nat. Mater.* **2022**, *21*, 1412–1418. [CrossRef]
20. Wang, S.; Zhang, L.; Li, J.; Zeng, Q.; Liu, X.; Chen, P.; Lai, W.-Y.; Zhao, T.; Zhang, L. A nanowire-nanoparticle double composite polymer electrolyte for high performance ambient temperature solid-state lithium batteies. *Electrochim. Acta* **2019**, *320*, 134560. [CrossRef]
21. Ding, J.; Chuy, C.; Holdcroft, S. Solid Polymer Electrolytes Based on Ionic Graft Polymers: Effect of Graft Chain Length on Nano-Structured, Ionic Networks. *Adv. Funct. Mater.* **2002**, *12*, 389–394. [CrossRef]
22. Bouchet, R.; Maria, S.; Meziane, R.; Aboulaich, A.; Lienafa, L.; Bonnet, J.-P.; Phan, T.; Bertin, D.; Gigmes, D.; Devaux, D. Single-ion BAB triblock copolymers as highly efficient electrolytes for lithium-metal batteries. *Nat. Mater.* **2013**, *12*, 452. [CrossRef]
23. Wang, S.; He, J.; Li, Q.; Wang, Y.; Liu, C.; Cheng, T.; Lai, W.-Y. Highly elastic energy storage device based on intrinsically su-per-stretchable polymer lithium-ion conductor with high conductivity. *Fund. Res.* **2022**. [CrossRef]
24. Wang, Z.; Wang, S.; Wang, A.; Liu, X.; Chen, J.; Zeng, Q.; Zhang, L.; Liu, W.; Zhang, L. Covalently linked metal-organic framework (MOF)-polymer all-solid-state electrolyte membranes for room temperature high performance lithium batteries. *J. Mater. Chem. A* **2018**, *6*, 17227–17234. [CrossRef]
25. Appetecchi, G.; Croce, F.; Hassoun, J.; Scrosati, B.; Salomon, M.; Cassel, F. Hot-pressed, dry, composite, PEO-based electrolyte membranes: I. Ionic conductivity characterization. *J. Power Sources* **2003**, *114*, 105–112. [CrossRef]
26. Niitani, T.; Shimada, M.; Kawamura, K.; Dokko, K.; Rho, Y.-H.; Kanamura, K. Synthesis of Li+ ion conductive PEO-PSt block co-polymer electrolyte with microphase separation structure. *Electrochem. Solid State Lett.* **2005**, *8*, A385. [CrossRef]
27. Wang, S.; Zhang, L.; Zeng, Q.; Liu, X.; Lai, W.-Y.; Zhang, L. Cellulose Microcrystals with Brush-Like Architectures as Flexible All-Solid-State Polymer Electrolyte for Lithium-Ion Battery. *ACS Sustain. Chem. Eng.* **2020**, *8*, 3200–3207. [CrossRef]
28. Balo, L.; Gupta, H.; Singh, V.; Singh, R. Flexible gel polymer electrolyte based on ionic liquid EMIMTFSI for rechargeable battery application. *Electrochim. Acta* **2017**, *230*, 123–131. [CrossRef]
29. Osada, I.; de Vries, H.; Scrosati, B.; Passerini, S. Ionic-Liquid-Based Polymer Electrolytes for Battery Applications. *Angew. Chem. Int. Ed.* **2016**, *55*, 500–513. [CrossRef]
30. Nakayama, A.; Kakugo, A.; Gong, J.; Osada, Y.; Takai, M.; Erata, T.; Kawano, S. High mechanical strength double-network hy-drogel with bacterial cellulose. *Adv. Funct. Mater.* **2004**, *14*, 1124–1128. [CrossRef]
31. Gong, J.; Katsuyama, Y.; Kurokawa, T.; Osada, Y. Double-network hydrogels with extremely high mechanical strength. *Adv. Mater.* **2003**, *15*, 1155–1158. [CrossRef]
32. Gong, J. Why are double network hydrogels so tough? *Soft Matter* **2010**, *6*, 2583–2590. [CrossRef]
33. Kamio, E.; Yasui, T.; Iida, Y.; Gong, J.; Matsuyama, H. Inorganic/Organic Double-Network Gels Containing Ionic Liquids. *Adv. Mater.* **2017**, *29*, 1704118. [CrossRef]
34. Yu, L.; Guo, S.; Lu, Y.; Li, Y.; Lan, X.; Wu, D.; Li, R.; Wu, S.; Hu, X. Highly Tough, Li-Metal Compatible Organic-Inorganic Dou-ble-Network Solvate Ionogel. *Adv. Energy Mater.* **2019**, *9*, 1900257. [CrossRef]
35. Wang, S.; Zeng, Q.; Wang, A.; Liu, X.; Chen, J.; Wang, Z.; Zhang, L. Constructing stable ordered ion channels for a solid electro-lyte membrane with high ionic conductivity by combining the advantages of liquid crystal and ionic liquid. *J. Mater. Chem. A* **2019**, *7*, 1069–1075. [CrossRef]
36. Zhang, L.; Wang, S.; Li, J.; Liu, X.; Chen, P.; Zhao, T.; Zhang, L. A nitrogen-containing all-solid-state hyperbranched polymer elec-trolyte for superior performance lithium batteries. *J. Mater. Chem. A* **2019**, *7*, 6801–6808. [CrossRef]
37. Wang, S.; Liu, X.; Wang, A.; Wang, Z.; Chen, J.; Zeng, Q.; Jiang, X.; Zhou, H.; Zhang, L. High-Performance All-Solid-State Polymer Electrolyte with Controllable Conductivity Pathway Formed by Self-Assembly of Reactive Discogen and Immobilized via a Facile Photopolymerization for a Lithium-Ion Battery. *ACS Appl. Mater. Interfaces* **2018**, *10*, 25273–25284. [CrossRef]
38. Chen, N.; Xing, Y.; Wang, L.; Liu, F.; Li, L.; Chen, R.; Wu, F.; Guo, S. "Tai Chi" philosophy driven rigid-flexible hybrid ionogel electrolyte for high-performance lithium battery. *Nano Energy* **2018**, *47*, 35–42. [CrossRef]
39. Wang, S.; Bai, M.; Liu, C.; Li, G.; Lu, X.; Cai, H.; Liu, C.; Lai, W.-Y. Highly stretchable multifunctional polymer ionic conductor with high conductivity based on organic-inorganic dual networks. *Chem. Eng. J.* **2022**, *440*, 135824. [CrossRef]

Disclaimer/Publisher's Note: The statements, opinions and data contained in all publications are solely those of the individual author(s) and contributor(s) and not of MDPI and/or the editor(s). MDPI and/or the editor(s) disclaim responsibility for any injury to people or property resulting from any ideas, methods, instructions or products referred to in the content.

Article

Characterization of PLA/PCL/Nano-Hydroxyapatite (nHA) Biocomposites Prepared via Cold Isostatic Pressing

Solechan Solechan [1,2,*], Agus Suprihanto [1], Susilo Adi Widyanto [1], Joko Triyono [3], Deni Fajar Fitriyana [4], Januar Parlaungan Siregar [5] and Tezara Cionita [6]

1. Department of Mechanical Engineering, Faculty of Engineering, Diponegoro University, Semarang 50275, Indonesia
2. Department of Mechanical Engineering, Universitas Muhammadiyah Semarang, Kampus Kasipah, Semarang 50254, Indonesia
3. Department of Mechanical Engineering, Sebelas Maret University, Surakarta 57126, Indonesia
4. Department of Mechanical Engineering, Universitas Negeri Semarang, Kampus Sekaran, Gunungpati, Semarang 50229, Indonesia
5. Faculty of Mechanical & Automotive Engineering Technology, Universiti Malaysia Pahang, Pekan 26600, Malaysia
6. Faculty of Engineering and Quantity Surveying, INTI International University, Nilai 71800, Malaysia
* Correspondence: solechan@unimus.ac.id

Citation: Solechan, S.; Suprihanto, A.; Widyanto, S.A.; Triyono, J.; Fitriyana, D.F.; Siregar, J.P.; Cionita, T. Characterization of PLA/PCL/ Nano-Hydroxyapatite (nHA) Biocomposites Prepared via Cold Isostatic Pressing. *Polymers* **2023**, *15*, 559. https://doi.org/10.3390/ polym15030559

Academic Editors: Jesús-María García-Martínez and Emilia P. Collar

Received: 30 December 2022
Revised: 19 January 2023
Accepted: 19 January 2023
Published: 21 January 2023

Copyright: © 2023 by the authors. Licensee MDPI, Basel, Switzerland. This article is an open access article distributed under the terms and conditions of the Creative Commons Attribution (CC BY) license (https:// creativecommons.org/licenses/by/ 4.0/).

Abstract: Hydroxyapatite has the closest chemical composition to human bone. Despite this, the use of nano-hydroxyapatite (nHA) to produce biocomposite scaffolds from a mixture of polylactic acid (PLA) and polycaprolactone (PCL) using cold isostatic pressing has not been studied intensively. In this study, biocomposites were created employing nHA as an osteoconductive filler and a polymeric blend of PLA and PCL as a polymer matrix for prospective usage in the medical field. Cold isostatic pressing and subsequent sintering were used to create composites with different nHA concentrations that ranged from 0 to 30 weight percent. Using physical and mechanical characterization techniques such as Fourier transform infrared spectroscopy (FTIR), X-ray diffraction (XRD), scanning electron microscopy (SEM), and density, porosity, tensile, and flexural standard tests, it was determined how the nHA concentrations affected the biocomposite's general properties. In this study, the presence of PLA, PCL, and nHA was well identified using FTIR, XRD, and SEM methods. The biocomposites with high nHA content showed intense bands for symmetric stretching and the asymmetric bending vibration of PO_4^{3-}. The incorporation of nHA into the polymeric blend matrix resulted in a rather irregular structure and the crystallization became more difficult. The addition of nHA improved the density and tensile and flexural strength of the PLA/PCL matrix (0% nHA). However, with increasing nHA content, the PLA/PCL/nHA biocomposites became more porous. In addition, the density, flexural strength, and tensile strength of the PLA/PCL/nHA biocomposites decreased with increasing nHA concentration. The PLA/PCL/nHA biocomposites with 10% nHA had the highest mechanical properties with a density of 1.39 g/cm^3, a porosity of 1.93%, a flexural strength of 55.35 MPa, and a tensile strength of 30.68 MPa.

Keywords: nano-hydroxyapatite; polylactic acid (PLA); polycaprolactone (PCL); biocomposites

1. Introduction

Bone fractures caused by cancer, traffic accidents, bone tumors, or aging are incapable of self-healing. These bone fractures require interventional therapy with implants or bone grafts in order to heal and regenerate. Bone tissue engineering has garnered interest due to its inherent advantages for the healing of bone fractures.

Important components of bone tissue engineering are scaffolds that can provide dynamic circumstances for cell growth. Scaffolds must be biocompatible and biodegradable, possess appropriate mechanical qualities and pore sizes, and have pores that are well-connected [1–3]. The success of an implanted biomaterial is determined by a number of criteria, including its shape and structural characterization, durability, mechanical loading, property of implant material, location of the implanted site, and host reaction [3]. The standard treatment for bone repair is autografting, but it has a number of drawbacks, including limited tissue availability, discomfort for the patient, morbidity at the donor site, the need for a second procedure, challenges in fabricating an anatomically shaped graft, and a failure rate of up to 50% for some sites. The creation of implants and scaffolds was urgently needed in order to address these drawbacks. Thus, the engineered scaffolds' goal is to rebuild bone tissue rather than just replace it [2].

Moreover, internal fixation devices made from metal may be utilized for the repair of bone fractures. Metallic materials such as stainless steel, CoCr alloys, Ti, and Ti alloys have a lot of good properties that mean they are utilized for bone implants. These properties include high fracture toughness, corrosion resistance, high strength, high hardness, and biocompatibility. However, the inadequate interfacial adhesion between metallic implants and tissue or bone results in the creation of a non-adhesive layer and movement at the implant tissue interface. This results in the failure of using metal implants to treat bone fractures. Furthermore, metallic implants have a substantially greater modulus of elasticity than bone, which can result in stress shielding. Consequently, osteoporosis, osteolysis, and secondary fractures will occur. During the implantation process, toxic effects induced by ions released from metallic implants represent a major problem [4–10]. The use of metal implants requires a second surgical procedure for implant removal, which can increase the cost of treatment [11]. This has prompted researchers to find substitutes for metal implants using other materials such as polymers, ceramics, and composites.

Presently, only a few polymer biomaterials are now available that are non-toxic, absorbable, and FDA-approved for use as scaffolding materials in medical applications. Among these, biomaterials with strong biocompatibility and biodegradability include polylactic acid (PLA) and polycaprolactone (PCL) [2,12]. PLA has a glass transition temperature (Tg) between 50 and 80 °C and a melting temperature (Tm) between 130 and 180 °C [13]. PCL has good solubility with other polymers, minimal viscosity, and hydrophobic characteristics, in which the molecular weight and crystallinity level affect the physical and mechanical characteristics. While the Tm ranges from 50 to 60 °C, the glass transition temperature (Tg) of PCL is roughly 60 °C. Take into account that PCL has a high crystallinity level (between 30 and 60%). However, the main limitations of PLA and PCL are their poor mechanical strength and low cell affinity limiting their application as bone scaffold materials [2,12].

PLA is a brittle material that degrades quickly in body fluids [14]. Therefore, the addition of PCL to PLA is used to minimize brittleness and extend the degradation time of the PLA. Subsequently, PLA and PCL are blended to make copolymers that have good qualities for tissue engineering, for instance, being biocompatible, biodegradable, and non-toxic [13,15,16]. Solechan et al. performed scaffold fabrication utilizing distinct PLA/PCL mixtures and observed the effects of PLA/PCL concentrations on their physical and mechanical properties. By increasing PLA content, the PLA/PCL blend porosity reduced, resulting in enhanced density and flexural strength [17]. An ideal scaffold ought to be osteoconductive (enabling pluripotent cells to develop into osteoblasts and supporting the proliferation of cells and capillaries to create bone), biocompatible, and biodegradable, as well as possess the proper biological qualities and mechanical strength [18]. To obtain the needed qualities, the incorporation of hydroxyapatite (HA) is able to overcome the hydrophobicity of the PLA/PCL blend, increase its mechanical properties, and stimulate osteoconduction, as well as osseointegration in the implanted scaffold [2,19,20].

Hydroxyapatite (HA) is known as the most common mineral found in bones and teeth. Because HA chemicals account for approximately 65% of bone, they are an interesting candidate for a synthetic bone composite. HA is a bioactive ceramic material that is extensively utilized in various biomedical applications, primarily as orthopedic implant materials and in the creation of dentistry materials [21–23].

Pijamit et al. utilize PLA/PCL/HA to be the biocomposite material for manufacturing 3D printing filaments. HA in the PLA/PCL/15HA mixture is used for producing the greatest compressive strength (82.72 ± 1.76 MPa). Furthermore, it was observed that HA also provided higher bone cell proliferation [16]. Fabrication of composite scaffolds made from PLA/PCL/HA by indirect 3D printing was studied by Hassanajili et al. According to their research, the scaffold with PLA/PCL 70/30 w/w and 35% HA had better osteoinduction, viability, and biocompatibility qualities [19]. Fitriyana et al. studied the effect of using HA on the mechanical, physical, and degradation properties of the composite materials using a matrix of PLA/PCL (80 wt%/20 wt%). According to their findings, the mechanical characteristics of the biocomposite got better as the HA concentration rose. However, the biocomposite degrades more quickly the greater the quantity of added HA content [20].

The use of additive manufacturing techniques to create scaffolds has been the subject of numerous investigations. The benefits of the additive manufacturing-based process for scaffold fabrication include closeness to the final dimensions, precision, and the capability of generating complex geometries, as well as low processing costs. The disadvantages that come along with this method are limited product size, relatively small dimensions, poor mechanical properties, the requirement of post-processing, which is expensive and time-consuming, residual stress, high surface roughness, frequently clogged nozzles, clumping, and the presence of delamination of the layers on the final product [24–26].

To overcome these problems, this research uses the cold isostatic pressing method to make biocomposites from PLA/PCL/nHA as a scaffold material. Prior to machining or sintering, powdered materials can be compacted via cold isostatic pressing to create a solid, uniform mass. The main advantage of cold isostatic pressing is the ability to produce products with more complex shapes. In addition, distortion and cracking due to non-uniform stresses are greatly reduced [27–29]. According to an investigation conducted by Abdallah et al., the use of the cold isostatic pressing method could improve the density, hardness, tensile strength, impact resistance, and ductility of the 93%W4.9%Ni-2.1%Fe alloy [30]. Cold isostatic pressing (CIP) has been widely used as an efficient processing process for the compaction of metal and ceramic powders. Compared to uniaxial pressing, CIP compression produces samples with a greater relative density, superior mechanical qualities, and a more uniform microstructure. According to some reports, after the CIP process, the ceramic powder can be compressed to a maximum relative density of 70%. After being subjected to high CIP pressure and sintering, the nanoparticles yield a relative density of up to 99.99 percent or more [31]. In the field of medical implants, porous metal structures are commonly utilized. By controlling the CIP and sintering process parameters, evenly porous metal components can be manufactured. Controlling porosity in CIP components necessitates a combination of parameters, including powder qualities, tool design, CIP process parameters, thermal processing conditions, and ingredient density throughout the process [32–34].

Al Bakri et al. found that the pressing process also affects the characteristics of the zirconia toughness alumina (ZTA) composite that is manufactured. CIP provides superior mechanical qualities compared to uniaxial pressing. Compared to the uniaxial pressing method, composites compacted using the CIP method exhibit superior characteristics at lower sintered temperatures [34]. The processes involved in the production of manganates are significantly affected by cold isostatic pressing. It is demonstrated that the compacting pressure has a bigger effect on the rate of a chemical reaction than on crystallization.

The results suggest that high hydrostatic pressures can be employed to reduce synthesis temperature and generate nanostructured ceramics and manganates with predetermined oxygen nonstoichiometry. The cold isostatic pressing affects manganate synthesis and increases the contact area, which mechanically activates the grains and amplifies the solid-phase sintering processes [35]. Akimov et al. showed that CIP has a big effect on the physical properties of many powders, such as stabilized zirconium dioxide, α-phase alumina powder, hydride-forming intermetallics $LaNi_{2.5}Co_{2.4}Al_{0.1}$ and $LaNi_5$, and manganese powder. In their research, they utilized the isostatic cold pressure technique to generate materials containing a mixture of amorphous boron powder, crystalline aluminum, and LaB_6–TiB_2 composites under 0.6 GPa of pressure and sintering at 1000 °C [36]. The decrease in crystallization temperature may be due to the fact that the CIP gives rise to crystallization nuclei. The sinterability and mechanical properties of green bodies subjected to cold isostatic pressing are significantly enhanced [37].

Cold isostatic pressing has not been intensively investigated as a method for producing scaffolds from PLA, PCL, or nHA biocomposites. This research was conducted to determine the effect of the concentration of HA used on the physical and mechanical properties of biocomposites with a PLA/PCL matrix prepared using the cold isostatic pressure method.

2. Materials and Methods

The properties of the polycaprolactone (PCL) and polylactic acid (PLA) used in this study are shown in Table 1. Meanwhile, nano-hydroxyapatite (n-HA) with particle size < 200 nm with 502.31 g/mol of molecular weight was obtained from Sigma-Aldrich Pte Ltd., Pasir Panjang, Singapore [38]. The PLA and PCL compositions used in this study were 80% and 20%, respectively.

Table 1. The properties of the polycaprolactone (PCL) and polylactic acid (PLA) [17].

Specifications	PLA	PCL
Form	Powder	Pellet
Manufacturer	Reprapper Tech Co., Kowloon, Hong Kong	Solvay Interox Limited, Warrington, UK
Density (g/cm^3)	1.24	1.1
Melting temperature (°C)	175–220	58–60
Grain size (µm)	5–10	-
Diameter (mm)	-	0.5

The percentages of nHA composition used in this study were 0, 10, 20, and 30 wt%. The PLA/PCL/nHA biocomposite formulation is presented in Table 2. In a laboratory ball mill, PCL, PLA, and nHA were blended for two hours at 80 revolutions per min (Bexco; Haryana, India). The finished mixture was then added to a mold of stainless steel 304 with 17 mm diameter and 3 mm thickness. The compacting procedure occurred afterward, which generated a green body under a pressure of 40 MPa. Based on the reference, at a pressure of 40 MPa, there is a tangential contact between HA particles, as evidenced by the good grain bonding found in sintered particles. Moreover, the relative density increases by increasing uniaxial pressure between 10 and 40 MPa. At pressures between 40 and 190 MPa, the relative density tends to be constant [39]. Increased pressure during hot compression resulted in an increased melt flow index (MFI), crystallinity, density, ultimate tensile strength (UTS), and Young's modulus of the PP-HA biocomposite. This happened because the mechanical bonding and surface locking of the HA and PP particles in the composite increased with increasing pressure, resulting in better mechanical properties and impact resistance [40].

Table 2. Label and composition of biocomposite specimens.

Polymeric Blends	Ratio (w/w)	HA (wt%)	Specimen Codes
PLA/PCL	80/20	0	0%
PLA/PCL	80/20	10	10%
PLA/PCL	80/20	20	20%
PLA/PCL	80/20	30	30%

The sintering technique was then performed on the green body formed at 150 °C for 2 h using a digital drying oven (D1570, made in Taiwan).

The developed PLA/PCL/nHA biocomposites were evaluated using FTIR, X-ray diffraction, scanning electron microscopy (SEM), and density, porosity, tensile, and flexural techniques.

In the PLA/PCL/nHA biocomposite, the functional groups constituting the material as well as the orientation of the molecular chain were identified using the Fourier transform infrared (FTIR) technique [41]. The functional groups in the PLA/PCL/nHA biocomposite were identified using a PerkinElmer Spectrum IR Version 10.6.1 spectrophotometer (USA). Aside from this, each spectrum was recorded in the range of 400 cm^{-1} to 4000 cm^{-1}. X-ray diffraction (XRD) analysis was a non-destructive technique for investigating crystalline materials that was used to identify the crystalline phases in a material by looking at its crystal structure [42]. The PLA/PCL/HA biocomposites samples were identified by X-ray diffraction using a Shimadzu XRD-7000 diffractometer at 40 kV with a current of 30 mA and Cu K radiation (λ = 0.15406 nm). Diffractograms were obtained using a scanning rate of 1°/min between (2θ) 10° and 90°. This was followed with a step of 0.02. Furthermore, using a scanning electron microscope (SEM) and energy dispersive X-ray spectroscopy (EDX) (JSM-6510, JEOL, Japan) at an accelerating voltage of 15 kV, the surface morphology and elemental composition of the PLA/PCL/HA biocomposites were examined [43]. To make composition photos, topography images, and shadow images, a high-sensitivity backscattered electron detector was mounted on the bottom of the objective lens.

The actual density, theoretical density, and void volume (%) of PLA/PCL/nHA biocomposites were determined using studies by Satapathy et al. (2017) [44] and Taib et al. (2018) [45]. Flexural tests on PLA/PCL/nHA biocomposites were carried out based on the American Society for Testing and Materials (ASTM) number D790-17 to assess flexural strength. These properties were measured using the three-point bending test on a rectangular-shaped sample with a dimension of 127 mm × 12.7 mm × 3 mm. The crosshead speed used in the flexural test was 2 mm/min at room temperature.

Tensile strength was measured using PLA/PCL/nHA biocomposites specimens in accordance with the American Society for Testing and Materials (ASTM) D3039. In this study, the specimens for the tensile test were rectangular in shape, with dimensions of 250 mm × 25 mm × 3 mm. Aluminum tabs were fastened at the ends of the specimen to provide proper grip, prevent gripping damage, and ensure deep failure at the gauge length. A 50 kN load cell and a clip-on-type MTS extensometer with a gauge length of 25 mm were used to measure load and strain. A loading rate of 2 mm/min was used for testing. The flexural and tensile tests in this study used the HT-2402 Series Computer Universal Testing Machine from Hung Ta Instrument Co., Ltd., Sammutprakarn, Thailand, with five replications, and the average data were analyzed.

3. Results and Discussion

Figure 1 shows the FTIR spectra of the biocomposite specimens. The biocomposite obtained in this study formed functional groups with almost similar wavenumbers. The appearance of C–O, C=O, and CH_3 peaks in the biocomposite specimens demonstrate that polylactic acid (PLA) was present. Meanwhile, the C–O, C=O, and CH_2 peaks indicate the presence of polycaprolactone (PCL) in the biocomposite specimens. The existence of PO_4^{3-} and O–H peaks in the biocomposite specimens indicates that nano-hydroxyapatite (nHA) is present [17,46,47]. The hydroxyapatite (HA) spectrum demonstrates that the PO_4^{3-} stretch band is around 1156–1000 cm^{-1} and the O–H bend is in the range of 950–910 cm^{-1}. Following the O–H bend, a band at 631 cm^{-1} showing the extension of the hydroxyl group (OH-) in nHA was also detected in all biocomposite samples. In this study, all of the biocomposite specimens showed a weak O–H stretch in the range of 3160–3640 cm^{-1} [46–50]. The results of the FTIR test in this study found that a peak between 1440 and 1475 cm^{-1} indicates a CH_2 asymmetric stretch. A peak between 2880 and 2975 cm^{-1} shows a CH_3 symmetrical stretch [51,52]. FTIR test results showed the presence of symmetrical bending of CH_3 in specimens of 0, 10, 20, and 30% at peaks of 1373, 1335, 1332, and 1336 cm^{-1}, respectively. Thus, the C=O and C–O stretches are measured at 1760–1670 cm^{-1} and 1100–1000 cm^{-1} [17,46].

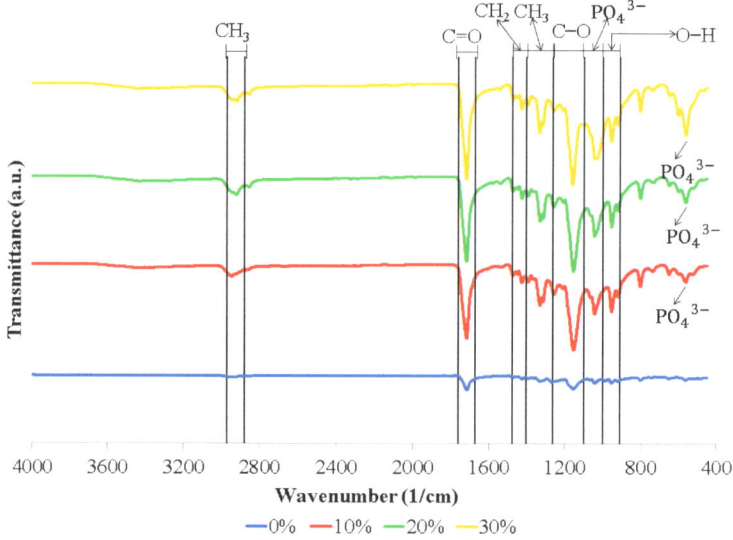

Figure 1. FTIR spectra of biocomposites with various concentrations of nHA.

The biocomposite specimens in this investigation demonstrated all of the peaks that were representative of nHA, PCL, and PLA. Furthermore, the findings of this investigation revealed no novel peaks in the spectrum. This suggests that the two polymers and nano-hydroxyapatite have a weak interaction and are fully incompatible. The results of this study are in accordance with the results of research conducted by Hassanajili et al., Shojaei et al., and Åkerlund et al. Based on the FTIR spectra, they found that there is a weak interaction between PLA, PCL, and HA, which indicates that they are completely incompatible. The carbonyl groups shifted, indicating that PLA, PCL, and HA interact with each other [19,46,47].

Figure 1 presents the FTIR spectrum of the biocomposite from PLA/PCL/nHA prepared via cold isostatic pressing. For the biocomposites obtained via isostatic cold pressing, the PCL, PLA, and nHA bands were well identified, with different band intensities being affected by the concentration of nHA used. Furthermore, biocomposites containing a high nHA content exhibited intense bands at 1047 and 551 cm^{-1}, which were associated with the symmetric stretching vibration of PO_4^{3-} and the asymmetric bending vibration of PO_4^{3-}. The findings in this study are similar to the results of research conducted by Bernardo et al. They made 3D filament from a biocomposite of PLA and HA. Their results showed that with a higher concentration of HA used, the FTIR spectra showed bands with stronger intensities at 1026 and 563 cm^{-1}, representing a symmetric stretching vibration of PO_4^{3-} and asymmetric bending vibration of PO_4^{3-}, respectively [53].

X-Ray diffraction (XRD) was utilized to identify the biocomposites' crystallized phases (Figure 2) within the 2θ range from 10° to 80°. The 2 theta values of 10.820; 16.841; 22.902; 25.879; 28.966; 31.773; 32.196; 32.902; 34.048; 39.818; 43.804; 46.711; 48.103; 49.468; 50.493; 51.283; 52.100; 53.143; 55.879; 61.660; 64.078; and 65.031 reflect the nHA peak based on the JCPDS card number 09-0432 for the stoichiometric peak of HA. The XRD peaks produced in this study have similarities with the results of research conducted by Herliansyah et al. The results of their research show the XRD graph on HA from bovine (BHA) with 2θ at 21, 22, 25, 28, 31, 32, 34, 35, 39, 41, 43, 45, 46, 48, and 49° [54]. Peaks at 2 theta values of 19.76,

22.74, and 28.82 suggested the presence of PLA. The peak that indicates PCL is denoted by the 2 theta value of 22.42 [17,55,56].

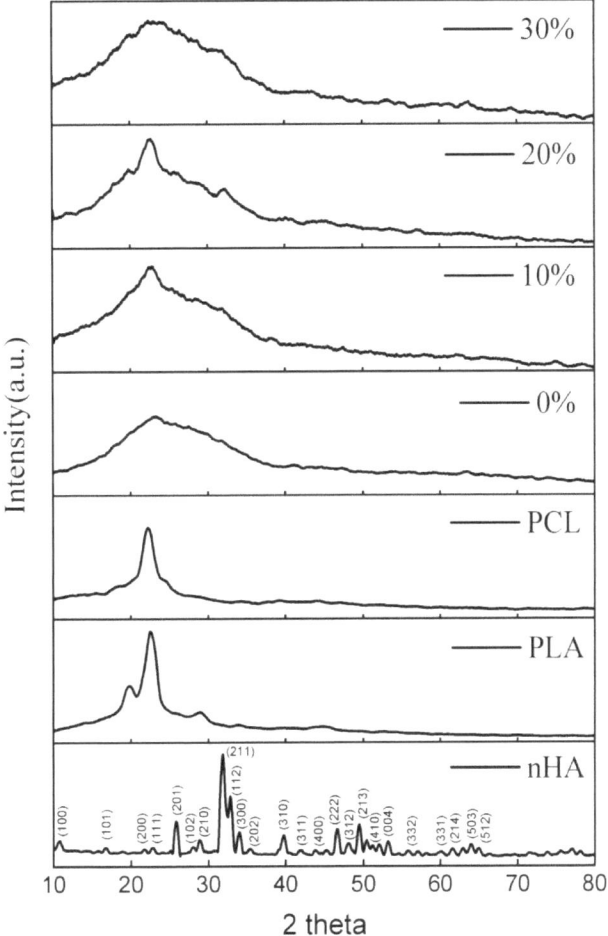

Figure 2. XRD patterns of the nHA, and PLA/PCL/HA biocomposites with various concentrations of nHA.

The lack of crystal peaks in the PLA/PCL blend (0% nHA) suggests that an amorphous structure is generated. When PCL is incorporated into PLA, a disordered structure is formed, making crystallization more difficult [17].

In this investigation, the PLA/PCL/nHA biocomposite exhibited amorphous phase dominance, as evidenced by the broadening of peaks between 2 thetas 10.00° and 40.00° for all nHA content variations (10%, 20%, and 30%). This occurred because the integration of nHA into the PCL and PLA matrix resulted in a less homogeneous structure and essentially no crystalline phase formation. The reason for this is that the addition of HA interferes with the arrangement of the PLA and PCL molecules, preventing crystal formation [57]. In addition, the hydrogen bonds between nHA and PCL/PCL inhibit the orderly arrangement of PCL and PLA molecular chains, thereby reducing crystallinity [58]. This investigation's findings are consistent with those reported by Pires et al. According to their research, the addition of 30 wt% glass prevented polymer matrix crystallization. Consequently,

PCL–bioglass composites have a lower crystallinity than pure PCL. This behavior is the result of the interaction between the filler and matrix interfaces. At high concentrations, bioglass inhibits the movement of polymer molecules. This results in a less crystalline or amorphous polymer [59].

The surface morphology of biocomposite samples has been examined using a scanning electron microscope (SEM). Figure 3 demonstrates that nHA was equally distributed throughout the biocomposite. The nHA (small white particles) is evenly distributed throughout the biocomposite. However, nHA agglomerations of diverse sizes were seen in biocomposites with higher nHA content (Figure 3b,c). This is because the biocomposite sample with a greater nHA content increases the surface energy between PCL and nHA, lowering the interfacial contact between PCL and nHA and causing the agglomeration of nHA particles in the polymer matrix [16,19]. The more nHA added resulted in higher Ca and P peaks on the EDX graph as shown in Figure 4.

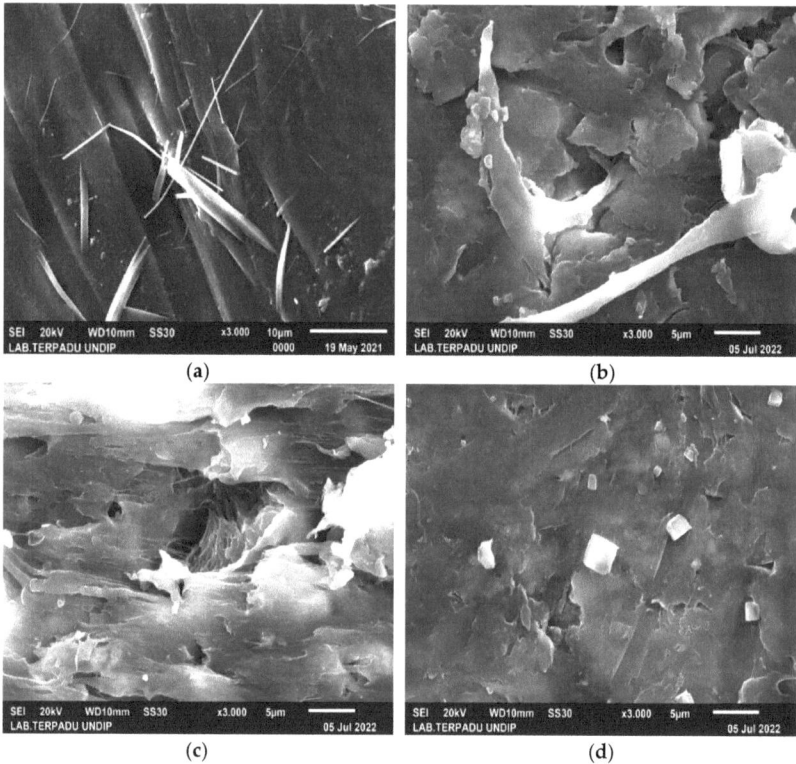

Figure 3. SEM images of PLA/PCL/nHA biocomposite specimens of (**a**) 0%; (**b**) 10%; (**c**) 20%; and (**d**) 30%, at 3000× magnification.

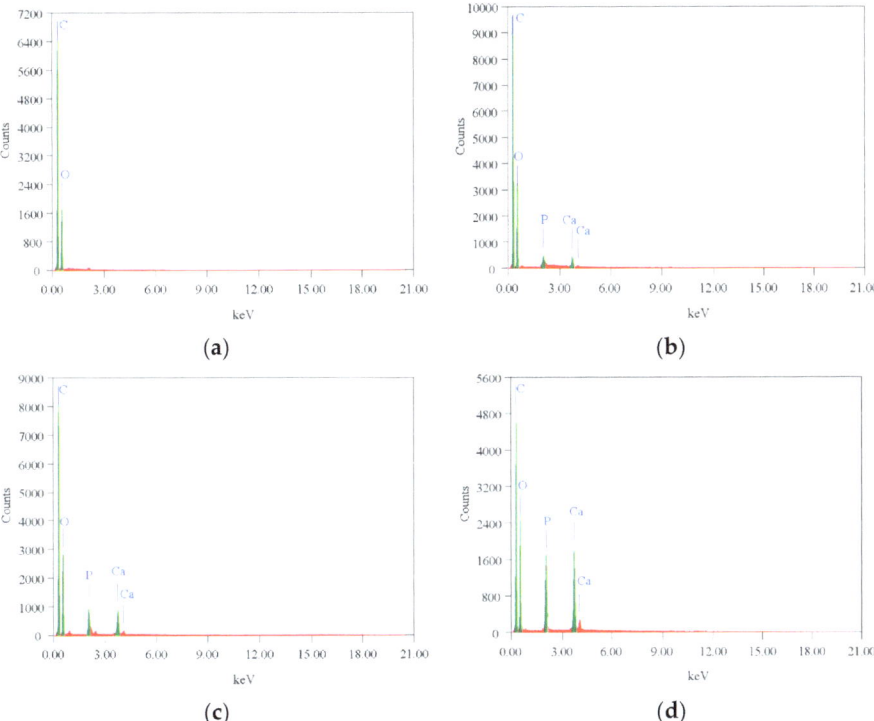

Figure 4. EDX profile of the PLA/PCL/nHA biocomposite specimens of (**a**) 0%; (**b**) 10%; (**c**) 20%; and (**d**) 30%.

However, in specimens with 0% nHA, the EDX graph only found C and O peaks. Aside from this, with more nHA added, the atomic (%) of Ca and P also increased (Table 3). The SEM and EDX test results support the XRD and FTIR test results which proved that nHA had been successfully incorporated into the PLA/PCL/nHA composite.

Table 3. Summary of EDX test results of PLA/PCL/nHA biocomposites.

Element	Atom (%)			
	0%	10%	20%	30%
C	75.73%	68.00%	67.77%	63.91%
O	24.27%	31.00%	30.39%	32.62%
Ca		0.76%	0.97%	1.88%
P		0.24%	0.87%	1.59%
Total	100%	100%	100.00%	100.00%

The studies of Cardoso et al. and Doyle et al. also showed similar results to this study. The agglomeration of nHA particles in the polymer matrix is caused by dispersion problems that occur during the mixing process in the raw materials [60,61]. According to Doyle et al., mixing nHA particles with chloroform can reduce agglomeration and produce a homogeneous dispersion of the nHA particles [50]. The unification of nHA into the polymer matrix not only improves the material's bioactivity but also increases surface roughness, which has the ability to alter cell adhesion and proliferation. Furthermore, nHA addition to the biocomposite material will result in the production of a porous surface. Adequate porosity of appropriate sizes and linkages between pores improves cell infiltration, migration, vascularization, oxygen and nutrient flow, and waste material elimination [46,62].

Figure 5 exhibits the effect of nHA concentration on biocomposite specimen density and porosity. Specimens without nHA had the highest porosity and lowest density of 11.76% and 1.14 g/cm^3, respectively. Adding nHA to the PLA/PCL matrix resulted in lower porosity than the sample without nHA (0%). This is because the nHA within the PLA/PCL matrix is strongly bound and is believed to be involved in the chemical changes that occur during biocomposite fabrication. During the process, nHA is adsorbed into the PLA/PCL matrix to fill voids and increases in density as the porosity of the biocomposite decreases [63]. The findings of this study are consistent with those of Kareem et al. and Kim et al. According to the findings of Kareem et al., adding 10% HA to the PLA matrix induced a decrease in porosity and an increase in density [64]. Meanwhile, Kim et al. found that adding 10% HA to the PCL matrix resulted in lower porosity than pure PCL porosity [65].

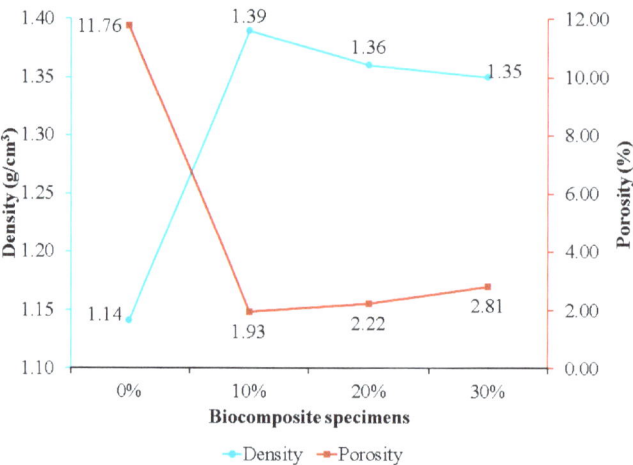

Figure 5. The density and porosity of PLA/PCL/nHA biocomposites.

The porosity of the biocomposite specimens increased as the concentration of nHA used increased. This occurred because the addition of nHA inhibited sintering for the matrix (PLA/PCL), resulting in the formation of pores in the biocomposite specimens [66]. Furthermore, biocomposite specimens containing 30% nHA exhibited greater porosity than biocomposite specimens using 10% and 20% nHA. The increase in porosity in the biocomposite specimens was due to the greater density of nHA compared to the matrix (PLA/PCL). With constant total nHA and matrix quality, the volume decreases as density increases with the addition of more nHA, resulting in the formation of more pores following solid–liquid phase separation [67]. The findings in this study are in accordance with the findings in research conducted by Fang et al. [66] and Casadei et al. [67]. Their research showed that the HA concentration addition increased the porosity of the scaffold biocomposite made of PLLA and HA.

The density of biocomposite specimens decreases as porosity rises [17,68]. The density of biocomposites obtained, formed from PLA, PCL, and nHA, is displayed in Figure 5. Biocomposites with a 0% nHA concentration produced the lowest density, 1.14 g/cm^3, while biocomposites with a 10% nHA concentration produced the highest density, 1.39 g/cm^3. The density of the biocomposite decreases as the nHA content increases from 10% to 30%. This is due to the wettability and clustering of nHA as the reinforcement particles [69]. The biocomposite specimen produced in this research has a density between 1.1 and 1.3 g/cm^3, which is nearly equivalent to that of human cortical bone density [70]. This explanation is consistent with Yousefpour et al.'s. In comparison with their results, increasing the HA concentration decreased the density of the Ce-TZP/Al$_2$O$_3$/HA bio-nanocomposite.

Furthermore, the density of the specimen with low HA is relatively identical to the density values noted in the theory. The increase in HA concentration resulted in the difference between the measured density and the theoretical density increasing [71].

The tensile strength of the PLA/PCL matrix (0% nHA) increased with the addition of nHA (Figure 6). This is due to the density increase that occurs with the addition of nHA. As the density of biocomposites increases, the ultimate tensile strength increases. As the density increases, the compatibility of the nHA matrix with the PLA/PCL improves, and the strength of the composite increases. However, the sample without nHA (0%) has low density due to high porosity. Higher porosity concentrates stress and reduces load-bearing capacity, thus reducing the strength of the material. The higher the compatibility, the more effectively the nHA can transfer stress between the PLA/PCL matrix. Therefore, nHA has a better stress concentration and can withstand higher stresses when stretched or pulled before failure [72]. The tensile strengths of the specimens 0%, 10%, 20%, and 30% were 16.75 MPa, 30.68 MPa, 27.57 MPa, and 26.07 MPa, respectively. The highest tensile strength in this study was found in biocomposite specimens with an nHA concentration of 10%. This occurs as a result of well-dispersed nHA particles, thereby extending the fracture propagation path, absorbing some of the energy, and increasing plastic deformation.

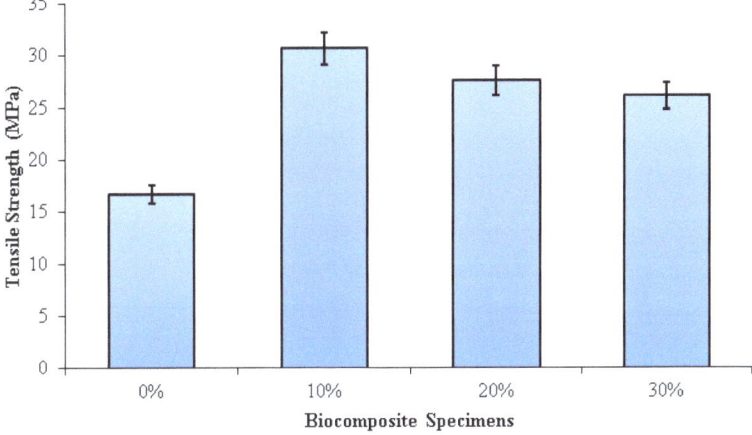

Figure 6. Tensile strength of PLA/PCL/nHA biocomposite specimens.

Consequently, the surface fracture strength and energy of the biocomposite specimen increase. Furthermore, as the concentration of nHA rises, the size of the voids form when the polymer matrix disintegrates from the nHA particles. When this happens, significant cracks begin to occur. In addition, the increased agglomeration of nHA particles resulting from uneven dispersion reduces the biocomposite's strength [73–76]. Aldabib et al. achieved the same findings in their study. Once the HA loading exceeded 5 wt%, the tensile strength declined. The use of a higher concentration of nHA resulted in agglomerations and an uneven distribution of nHA particles in the matrix. More agglomeration results in higher porosity. This phenomenon causes a reduction in the biocomposite's density and tensile strength [73].

Dehestani et al. found that as the content of HA increased, the tensile strength and ductility of iron–hydroxyapatite composites dropped. The higher the HA content, the more unequal the HA particle dispersion in the Fe matrix and the lower the tensile strength of the biocomposite. The magnitude of the decrease in mechanical properties obviously depends not only on the content of HA but also on the distribution of HA dispersed in the

Fe matrix and the particle size of the HA used [74]. According to research conducted by Ma et al., increasing the amount of HA from 0% to 40% increased the elastic modulus and decreased the tensile strength.

When the HA concentration is 30 percent or less, the tensile strength of the HA/PEEK composite is greater than that of cortical bone (50 MPa). The 40% HA/PEEK composite's tensile strength, however, was less than 50 MPa, rendering it incompatible with cortical bone [61]. According to research by Kang et al., the elastic modulus of the composite increased from 2.36 GPa to 2.79 GPa as the HA content increased from 10% to 30%, while the tensile strength decreased from 95 MPa to 74 MPa. The homogeneity of various particle sizes and dispersions has a significant impact on mechanical properties. The 10% HA concentration improved dispersion in the PEEK matrix and enhanced the composite's tensile strength [76].

Figure 7 depicts the three-point bending test results on PLA/PCL/nHA biocomposites. The flexural strength of the PLA/PCL matrix was significantly enhanced by the addition of nHA (10%, 20%, and 30%). Increased adhesion between the PLA/PCL matrix and nHA results in greater stress transfer from the polymer matrix to nHA. The presence of nHA in the PLA/PCL blend facilitates the formation of a more rigid bond, which contributes to the improvement of flexural strength.

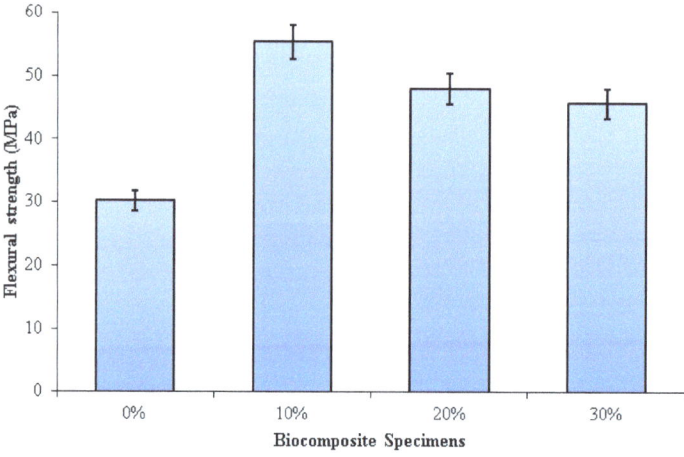

Figure 7. Flexural strength of PLA/PCL/nHA biocomposite specimens.

Flexural strengths of 0%, 10%, 20%, and 30% of specimens were 30.21 MPa, 55.35 MPa, 47.99 MPa, and 45.67 MPa, respectively. The highest flexural strength (55.35 MPa) was obtained in biocomposite specimens containing 10% nHA. The flexural strength of the biocomposite specimens decreases as the nHA concentration increases from 10% to 30%. A rise in nHA concentration results in significant nHA agglomeration in the matrix. Consequently, a propagating fracture may form as a result of stress concentration, which subsequently readily results in brittle failure. Furthermore, the addition of nHA with a concentration of 10% gave a suitable stiff phase in the matrix made of a mixture of PCL and PLA. The use of nHA with the right concentration can limit deformation and mobility in the matrix, resulting in high flexural strength [77].

Aldabib et al. explained that the homogeneous distribution of nHA particles within the biocomposite can be linked to the increased flexural strength of the biocomposite specimen at lower nHA concentrations. The flexural strength increases as the dispersion of nHA particles improves [73]. As previously explained, a higher concentration of nHA used results in an increase in the porosity of the biocomposite specimen. Due to this, mechanical properties such as tensile strength and flexural strength diminish as well as density. Thus, a higher concentration of nHA used makes a decrease in the flexural strength of the biocomposite specimen. Based on the study of Yadav et al., the cause of the decrease in flexural strength in dental restorative composite specimens with an inclination in nHAPs filler could be due to the presence of harder and stiffer ceramic particles resulting in brittle composites.

Furthermore, the higher concentration of nHA resulted in the formation of a lot of pore content in the composite specimen [78]. Comparably similar findings were reported by Nawang et al. It was discovered that increasing the amount of filler decreased flexural strength [79]. The liquid absorption and the amount of contraction stress decreased as the hydroxyapatite added to the polymer matrix was increased. Thus, the resulting bond strength and flexural strength of the composite decreased [80].

The investigation findings are aligned with Ferri et al.'s research. Their findings show that the higher the concentration of HA used, the lower the flexural strength of the PLA/HA composite. Flexural strength is reduced as a result of the biocomposites' greater stiffness produced by the increased HA concentration. Furthermore, when HA concentration increases, particle aggregation becomes more possible, and the nucleating effect becomes less evident. As a result, the cracking probability is increased because HA aggregates perform as crack initiators [81]. According to Shyang et al. [67] and Bilic-Prcic et al. [68], an increase in hydroxyapatite concentrations results in a decrease in flexural strength.

4. Conclusions

The PLA/PCL/nHA biocomposites were successfully prepared via the cold isostatic pressing method. The biocomposites obtained via isostatic cold pressing, PCL, PLA, and nHA bands, were well identified using the FTIR test. The different band intensities are affected by the concentration of nHA used. The biocomposites with high nHA content showed intense bands at 1047 and 551 cm^{-1}, which were associated with a symmetric stretching vibration of PO_4^{3-} and asymmetric bending vibration of PO_4^{3-}, respectively. The PLA/PCL/nHA biocomposites lack crystal peaks, indicating that the resulting structure is amorphous. A broadening of the peak between 2 thetas of 10.00° and 40.00° on the XRD diffractogram indicated an increase in the distance between PLA layers. The spacing between PLA layers increased, indicating a more amorphous phase in these PLA/PCL/nHA biocomposites, whereas the addition of nHA led to a less uniform structure and made crystallization more difficult. The nHA (small white particles) was evenly distributed throughout the biocomposite. However, nHA agglomerations of diverse sizes were seen in biocomposites with higher nHA content.

The addition of nHA improves the density and tensile and flexural strength of the PLA/PCL matrix (0% nHA). However, by increasing nHA content, the PLA/PCL/nHA biocomposites became more porous. In addition, the density of the PLA/PCL/nHA biocomposites decreased linearly as the nHA concentration increased. The PLA/PCL/nHA biocomposites with 10 wt% nHA exhibited the highest density (1.39 g/cm^3) and the smallest porosity (1.93%).

The flexural and tensile strength of the PLA/PCL/nHA biocomposite decreased with decreasing density. The tensile strength of biocomposite specimens decreased when the nHA concentration exceeded 10 wt% because the nHA particles were not well dispersed. Furthermore, the increase in the concentrations of nHA led to a decrease in flexural strength. Flexural strength is reduced as a result of the biocomposites' greater stiffness produced by the increased nHA concentration. The highest tensile and flexural strength were found in PLA/PCL/nHA biocomposites with 10 wt% nHA, with a tensile and flexural strength of 30.68 MPa and 55.35 MPa, respectively.

Author Contributions: S.S.: writing—original draft, investigation; A.S.: project administration, funding acquisition; S.A.W.: writing—review and editing, funding acquisition; J.T.: methodology, resources; D.F.F.: formal analysis, investigation; J.P.S.: supervision, formal analysis; T.C.: investigation, formal analysis. All authors have read and agreed to the published version of the manuscript.

Funding: This study was funded by the Universitas Muhammadiyah Semarang (UNIMUS), Central Java, Indonesia, with Grant No. 2077/UNIMUS/KP/2021.

Institutional Review Board Statement: Not applicable.

Data Availability Statement: Data are contained within the article.

Acknowledgments: The authors would like to express their gratitude to the Universitas Muhammadiyah Semarang (UNIMUS) for the research grant in the year 2021. The material characterization was partially supported by a research fund from the Department of Mechanical Engineering, Faculty of Engineering, Diponegoro University, 2022.

Conflicts of Interest: The authors declare no conflict of interest.

References

1. Cho, Y.S.; Gwak, S.-J.; Cho, Y.-S. Fabrication of Polycaprolactone/Nano Hydroxyapatite (PCL/nHA) 3D Scaffold with Enhanced In Vitro Cell Response via Design for Additive Manufacturing (DfAM). *Polymers* **2021**, *13*, 1394. [CrossRef] [PubMed]
2. Bernardo, M.P.; da Silva, B.C.R.; Hamouda, A.E.I.; de Toledo, M.A.S.; Schalla, C.; Rütten, S.; Goetzke, R.; Mattoso, L.H.C.; Zenke, M.; Sechi, A. PLA/Hydroxyapatite scaffolds exhibit in vitro immunological inertness and promote robust osteogenic differentiation of human mesenchymal stem cells without osteogenic stimuli. *Sci. Rep.* **2022**, *12*, 2333. [CrossRef] [PubMed]
3. Prasadh, S.; Raguraman, S.; Wong, R.; Gupta, M. Current Status and Outlook of Temporary Implants (Magnesium/Zinc) in Cardiovascular Applications. *Metals* **2022**, *12*, 999. [CrossRef]
4. Mosas, K.K.A.; Chandrasekar, A.R.; Dasan, A.; Pakseresht, A.; Galusek, D. Recent Advancements in Materials and Coatings for Biomedical Implants. *Gels* **2022**, *8*, 323. [CrossRef]
5. Pisecky, L.; Luger, M.; Klasan, A.; Gotterbarm, T.; Klotz, M.C.; Hochgatterer, R. Bioabsorbable implants in forefoot surgery: A review of materials, possibilities and disadvantages. *EFORT Open Rev.* **2021**, *6*, 1132–1139. [CrossRef]
6. Katti, K.S.; Verma, D.; Katti, D.R. 4—Materials for joint replacement. In *Woodhead Publishing Series in Biomaterials*; Revell, P.A., Ed.; Woodhead Publishing: Sawston, UK, 2008; pp. 81–104. ISBN 978-1-84569-245-2.
7. Beig, B.; Liaqat, U.; Douna, I.; Zahoor, M. Coatings Current Challenges and Innovative Developments in Hydroxyapatite-Based Coatings on Metallic Materials for Bone Implantation: A Review. *Coatings* **2020**, *10*, 1249. [CrossRef]
8. Davis, R.; Singh, A.; Jackson, M.J.; Coelho, R.T.; Prakash, D.; Charalambous, C.P.; Ahmed, W.; da Silva, L.R.R.; Lawrence, A.A. A comprehensive review on metallic implant biomaterials and their subtractive manufacturing. *Int. J. Adv. Manuf. Technol.* **2022**, *120*, 1473–1530. [CrossRef]
9. Priyadarshini, B.; Rama, M.; Chetan; Vijayalakshmi, U. Bioactive coating as a surface modification technique for biocompatible metallic implants: A review. *J. Asian Ceram. Soc.* **2019**, *7*, 397–406. [CrossRef]
10. Darsan, R.S.; Retnam, B.S.J.; Sivapragash, M. Material Characteristic Study and Fabrication of Hydroxyapatite (HA) with Poly (Lactide/Lactic) Acids (PLA) for Orthopaedic Implants Department of Automobile Engineering, Noorul Islam Center for Higher Education. *Middle-East J. Sci. Res.* **2017**, *25*, 1491–1500. [CrossRef]
11. Ma'ruf, M.T.; Siswomihardjo, W.; Soesatyo, M.H.N.E.; Tontowi, A. Effect of glutaraldehyde as a crosslinker on mechanical characteristics of catgut reinforced polyvinyl alcohol-hydroxyapatite composite as bone-fracture fixation material. *ARPN J. Eng. Appl. Sci.* **2015**, *10*, 6359–6364.
12. Washington, K.E.; Kularatne, R.N.; Karmegam, V.; Biewer, M.C.; Stefan, M.C. Recent advances in aliphatic polyesters for drug delivery applications. *Wiley Interdiscip. Rev. Nanomed. Nanobiotechnol.* **2017**, *9*, e1446. [CrossRef]
13. Moura, N.K.; Siqueira, I.A.W.B.; Machado, J.P.; Kido, H.W.; Avanzi, I.R.; Rennó, A.C.; Trichês, E.D.; Passador, F.R. Production and Characterization of Porous Polymeric Membranes of PLA/PCL Blends with the Addition of Hydroxyapatite. *J. Compos. Sci.* **2019**, *3*, 45. [CrossRef]
14. Pratama, J.; Cahyono, S.I.; Suyitno, S.; Muflikhun, M.A.; Salim, U.A.; Mahardika, M.; Arifvianto, B. A Review on Reinforcement Methods for Polymeric Materials Processed Using Fused Filament Fabrication (FFF). *Polymers* **2021**, *13*, 4022. [CrossRef]
15. Torres, E.; Dominguez-Candela, I.; Castello-Palacios, S.; Vallés-Lluch, A.; Fombuena, V. Development and characterization of polyester and acrylate-based composites with hydroxyapatite and halloysite nanotubes for medical applications. *Polymers* **2020**, *12*, 1703. [CrossRef]
16. Pitjamit, S.; Thunsiri, K.; Nakkiew, W.; Wongwichai, T. The Possibility of Interlocking Nail Fabrication from FFF 3D Printing PLA/PCL/HA Composites Coated by Local Silk Fibroin for Canine Bone Fracture Treatment. *Materials* **2020**, *13*, 1564. [CrossRef]
17. Solechan, S.; Suprihanto, A.; Widyanto, S.A.; Triyono, J.; Fitriyana, D.F.; Siregar, J.P.; Cionita, T. Investigating the Effect of PCL Concentrations on the Characterization of PLA Polymeric Blends for Biomedical Applications. *Materials* **2022**, *15*, 7396. [CrossRef]
18. Zimmerling, A.; Yazdanpanah, Z.; Cooper, D.M.L.; Johnston, J.D.; Chen, X. 3D printing PCL/nHA bone scaffolds: Exploring the influence of material synthesis techniques. *Biomater. Res.* **2021**, *25*, 3. [CrossRef]
19. Hassanajili, S.; Pour, A.; Oryan, A.; Talaei-Khozani, T. Preparation and characterization of PLA/PCL/HA composite scaffolds using indirect 3D printing for bone tissue engineering. *Mater. Sci. Eng. C* **2019**, *104*, 109960. [CrossRef]

20. Fitriyana, D.F.; Nugraha, F.W.; Laroybafih, M.B.; Ismail, R.; Bayuseno, A.P.; Muhamadin, R.C.; Ramadan, M.B.; Qudus, A.R.A.; Siregar, J.P. The effect of hydroxyapatite concentration on the mechanical properties and degradation rate of biocomposite for biomedical applications. *IOP Conf. Ser. Earth Environ. Sci.* **2022**, *969*, 12045. [CrossRef]
21. Fitriyana, D.F.; Ismail, R.; Santosa, Y.I.; Nugroho, S.; Hakim, A.J.; Syahreza Al Mulqi, M. Hydroxyapatite Synthesis from Clam Shell Using Hydrothermal Method: A Review. In Proceedings of the 2019 International Biomedical Instrumentation and Technology Conference (IBITeC), Yogyakarta, Indonesia, 23–24 October 2019; pp. 7–11. [CrossRef]
22. Pokhrel, S. Hydroxyapatite: Preparation, Properties and Its Biomedical Applications. *Adv. Chem. Eng. Sci.* **2018**, *08*, 225–240. [CrossRef]
23. Fernando, S.; McEnery, M.; Guelcher, S.A. 16—Polyurethanes for bone tissue engineering. In *Advances in Polyurethane Biomaterials*; Cooper, S.L., Guan, J., Eds.; Woodhead Publishing: Sawston, UK, 2016; pp. 481–501. ISBN 978-0-08-100614-6.
24. Wu, Y.; Lu, Y.; Zhao, M.; Bosiakov, S.; Li, L. A Critical Review of Additive Manufacturing Techniques and Associated Biomaterials Used in Bone Tissue Engineering. *Polymers* **2022**, *14*, 2117. [CrossRef] [PubMed]
25. Vranić, A.; Bogojević, N.; Ćirić-Kostić, S.; Croccolo, D.; Olmi, G. Advantages and drawbacks of additive manufacturing. *IMK-14 Istraz. Razvoj* **2017**, *23*, 57–62. [CrossRef]
26. Hasanov, S.; Alkunte, S.; Rajeshirke, M.; Gupta, A.; Huseynov, O.; Fidan, I.; Alifui-Segbaya, F.; Rennie, A. Review on Additive Manufacturing of Multi-Material Parts: Progress and Challenges. *J. Manuf. Mater. Process.* **2022**, *6*, 4. [CrossRef]
27. Snead, L.L.; Hoelzer, D.T.; Rieth, M.; Nemith, A.A.N. Chapter 13—Refractory Alloys: Vanadium, Niobium, Molybdenum, Tungsten. In *Structural Alloys for Nuclear Energy Applications*; Odette, G.R., Zinkle, S., Eds.; Elsevier: Amsterdam, The Netherlands, 2019; pp. 585–640. ISBN 978-0-12-397046-6.
28. Morris, K.J. Cold Isostatic Pressing. In *Concise Encyclopedia of Advanced Ceramic Materials*; Brook, R.J., Ed.; Pergamon: Oxford, UK, 1991; pp. 84–88. ISBN 978-0-08-034720-2.
29. Ruys, A. Processing, Structure, and Properties. In *Alumina Ceramics: Biomedical and Clinical Applications: Woodhead Publishing Series in Biomaterials*; Woodhead Publishing: Sawston, UK, 2019; Chapter 4; pp. 71–121. ISBN 9780081024423.
30. Abdallah, A.; Fayed, A.; Abdo, G.; Sallam, M. Effect of Cold Isostatic Pressing On The Physical and Mechanical Properties of Tungsten Heavy Alloys. *J. Eng. Sci. Mil. Technol.* **2017**, *17*, 1–12. [CrossRef]
31. Peng, Y.; Liu, J.Z.; Wang, K.; Cheng, Y.-B. Influence of Parameters of Cold Isostatic Pressing on TiO_2 Films for Flexible Dye-Sensitized Solar Cells. *Int. J. Photoenergy* **2011**, *2011*, 410352. [CrossRef]
32. Al Bakri, A.M.M.; Ahmad Fauzi, M.N.; Kamarudin, H.; Norazian, M.N.; Salleh, M.A.A.; Alida, A. Mechanical properties of ZTA composite using cold isostatic pressing and uniaxial pressing. *Adv. Mater. Res.* **2013**, *740*, 728–733. [CrossRef]
33. Prilipko, S.Y.; Timchenko, V.M.; Akimov, G.Y.; Tkach, V.I. Effect of cold isostatic pressing on the synthesis and particle size of lanthanum manganate. *Powder Metall. Met. Ceram.* **2008**, *47*, 284–287. [CrossRef]
34. Akimov, G.Y.; Soloviova, T.A.; Loboda, P.I.; Prilipko, S.Y. Mechanical activation of crystallization of amorphous boron and synthesis of Al3Ti under cold isostatic pressing of B–Al–(LaB6–TiB2) powder. *J. Superhard Mater.* **2017**, *39*, 343–348. [CrossRef]
35. Ramesh, S.; Christopher, P.; Tan, C.Y.; Teng, W.D. The effect of cold isostatic pressing on the sinterability of synthesized ha. *Biomed. Eng. Appl. Basis Commun.* **2004**, *16*, 199–204. [CrossRef]
36. Attia, U.M. Cold-isostatic pressing of metal powders: A review of the technology and recent developments. *Crit. Rev. Solid State Mater. Sci.* **2021**, *46*, 587–610. [CrossRef]
37. Cuesta, I.I.; Martínez-Pañeda, E.; Díaz, A.; Alegre, J.M. Cold isostatic pressing to improve the mechanical performance of additively manufactured metallic components. *Materials* **2019**, *12*, 2495. [CrossRef]
38. *Nano-Hydroxyapatite Safety Data Sheet (SDS) No 702153*; Sigma-Aldrich Pte Ltd.: Singapore, 2022.
39. 39. Indra, A.; Putra, A.B.; Handra, N.; Fahmi, H.; Nurzal; Asfarizal; Perdana, M.; Anrinal; Subardi, A.; Affi, J.; et al. Behavior of sintered body properties of hydroxyapatite ceramics: Effect of uniaxial pressure on green body fabrication. *Mater. Today Sustain.* **2022**, *17*, 100100. [CrossRef]
40. Younesi, M.; Bahrololoom, M.E. Effect of temperature and pressure of hot pressing on the mechanical properties of PP–HA bio-composites. *Mater. Des.* **2009**, *30*, 3482–3488. [CrossRef]
41. Ismail, R.; Fitriyana, D.F.; Santosa, Y.I.; Nugroho, S.; Hakim, A.J.; Al Mulqi, M.S.; Jamari, J.; Bayuseno, A.P. The potential use of green mussel (Perna Viridis) shells for synthetic calcium carbonate polymorphs in biomaterials. *J. Cryst. Growth* **2021**, *572*, 126282. [CrossRef]
42. Ismail, R.; Cionita, T.; Shing, W.L.; Fitriyana, D.F.; Siregar, J.P.; Bayuseno, A.P.; Nugraha, F.W.; Muhamadin, R.C.; Junid, R.; Endot, N.A. Synthesis and Characterization of Calcium Carbonate Obtained from Green Mussel and Crab Shells as a Biomaterials Candidate. *Materials* **2022**, *15*, 5712. [CrossRef]
43. Fitriyana, D.; Suhaimi, H.; Sulardjaka, S.; Noferi, R.; Caesarendra, W. Synthesis of Na-P Zeolite from Geothermal Sludge. In *NAC 2019: Proceedings of the 2nd International Conference on Nanomaterials and Advanced Composites*; Springer Proceedings in Physics; Springer: Singapore, 2020; pp. 51–59. ISBN 978-981-15-2293-2.
44. Nanda, B.P.; Satapathy, A. Processing and characterization of epoxy composites reinforced with short human hair. *J. Phys. Conf. Ser.* **2017**, *178*, 012012. [CrossRef]
45. Taib, M.N.A.M.; Julkapli, N.M. *Dimensional Stability of Natural Fiber-Based and Hybrid Composites*; Elsevier Ltd.: Amsterdam, The Netherlands, 2018; ISBN 9780081022924.

46. Åkerlund, E.; Diez-escudero, A.; Grzeszczak, A. The Effect of PCL Addition on 3D-Printable PLA / HA Composite Filaments for the Treatment of Bone Defects. *Polymers* **2022**, *14*, 3305. [CrossRef]
47. Shojaei, S.; Nikuei, M.; Goodarzi, V.; Hakani, M.; Khonakdar, H.A.; Saeb, M.R. Disclosing the role of surface and bulk erosion on the viscoelastic behavior of biodegradable poly(ε-caprolactone)/poly(lactic acid)/hydroxyapatite nanocomposites. *J. Appl. Polym. Sci.* **2019**, *136*, 47151. [CrossRef]
48. Mushtaq, A.; Mukhtar, H.; Shariff, A.M. FTIR study of enhanced polymeric blend membrane with amines. *Res. J. Appl. Sci. Eng. Technol.* **2014**, *7*, 1811–1820. [CrossRef]
49. Chukanov, N.V.; Chervonnyi, A.D. Some General Aspects of the Application of IR Spectroscopy to the Investigation of Minerals. In *Springer Mineralogy*; Springer: Berlin/Heidelberg, Germany, 2016; pp. 1–49. ISBN 9783319253497.
50. Kędzierska-Matysek, M.; Matwijczuk, A.; Florek, M.; Barłowska, J.; Wolanciuk, A.; Matwijczuk, A.; Chruściel, E.; Walkowiak, R.; Karcz, D.; Gładyszewska, B. Application of FTIR spectroscopy for analysis of the quality of honey. *BIO Web Conf.* **2018**, *10*, 02008. [CrossRef]
51. Wahab, R.; Mustafa, M.T.; Fauzi, N.; Samsi, H. Thermal Degradation Analysis on 4-year-old Culms of Cultivated Tropical Bamboo Bambusa Vulgaris. *J. Agric. Stud.* **2017**, *5*, 50. [CrossRef]
52. Ferati, F. Structural Information from Ratio Bands in the FTIR Spectra of Long Chain and Branched Alkanes in Petrodiesel Samples. *J. Environ. Treat. Tech.* **2020**, *8*, 1140–1143.
53. Bernardo, M.P.; da Silva, B.C.R.; Mattoso, L.H.C. Development of three-dimensional printing filaments based on poly(lactic acid)/hydroxyapatite composites with potential for tissue engineering. *J. Compos. Mater.* **2021**, *55*, 2289–2300. [CrossRef]
54. Herliansyah, M.K.; Hamdi, M.; Ide-Ektessabi, A.; Wildan, M.W.; Toque, J.A. The influence of sintering temperature on the properties of compacted bovine hydroxyapatite. *Mater. Sci. Eng. C* **2009**, *29*, 1674–1680. [CrossRef]
55. Sun, H.; Yu, B.; Han, J.; Kong, J.; Meng, L.; Zhu, F. Microstructure, thermal properties and rheological behavior of PLA/PCL blends for melt-blown nonwovens. *Polymer* **2014**, *38*, 477–483. [CrossRef]
56. Hasan, A.; Soliman, S.; El Hajj, F.; Tseng, Y.T.; Yalcin, H.C.; Marei, H.E. Fabrication and in Vitro Characterization of a Tissue Engineered PCL-PLLA Heart Valve. *Sci. Rep.* **2018**, *8*, 8187. [CrossRef]
57. Jing, X.; Mi, H.-Y.; Turng, L.-S. Comparison between PCL/hydroxyapatite (HA) and PCL/halloysite nanotube (HNT) composite scaffolds prepared by co-extrusion and gas foaming. *Mater. Sci. Eng. C* **2017**, *72*, 53–61. [CrossRef]
58. Li, Y.; Yu, Z.; Ai, F.; Wu, C.; Zhou, K.; Cao, C.; Li, W. Characterization and evaluation of polycaprolactone/hydroxyapatite composite scaffolds with extra surface morphology by cryogenic printing for bone tissue engineering. *Mater. Des.* **2021**, *205*, 109712. [CrossRef]
59. Pires, L.S.O.; Fernandes, M.H.F.V.; de Oliveira, J.M.M. Crystallization kinetics of PCL and PCL–glass composites for additive manufacturing. *J. Therm. Anal. Calorim.* **2018**, *134*, 2115–2125. [CrossRef]
60. Doyle, S.E.; Henry, L.; McGennisken, E.; Onofrillo, C.; Di Bella, C.; Duchi, S.; O'Connell, C.D.; Pirogova, E. Characterization of polycaprolactone nanohydroxyapatite composites with tunable degradability suitable for indirect printing. *Polymers* **2021**, *13*, 295. [CrossRef]
61. Cardoso, G.B.C.; Ramos, S.L.F.; Rodas, A.C.D.; Higa, O.Z.; Zavaglia, C.A.C.; Arruda, A.C.F. Scaffolds of poly (ε-caprolactone) with whiskers of hydroxyapatite. *J. Mater. Sci.* **2010**, *45*, 4990–4993. [CrossRef]
62. Abbasi, N.; Hamlet, S.; Love, R.M.; Nguyen, N.-T. Porous scaffolds for bone regeneration. *J. Sci. Adv. Mater. Devices* **2020**, *5*, 1–9. [CrossRef]
63. Wulandari, E.; Wardani, F.R.A.; Fatimattuzahro, N.; Dewanti, I.D.A.R. Addition of gourami (Osphronemus goramy) fish scale powder on porosity of glass ionomer cement. *Dent. J.* **2022**, *55*, 33–37. [CrossRef]
64. Kareem, M.M.; Tanner, K.E. Optimising micro-hydroxyapatite reinforced poly(lactide acid) electrospun scaffolds for bone tissue engineering. *J. Mater. Sci. Mater. Med.* **2020**, *31*, 38. [CrossRef]
65. Kim, J.W.; Shin, K.H.; Koh, Y.H.; Hah, M.J.; Moon, J.; Kim, H.E. Production of poly(ε-caprolactone)/hydroxyapatite composite scaffolds with a tailored macro/micro-porous structure, high mechanical properties, and excellent bioactivity. *Materials* **2017**, *10*, 1123. [CrossRef]
66. Fang, Z.; Feng, Q. Improved mechanical properties of hydroxyapatite whisker-reinforced poly(l-lactic acid) scaffold by surface modification of hydroxyapatite. *Mater. Sci. Eng. C* **2014**, *35*, 190–194. [CrossRef]
67. Casadei, A.P.M.; Dingee, F.; da Silva, T.E.; Prette, A.L.G.; Rambo, C.R.; Fredel, M.C.; Duek, E.A.R. Manufacturing of Porous PPLA-HA Composite Scaffolds by Sintering for Bone Tissue Engineering. In Proceedings of the the 33rd International Conference on Advanced Ceramics and Composites, Daytona Beach, FL, USA, 18–23 January 2009; pp. 169–177. [CrossRef]
68. Liu, H.; Luo, G.; Wei, H.; Yu, H. Strength, Permeability, and Freeze-Thaw Durability of Pervious Concrete with Different Aggregate Sizes, Porosities, and Water-Binder Ratios. *Appl. Sci.* **2018**, *8*, 1217. [CrossRef]
69. Ünal, T.G.; Diler, E.A. Properties of AlSi9Cu3 metal matrix micro and nano composites produced via stir casting. *Open Chem.* **2018**, *16*, 726–731. [CrossRef]
70. Kroemer, K.H.E.; Kroemer, H.J.; Kroemer-Elbert, K.E. *Engineering Physiology: Bases of Human Factors Engineering/Ergonomics*, 4th ed.; Springer: Berlin/Heidelberg, Germany, 2010.
71. Yousefpour, M.; Askari, N.; Abdollah-Pour, H.; Amanzadeh, A.; RIAHI, N. Investigation on biological properties of dental implant by Ce-TZP/Al2O3/ha bio-nano-composites. *Dig. J. Nanomater. Biostruct.* **2011**, *6*, 675–681. [CrossRef]

72. Chiang, T.C.; Hamdan, S.; Osman, M.S. Effects of density of sago/urea formaldehyde particleboard towards its thermal stability, mechanical and physical properties. *J. Teknol.* **2016**, *78*, 187–197. [CrossRef]
73. Aldabib, J.M.; Ishak, Z.A.M. Effect of hydroxyapatite filler concentration on mechanical properties of poly (methyl methacrylate) denture base. *SN Appl. Sci.* **2020**, *2*, 732. [CrossRef]
74. Dehestani, M.; Adolfsson, E.; Stanciu, L.A. Mechanical properties and corrosion behavior of powder metallurgy iron-hydroxyapatite composites for biodegradable implant applications. *Mater. Des.* **2016**, *109*, 556–569. [CrossRef]
75. Ma, R.; Guo, D. Evaluating the bioactivity of a hydroxyapatite-incorporated polyetheretherketone biocomposite. *J. Orthop. Surg. Res.* **2019**, *14*, 23. [CrossRef] [PubMed]
76. Kang, J.; Zheng, J.; Hui, Y.; Li, D. Mechanical Properties of 3D-Printed PEEK/HA Composite Filaments. *Polymers* **2022**, *14*, 4293. [CrossRef]
77. Verma, N.; Zafar, S.; Talha, M. Influence of nano-hydroxyapatite on mechanical behavior of microwave processed polycaprolactone composite foams. *Mater. Res. Express* **2019**, *6*, 085336. [CrossRef]
78. Yadav, S.; Gangwar, S. The effectiveness of functionalized nano-hydroxyapatite filler on the physical and mechanical properties of novel dental restorative composite. *Int. J. Polym. Mater. Polym. Biomater.* **2020**, *69*, 907–918. [CrossRef]
79. Nawang, R.; Hussein, M.Z.; Matori, K.A.; Abdullah, C.A.C.; Hashim, M. Physicochemical properties of hydroxyapatite/montmorillonite nanocomposite prepared by powder sintering. *Results Phys.* **2019**, *15*, 102540. [CrossRef]
80. Hapsari, D.N.; Wardani, S.C.; Firdausya, W.A.; Amaturrohman, K.; Wiratama, H.P. The effect of addition of hydroxyapatite from skipjack tuna (katsuwonus pelamis) fish bone flour to the transverse, impact, and tensile strength of heat cured acrylic resin. *J. Dentomaxillofac. Sci.* **2020**, *5*, 94. [CrossRef]
81. Ferri, J.M.; Jordá, J.; Montanes, N.; Fenollar, O.; Balart, R. Manufacturing and characterization of poly(lactic acid) composites with hydroxyapatite. *J. Thermoplast. Compos. Mater.* **2018**, *31*, 865–881. [CrossRef]

Disclaimer/Publisher's Note: The statements, opinions and data contained in all publications are solely those of the individual author(s) and contributor(s) and not of MDPI and/or the editor(s). MDPI and/or the editor(s) disclaim responsibility for any injury to people or property resulting from any ideas, methods, instructions or products referred to in the content.

Communication

One-Step Low Temperature Synthesis of CeO$_2$ Nanoparticles Stabilized by Carboxymethylcellulose

Vasily V. Spiridonov [1,*], Andrey V. Sybachin [1], Vladislava A. Pigareva [1], Mikhail I. Afanasov [1], Sharifjon A. Musoev [2], Alexander V. Knotko [2] and Sergey B. Zezin [1]

[1] Department of Chemistry, Lomonosov Moscow State University, Leninskie Gory 1-3, 119991 Moscow, Russia; sybatchin@mail.ru (A.V.S.)
[2] Faculty of Materials Science, Lomonosov Moscow State University, Leninskie Gory 1-73, 119991 Moscow, Russia
* Correspondence: vasya_spiridonov@mail.ru

Abstract: An elegant method of one-pot reaction at room temperature for the synthesis of nanocomposites consisting of cerium containing nanoparticles stabilized by carboxymethyl cellulose (CMC) macromolecules was introduced. The characterization of the nanocomposites was carried out with a combination of microscopy, XRD, and IR spectroscopy analysis. The type of crystal structure of inorganic nanoparticles corresponding to CeO$_2$ was determined and the mechanism of nanoparticle formation was suggested. It was demonstrated that the size and shape of the nanoparticles in the resulting nanocomposites does not depend on the ratio of the initial reagents. Spherical particles with a mean diameter 2–3 nm of were obtained in different reaction mixtures with a mass fraction of cerium from 6.4 to 14.1%. The scheme of the dual stabilization of CeO$_2$ nanoparticles with carboxylate and hydroxyl groups of CMC was proposed. These findings demonstrate that the suggested easily reproducible technique is promising for the large-scale development of nanoceria-containing materials.

Keywords: carboxymethyl cellulose; cerium oxide; nanocomposite; cerium nanoparticles; stabilization; microscopy

1. Introduction

Cerium oxide nanoparticles (nanoceria) are of interest in a wide range of applications, especially in the field of biomedicine. Many chemical methods for the synthesis of nanoceria are reported in the literature. Precipitation methods such as co-precipitation, chemical precipitation, microwave, sonochemical, hydrothermal, reverse-co-precipitation, and microwave–hydrothermal methods are basic procedures of synthesis [1–14]. The key disadvantage of these methods for obtaining cerium-containing nanoparticles is the requirement to conduct the reactions at high temperatures. That, in turn, leads to the formation of particles of arbitrary sizes and shapes. As a result, no control of the geometric and morphological parameters of nanoparticles can be achieved. Moreover, the problem of stabilization of cerium-containing nanoparticles from aggregation and precipitation also remains relevant. The most commonly used approach to stabilize cerium oxide nanoparticles is modification of their surfaces with surfactants [15]. A significant drawback of this approach is the impossibility of complete purification of the formed nanocomposites from surfactants and, as a result, limitation of the use of such materials for biomedical purposes. Eco-friendly examples of green synthesis of nanoceria were reported as well, for example, the synthesis of nanoceria using nutrient *Salvia macrosiphon Boiss* seeds [16]. However, such approaches did not find wide distribution.

A high degree of biocompatibility, low toxicity, and catalytic activity of nanodispersed cerium dioxide make it possible to consider it as a promising material for biomedical applications as antioxidant, anticancer, and antibacterial agents [17–19]. That is why cerium oxide nanoparticles, which have biomimetic and antioxidant activity, are now increasingly being used in medicine.

The influence of nanocrystalline cerium dioxide in the protection of living cells from oxidative stress is of particular interest. The uniqueness of cerium dioxide nanoparticles is due to the fact that they can exist in different oxidation states, Ce^{3+} and Ce^{4+}, which distinguishes them from most other rare earth metals, which exist predominantly in the trivalent state [20,21]. The biological activity of cerium dioxide nanoparticles is determined by oxygen nonstoichiometry, which depends on the size of the nanoparticle and the nature of the surface ligand. It has been shown that cerium-containing nanoparticles can act as superoxide dismutase, catalases, oxidases, and oxidoreductases [22]. In this case, the efficiency of radical neutralization is proportional to the concentration of Ce^{3+} ions on the surface of nanoparticles. In addition, the size and state of the surface of CeO_2 particles determine the possibility of inactivation of superoxide free radicals, preventing oxidative stress in cells [22]. The most effective anti-oxidative-stress activity cerium oxide nanoparticles demonstrate has been shown in alkaline and neutral (physiological) media [23,24].

Along with the complicated preparation procedures of cerium-containing nanoparticles, there exists the important problem of their stabilization against aggregation (agglomeration). Different approaches to the stabilization of nanoparticles, such as gold, silver, and iron oxides as well as cerium oxide, are widely discussed in the literature. For example, various stabilizers have been found to be effective in preventing the agglomeration of nanosized particles, including thiols, carboxylic acids, surfactants, and polymers [25–30]. In general, some stabilizers are not environmentally friendly, while others are prohibitively expensive. The key requirements for the stabilizers for the nanocomposites proposed for biomedical applications are the following: the stabilizer must be capable of specific interactions with nanoparticles to inhibit their growth and they should be harmless to living organisms and the environment. As well, it is necessary to use a stabilizer that provides a longer effective stabilization, which will facilitate the shelf-life of nanoparticles.

In this work, for the preparation of stable cerium oxide (CeO_2) nanoparticles, sodium salt of carboxymethyl cellulose was used. CMC is water soluble and commonly used in the food industry [30]. In addition, CMC has been successfully used as an effective stabilizer in the preparation processes of superparamagnetic iron oxide nanoparticles and Ag nanoparticles [31,32]. CMC is environmental friendly and harmless to living organisms. It is extremely important that CMC is a polyelectrolyte and contains carboxylate groups along with hydroxyl groups in monomer units [33]. Polyelectrolytes are promising components for the effective stabilization of cerium-containing nanoparticles. The efficiency of preventing aggregation can be achieved due to the ability of charged groups in the main chain to interact electrostatically with the surface of nanoparticles. In addition, the natural origin of CMC stabilizer will make it possible to obtain biocompatible nanocomposite materials.

In this communication, we propose a facile original method for the preparation of fine cerium oxide nanoparticles, which was carried out by the reduction of the Ce^{4+} complex salt with sodium borohydride in the presence of CMC at room temperature under aerobic conditions. The suggested technique allows one to easily separate the resulting product from the non-reacted initial compounds and side products with dialysis. The obtained nanocomposite could be stored as a colloid stable solution or as a lyophilized powder without losing its quality.

2. Materials and Methods
2.1. Materials

The following reagents were used in this work: sodium salt of CMC with molecular mass 90.000 Da DS 0.7 (Merck, Rahway, NJ, USA, Analytical grade), ammonium cerium nitrate, $(NH_4)_2Ce(NO_3)_6$ (Reakhim, Russia, Special Purification grade), and sodium borohydride (99%, Pulver, Belgium, Special Purification grade).

2.2. Nanoparticle Synthesis Procedure

The synthesis of composite materials based on the sodium salt of carboxymethyl cellulose and cerium oxide nanoparticles was carried out according to the following procedure. To 2.5 mL of a 2 wt.% solution of CMC, 5 mL of the solution of $(NH_4)_2Ce(NO_3)_6$ was added dropwise under vigorous stirring. The masses of $(NH_4)_2Ce(NO_3)_6$ in solutions were varied from 6.3 to 17.6 mg. Then, 0.5 mL solution containing 3 mg to 5 mg of $NaBH_4$, respectively, was added to the reaction mixture under intensive stirring. The resulting volume of the reaction mixture was 8 mL. The obtained solution was stirred for 12 h. Then, purification from low-molecular-weight components was carried out using dialysis (in dialysis bags, Sigma, MWCO~12 kDa) against water for 24 h. The resulting products were lyophilized.

2.3. Research Methods for Cerium-Containing Nanocomposites

2.3.1. UV Spectroscopy

The determination of the cerium content in the composites was carried out using the UV spectroscopy method. The measurements were carried out on a Specord M40 device from Carl Zeiss (Jena, Germany) in the spectral range from 280 to 500 nm [34,35]. Sample solutions were prepared to record the UV spectra. A calibration graph was built according to the method given in [15]. To construct this calibration graph, weighed amounts of cerium ammonium nitrate (0.5 mg, 1 mg, 1.5 mg, and 2 mg) were dissolved in 100 μL of concentrated H_2SO_4. Then, 10 mL of an aqueous solution containing 0.1% wt. of silver nitrate and 0.2 g of ammonium persulfate was added to the solutions. After that, the UV spectra of the obtained solutions were recorded in the wavelength range from 200 to 500 nm and the absorption intensity was measured at a wavelength of 310 nm (D). Absorption spectra of the solutions containing cerium ions of various concentrations and a calibration curve are presented in Supplementary Materials Figure S1.

2.3.2. Transmission Electron Microscopy (TEM)

Structural studies were carried out with transmission (tunneling) electron microscopy (TEM) on a JEM-100B setup (JEML) equipped with an attachment for X-ray phase analysis [35]. The samples were prepared by applying a drop of an aqueous solution containing the test substance to a copper grid with further drying in an air atmosphere. The samples were examined without preliminary contrasting. The series of images was analyzed using ImageJ software. The obtained results were used to calculate the sizes of the nanoparticles fractions and their differential share.

2.3.3. X-ray Diffraction Analysis (XRD)

X-ray studies were carried out on a Rigaku D/Max2500 diffractometer with a rotating anode (Japan). The survey was carried out in reflection mode (Bragg–Brentano geometry) using CuKα1 radiation (cf. wavelength λ = 1.54183 Å). The generator operation parameters were: accelerating voltage 40 kV, tube current 200 mA. The survey was carried out in quartz cuvettes without averaging rotation. Solvents were not used to fix the powder samples. The recording parameters were: angle interval 2θ = 2°–60°, step (in 2θ) 0.02°, spectrum recording rate 5°/min [34]. Silicon powder was used as an internal standard for correction.

Qualitative analysis of the obtained radiographs was performed using the WinXPOW software package (version 1.06) using the ICDDPDF-2 database.

2.3.4. IR Spectroscopy

The nature of the interaction between the samples was studied using IR spectroscopy on a Specord M80 instrument (Carl Zeiss, Jena, Germany). Studies were carried out in the absorption mode by preparing tablets from KBr (matrix) [35].

2.3.5. Raster (Scanning) Electron Microscopy (SEM)

The microstructure of the samples was studied using a scanning electron microscope with a LEO SUPRA 50VP field emission source (Carl Zeiss, Jena, Germany). For the study, the samples were glued on a metal table using a conductive carbon adhesive tape and a layer of carbon or chromium was deposited on them (sputtering unit Quorum 150T—(UK), voltage 1000–1200 V, current strength 5–10 mA, deposition time 5–10 min [34]). The accelerating voltage of the electron gun was 5–20 kV. Images were obtained in secondary electrons (detector SE2) at magnifications up to 100,000× and recorded in digitized form on a computer.

3. Results

An original technique was used for the first time to obtain CMC–CeO$_2$ nanocomposites. The CMC–CeO$_2$ nanocomposites were obtained by treating CMC with solutions of (NH$_4$)$_2$Ce(NO$_3$)$_6$ and a reducing agent (NaBH$_4$) at room temperature. A feature of the proposed approach is the formation of cerium oxide in an aqueous medium simultaneously in the presence of a strong reducing agent and atmospheric oxygen. The need to use a reducing agent is due to the chemical specificity of this process, which, apparently, includes the following stages [36,37]:

$$Ce^{4+} + BH_4^- \rightarrow Ce^{3+} + B(OH)_3 + H_2 \quad (1)$$

$$NaBH_4 + 4H_2O \rightarrow NaB(OH)_4 + 4H_2 \quad (2)$$

$$NaB(OH)_4 \rightarrow NaOH + B(OH)_3 \quad (3)$$

$$Ce^{3+} + 3OH^- \rightarrow Ce(OH)_3 \quad (4)$$

$$4Ce(OH)_3 + O_2 \rightarrow 4CeO_2 + 6H_2O \quad (5)$$

During the synthesis, the concentration of CMC was kept constant, while the concentration of the cerium salt, the source of cerium (IV) ions, changed six-fold (see details in Table 1). This made it possible to trace the effect of the molar ratio of components in the reaction system on the composition of the final product.

Table 1. Composition of the reaction mixture for the synthesis of cerium-containing nanoparticles and Ce^{4+} yield.

[CMC], Base-Mole/L	(NH$_4$)$_2$Ce(NO$_3$)$_6$, mM	[CMC]/[Ce^{4+}]	Ce^{4+} in Composite, wt.% *
0.08	0.002	20:1	6.4
0.08	0.0053	15:1	7.1
0.08	0.008	10:1	9.0
0.08	0.011	7:1	11.0
0.08	0.016	5:1	14.1

* Determined with UV spectroscopy.

The TEM method was used to visualize the inorganic particles included in the composition of the obtained substances (Figures 1–4). The figures present TEM images of samples obtained at various [CMC]/[Ce^{4+}] ratios in the reaction mixture.

All presented TEM images demonstrated the presence of dark contrasting nanometer-sized spherical particles. On dark-field TEM images of the samples, nanometer-sized particles appeared as bright spots. This indicates that they have a crystalline structure and represent themselves as sources of diffraction. The presence of a crystalline structure in nanoparticles was confirmed with the electron diffraction patterns, which could be described as a set of diffuse Bragg reflections.

Using TEM images, the sizes of spherical nanoparticles in the composites were calculated. The resulting mean values in all the studied samples were in the range 2–3 nm (for the initial TEM images please see Supplementary Materials Figure S2). Thus, the ratio of components in the reaction mixture did not affect the shape and size of nanoparticles in the obtained products.

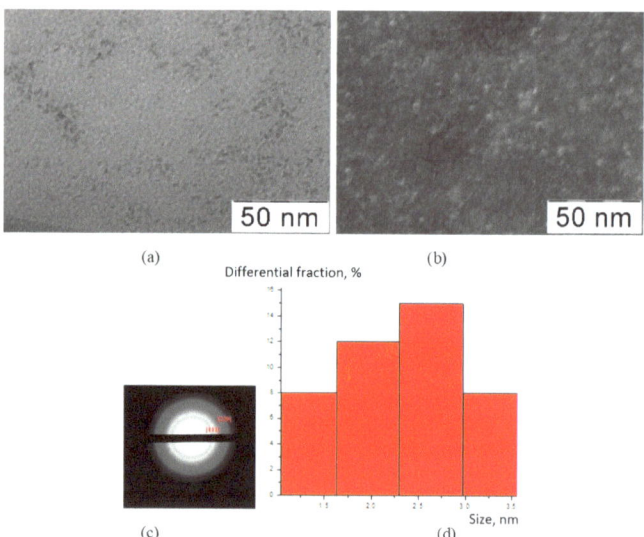

Figure 1. TEM image (**a**), TEM image in dark field (**b**), diffractogram (**c**), nanoparticle size distribution (**d**) of nanoceria-containing composites of CMC with 7.1 wt.% Ce^{4+}.

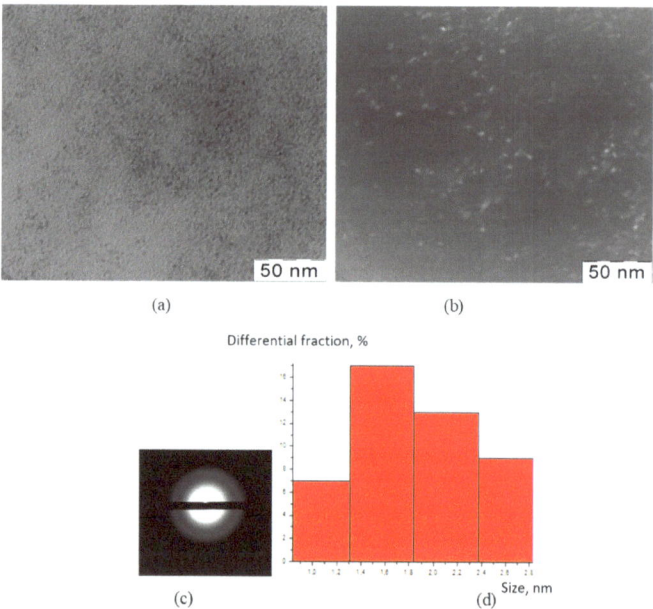

Figure 2. TEM image (**a**), TEM image in dark field (**b**), diffractogram (**c**), nanoparticle size distribution (**d**) of nanoceria-containing composites of CMC with 9.0 wt.% Ce^{4+}.

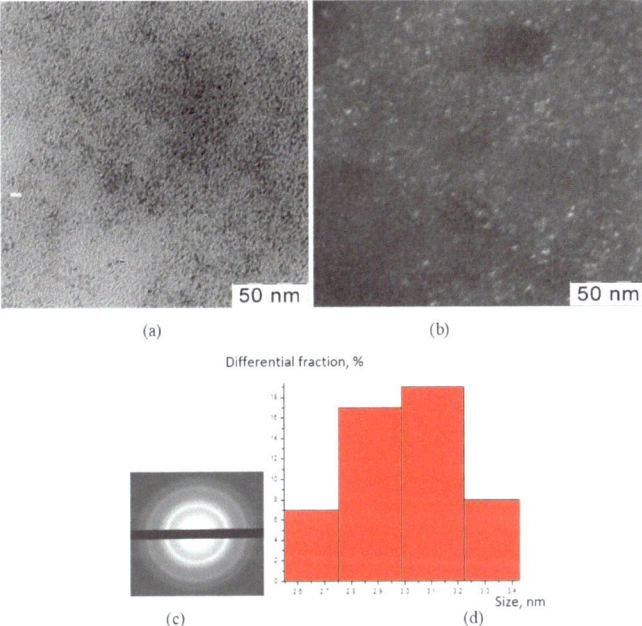

Figure 3. TEM image (**a**), TEM image in dark field (**b**), diffractogram (**c**), nanoparticle size distribution (**d**) of nanoceria-containing composites of CMC with 11.0 wt.% Ce^{4+}.

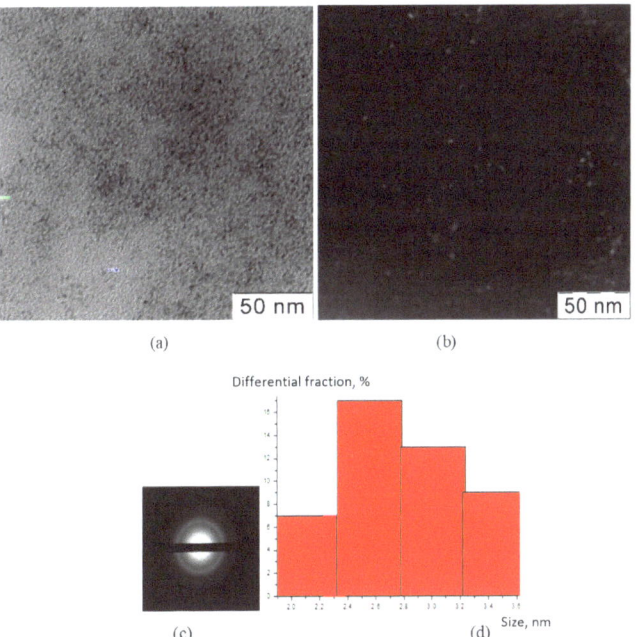

Figure 4. TEM image (**a**), TEM image in dark field (**b**), diffractogram (**c**), nanoparticle size distribution (**d**) of nanoceria-containing composites of CMC with 14.1 wt.% Ce^{4+}.

The identification of the crystal structure of the inorganic phase in cerium-containing products based on CMC was carried out using the XRD method (Figure 5).

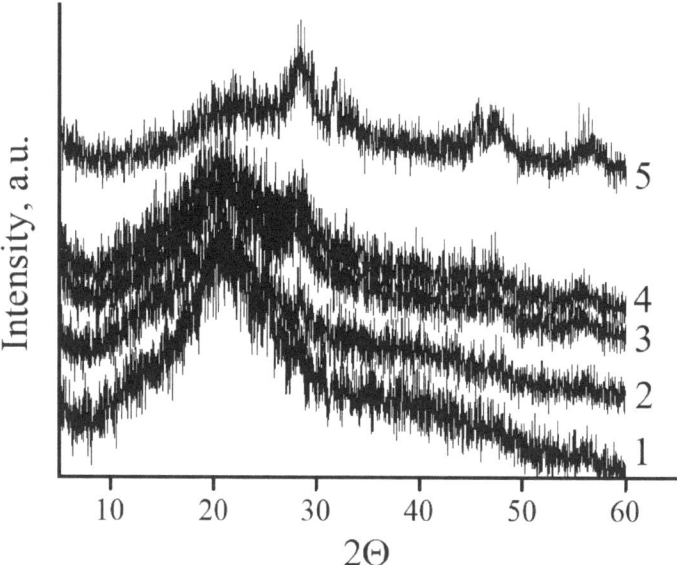

Figure 5. XRD patterns of cerium-containing products based on CMC: pure CMC (**1**); composite with 7.1 wt.% Ce^{4+} (**2**); composite with 9 wt.% Ce^{4+} (**3**); composite with 11.0 wt.% Ce^{4+} (**4**); composite with 14.1 wt.% Ce^{4+} (**5**).

All diffractograms presented in Figure 6 curves 2–6 had peaks at angles $2\theta = 28.4°$, $34.38°$, $47.6°$, and $57°$. Broad reflections indicate the presence of a small particle size of the inorganic phase. The diffraction pattern of the original CMC (Figure 6 curve 1) did not contain these peaks.

The microstructure of cerium-containing composites based on CMC was studied using SEM (Figure 6).

An image of the plane surface was obtained with SEM for the initial CMC. The appearance of CeO_2 nanoparticles in the content of composites resulted in the formation of a porous structure. The pore sizes were practically independent of the mass content of CeO_2 nanoparticles and varied from 4 to 10 µm.

The nature of interactions between CMC macromolecules and cerium oxide nanoparticles was studied using IR spectroscopy (Figure 7).

The CMC spectrum contained bands at wavelengths $\nu = 1688$ cm^{-1} and $\nu = 1410$ cm^{-1}, related to antisymmetric and symmetric stretching vibrations of the carboxyl groups of the polyanion, respectively. It should be noted that a wide peak in the range of 1690–1740 cm^{-1}, corresponding to the stretching vibrations of the C=O group in the composition of the carboxyl groups of the polyanion, was observed in the spectrum. It is known that the presence of this peak indicates the formation of a system of intra- and intermolecular hydrogen bonds with its participation [34,38]. In addition, in the CMC spectrum there was a band at $\nu = 1220$ cm^{-1}, associated with out-of-plane bending vibrations of the hydroxyl groups of the polysaccharide.

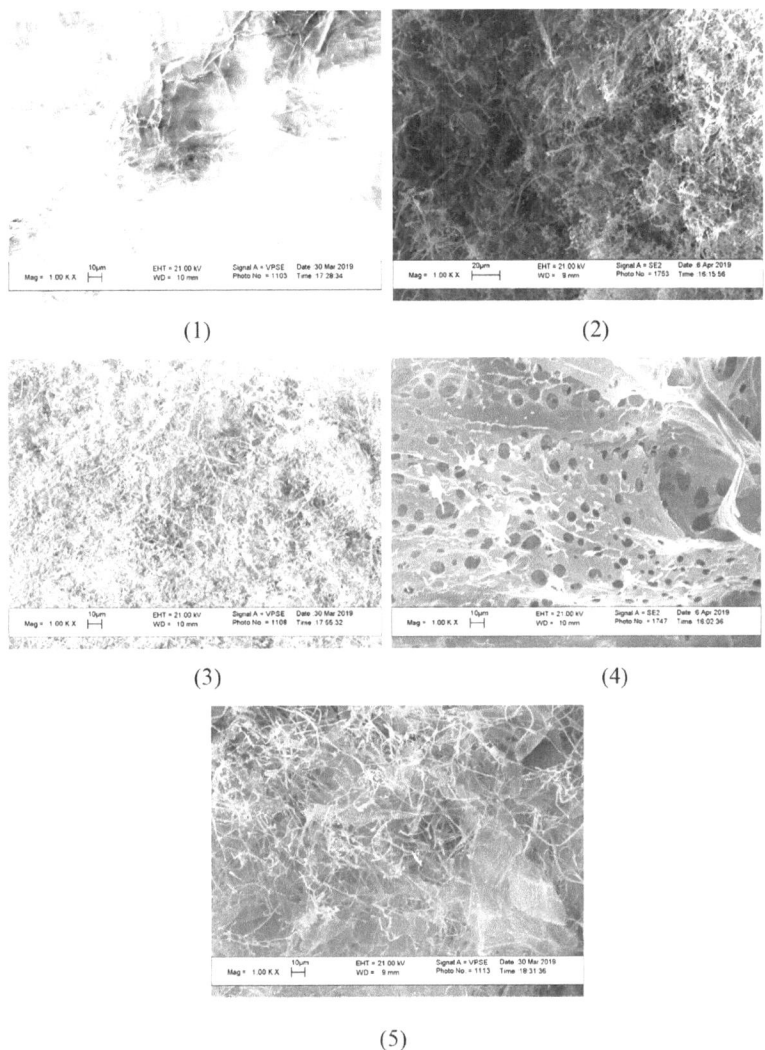

Figure 6. SEM image of cerium-containing products based on CMC: composite with 6.4 wt.% Ce^{4+} (**1**); composite with 7.1 wt.% Ce^{4+} (**2**); composite with 9 wt.% Ce^{4+} (**3**); composite with 11.0 wt.% Ce^{4+} (**4**); composite with 14.1 wt.% Ce^{4+} (**5**).

In the spectra of the nanocomposites, as the cerium content increased, the absorption intensity decreased in the range of 1690–1740 cm^{-1}. This phenomenon is attributed to a violation of the internal structure of the CMC, accompanied by the destruction of the system of hydrogen bonds and the formation of electrostatic contacts between the carboxyl groups of the polysaccharide and the surface of the nanoparticles. In the spectra of the nanocomposites, a shift of the band at ν = 1688 cm^{-1} to the region of lower wave numbers up to ν = 1668 cm^{-1} was observed. The shift of this band is additional evidence that some of the carboxyl groups of the polyanion were involved in an electrostatic interaction with the surface of the cerium oxide nanoparticles. Simultaneously, there is a decrease in the intensity of the band at ν = 1220 cm^{-1}. The observed phenomenon indicates that some of the hydroxyl groups of the polyanion did not realize out-of-plane bending vibrations.

These hydroxyl groups take part in the interaction with the particles of the inorganic phase, forming coordination bonds with cerium ions on the surface of the nanoparticles.

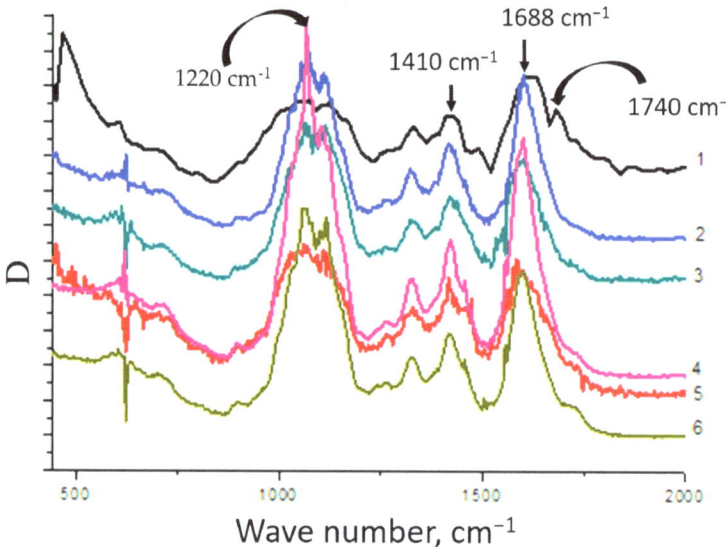

Figure 7. IR spectra of cerium-containing products based on CMC: pure CMC (**1**), composite with 6.4 wt.% Ce^{4+} (**2**); composite with 7.1 wt.% Ce^{4+} (**3**); composite with 9 wt.% Ce^{4+} (**4**); composite with 11.0 wt.% Ce^{4+} (**5**); composite with 14.1 wt.% Ce^{4+} (**6**).

Thus, nanoparticles are included in the composition of the polymer matrix due to the implementation of interactions between surface cerium ions and functional groups of carboxymethyl cellulose. Carboxyl groups of the polysaccharide form electrostatic contacts with nanoparticles. In addition, the formation of the composite results in the destruction of the system of hydrogen bonds with the participation of carboxyl groups of the initial CMC. Due to this effect, the formation of a system of coordination bonds between the hydroxyl groups of polysaccharides and cerium ions on the surface of nanoparticles is possible. That, in turn, is an additional factor that increases the stability of nanoparticles and their aggregative stability. The scheme of the stabilized nanoparticles is presented in Figure 8.

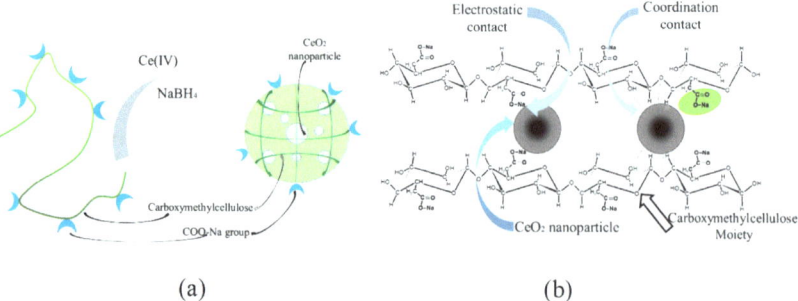

Figure 8. Scheme of formation of the CMC–CeO_2 composite (**a**); Electrostatic and coordination bond formation between polysaccharide and surface of cerium oxide nanoparticles (**b**).

4. Discussion

A one-pot synthesis of fine dispersed CeO_2 nanoparticles could be carried out at room temperature by direct synthesis in a solution of CMC. The formation of nanoparticles occurs in several stages. First, cerium (IV) from the dissolved $(NH_4)_2Ce(NO_3)_6$ undergoes reduction to cerium (III) in the form of cerium hydroxide, which undergoes oxidation by air oxygen dissolved in water. The growth of the nanoparticles is restricted by a stabilizing agent—CMC. Using different CMC-to-cerium molar ratios, a series of nanocomposites was obtained. The morphology and the size of the inorganic nanoparticles stabilized with CMC was studied using TEM. For the all studied compositions of the reaction mixtures the resulting nanoparticles had a spherical shape and a mean diameter of 2–3 nm. Thus, CMC restricts the growth of nanoparticles, providing effective stabilization of the inorganic surface with carboxylate and hydroxyl functional groups. It should be stressed that not only electrostatic but coordination contacts as well provide the stabilization of the nanoparticles. The SEM experiments demonstrated a violation of the continuous film structure of the CMC after the formation of the nanocomposites. In the initial CMC film the intra- and intermolecular hydrogen bonds of the polymer supply the smooth, non-porous structure. The formation of the nanoparticles and their incorporation in the polymer matrix result in destruction of the internal structure of the polymer film due to a change in the function of coordination bonds from stabilization of the film to stabilization of the nanoparticles.

The nature and the structural type of the nanoparticles formed were determined using XRD and TEM analyses. Dark field TEM images of the nanocomposites allows one to estimate the crystalline nature of the inorganic phase. Diffractograms of these particles confirm the crystalline structure of the nanoparticles. However, due to the small size of inorganic particles no detailed information about the structure could be obtained with diffractometer-integrated TEM. The XRD analysis confirmed that the nanoparticles represent themselves as crystals of CeO_2. For the CeO_2 powder the characteristic peaks should be observed at angles $2\theta = 28.4°$, $34.38°$, $47.6°$, and $57°$. The obtained values of 2θ correspond to the following parameters of the CeO_2 crystalline lattice: 110; 004; 220; 214; 311. With the increase in the inorganic phase share in the nanocomposites, the reflections corresponding to the crystalline lattice of CeO_2 become more intensive. No additional peaks arise with the increase in the Ce(IV)-to-CMC ratio in the nanocomposite, confirming that the structure of inorganic nanoparticles does not depend on ratio of the components in the reaction mixture.

To support the proposed nature of the stabilization of CeO_2 nanoparticles with groups of CMC, IR spectroscopy was applied. The initial spectrum of the CMC contains characteristic peaks at 1740 cm^{-1}, 1688 cm^{-1}, 1420 cm^{-1}, and 1220 cm^{-1} reflecting carboxyl, carboxylate, and hydroxyl groups. The formation of nanocomposites results in the vanishing of the 1740 cm^{-1} peak, which reflects the formation of intra- and intermolecular hydrogen bonds. At the same time, the intensive peaks reflecting carboxylate groups were retained in the spectra of the nanocomposites. Hence, the stabilization of the nanoparticles is supplied by electrostatic interactions. Transformation of the 1220 cm^{-1} peak from a wide to a sharp shape with the increase in the cerium content reflects the incorporation of the hydroxyl groups of CMC in interactions with the surface of the inorganic particles.

We suggest that geometrically surrounded nanoparticles of cerium oxide are stabilized with both carboxylate and hydroxyl groups of CMC. The first one ensures electrostatic stabilization while the second one is responsible for the coordination contacts. The dual action of these groups results in effective stabilization of the nanoparticles with a relatively small size.

5. Conclusions

Nanocomposites consisting of CeO_2 nanoparticles stabilized by carboxymethyl cellulose macromolecules were obtained. In the resulting nanocomposites, the content of the inorganic component was determined with UV spectrophotometry. The nanocomposites were also characterized by TEM, SEM, XRD, and IR spectroscopy. It has been found that varying the ratio of the components of the reaction mixture in the course of synthesis makes it possible to obtain nanocomposites with different contents of the inorganic phase while the size and morphology of the nanoparticles did not depend on the reaction mixture composition. The nanoparticles had a spherical shape with an average diameter of 2–3 nm. It was demonstrated that the type of crystal lattice of the nanoparticles corresponds to the structure of CeO_2. The type of crystal structure did not depend on the ratio of the components in the reaction mixture as well. With the use of a combination of SEM and IR spectroscopy, it has been demonstrated that the stabilization of the nanoparticles is attributed to both electrostatic and coordination contacts of the inorganic surface with the macromolecules. These findings could be useful for the further development and application of cerium oxide nanoparticles.

Supplementary Materials: The following supporting information can be downloaded at: https://www.mdpi.com/article/10.3390/polym15061437/s1, Figure S1: UV spectra of $(NH_4)_2Ce(NO_3)_6$ solutions of various concentrations (a): 0.05 mg/mL (1); 0.1 mg/mL (2); 0.15 mg/mL (3); 0.2 mg/mL (4). Calibration plot of absorbance of $(NH_4)_2Ce(NO_3)_6$ solutions versus concentration at 310 nm (b).; Figure S2: TEM images of nanoceria-containing composites of CMC with 7.1 wt.% Ce4+ (a); 9.0 wt.% Ce4+ (b); 11.0 wt.% Ce4+ (c); 14.1 wt.% Ce4+ (d).

Author Contributions: Conceptualization, V.V.S.; methodology, M.I.A., A.V.K. and S.B.Z.; validation, V.V.S.; formal analysis, V.V.S.; investigation, V.V.S., S.A.M., V.A.P., S.B.Z. and A.V.S.; resources, V.V.S.; data curation, V.V.S.; writing—original draft preparation, V.V.S.; writing—review and editing, A.V.S.; visualization, V.V.S.; supervision, V.V.S.; project administration, V.V.S.; funding acquisition, V.V.S. All authors have read and agreed to the published version of the manuscript.

Funding: This research was funded by the Ministry of Science and Higher Education of the Russian Federation, project No. 13.1902.21.0017.

Institutional Review Board Statement: Not applicable.

Data Availability Statement: Not applicable.

Conflicts of Interest: The authors declare no conflict of interest.

References

1. Liu, Y.H.; Zuo, J.C.; Ren, X.F. Synthesis and character of cerium oxide (CeO_2) nanoparticles by the precipitation method. *Metalurgija* **2019**, *53*, 463–465.
2. Farahmandjou, M.; Zarinkamar, M.; Firoozabadi, T.P. Synthesis of cerium oxide (CeO_2) nanoparticles using simple CO-precipitation method. *Rev. Mex. Fis.* **2016**, *62*, 496–499.
3. Ketzial, J.; Jasmine, A.; Nesaraj, S. Synthesis of CeO_2 nanoparticles by chemical precipitation and the effect of a surfactant on the distribution of particle sizes. *J. Ceram. Process. Res.* **2011**, *12*, 74–79.
4. Soren, S.; Jena, S.R.; Samanta, L.; Parhi, P. Antioxidant Potential and Toxicity Study of the Cerium Oxide Nanoparticles Synthesized by Microwave-Mediated Synthesis. *Appl. Biochem. Biotechnol.* **2015**, *177*, 148–161. [CrossRef] [PubMed]
5. Pinjari, D.V.; Pandit, A.B. Room temperature synthesis of crystalline CeO_2 nanopowder: Advantage of sonochemical method over conventional method. *Ultrason. Sonochem.* **2011**, *18*, 1118–1123. [CrossRef]
6. Masui, T.; Hirari, H.; Imanaka, N.; Adachi, G.; Sakata, T.; Mori, H. Synthesis of cerium oxide nanoparticles by hydrothermal crystallization with citric acid. *J. Mater. Sci. Lett.* **2002**, *21*, 489–491. [CrossRef]
7. Masui, T.; Hirari, H.; Hamada, R.; Imanaka, N.; Adachi, G.; Sakata, T.; Mori, H. Synthesis and Characterization of Cerium Oxide Nanoparticles Coated With Turbostratic Boron Nitride. *J. Mater. Chem.* **2003**, *13*, 622–627. [CrossRef]
8. Jalilpour, M.; Fathalilou, M. Effect of aging time and calcination temperature on the cerium oxide nanoparticles synthesis via reverse co-precipitation method. *Int. J. Phys. Sci.* **2012**, *7*, 944–948.
9. Gao, F.; Lu, Q.; Komarneni, S. Fast synthesis of cerium oxide nanoparticles and nanorods. *J. Nanosci. Nanotechnol.* **2006**, *6*, 3812–3819. [CrossRef] [PubMed]
10. Hosseini, M.; Amjadi, I.; Mohajeri, M.; Mozafari, M. Sol–Gel Synthesis, Physico-Chemical and Biological Characterization of Cerium Oxide/Polyallylamine Nanoparticles. *Polymers* **2020**, *12*, 1444. [CrossRef]
11. Pujar, M.S.; Hunagund, S.M.; Desai, V.R.; Patil, S.; Sidarai, A.H. One-step synthesis and characterizations of cerium oxide nanoparticles in an ambient temperature via Co-precipitation method. *AIP Conf. Proc.* **2018**, *1942*, 050026.

12. Kockrick, E.; Schrage, C.; Grigas, A.; Geiger, D.; Kaskel, S. Synthesis and catalytic properties of microemulsion-derived cerium oxide nanoparticles. *J. Solid State Chem.* **2008**, *181*, 1614–1620. [CrossRef]
13. Annis, J.W.; Fisher, J.M.; Thompsett, D.; Walton, R.I. Solvothermal Synthesis Routes to Substituted Cerium Dioxide Materials. *Inorganics* **2021**, *9*, 40. [CrossRef]
14. Tumkur, P.P.; Gunasekaran, N.K.; Lamani, B.R.; Nazario Bayon, N.; Prabhakaran, K.; Hall, J.C.; Ramesh, G.T. Cerium Oxide Nanoparticles: Synthesis and Characterization for Biosafe Applications. *Nanomanufacturing* **2021**, *1*, 176–189. [CrossRef]
15. Bumajdad, A.; Eastoe, J.; Mathew, A. Cerium oxide nanoparticles prepared in self-assembled systems. *Adv. Colloid Interface Sci.* **2009**, *147–148*, 56–66. [CrossRef]
16. Elahi, B.; Mirzaee, M.; Darroudi, M.; Oskuee, R.K.; Sadri, K.; Amiri, M.S. Preparation of cerium oxide nanoparticles in Salvia Macrosiphon Boiss seeds extract and investigation of their photo-catalytic activities. *Ceram. Int.* **2019**, *45*, 4790–4797. [CrossRef]
17. Thakur, N.; Manna, P.; Das, J. Synthesis and biomedical applications of nanoceria, a redox active nanoparticle. *J. Nanobiotechnol.* **2019**, *17*, 84. [CrossRef]
18. Saifi, M.A.; Seal, S.; Godugu, C. Nanoceria, the versatile nanoparticles: Promising biomedical applications. *J. Control. Release* **2021**, *338*, 164–189. [CrossRef]
19. Rajeshkumar, S.; Naik, P. Synthesis and biomedical applications of Cerium oxide nanoparticles—A Review. *Biotechnol. Rep.* **2018**, *17*, 1–5. [CrossRef] [PubMed]
20. Lord, M.S.; Berret, J.F.; Singh, S.; Vinu, A.; Karakoti, A.S. Redox Active Cerium Oxide Nanoparticles: Current Status and Burning Issues. *Small* **2021**, *17*, 2102342. [CrossRef]
21. Habib, S.; Fayyad, E.; Nawaz, M.; Khan, A.; Shakoor, R.A.; Kahraman, R.; Abdullah, A. Cerium Dioxide Nanoparticles as Smart Carriers for Self-Healing Coatings. *Nanomaterials* **2020**, *10*, 791. [CrossRef] [PubMed]
22. Ferreira, C.A.; Ni, D.; Rosenkrans, Z.T.; Cai, W. Scavenging of reactive oxygen and nitrogen species with nanomaterials. *Nano Res.* **2018**, *11*, 4955–4984. [CrossRef] [PubMed]
23. Nelson, B.; Johnson, M.; Walker, M.; Riley, K.; Sims, C. Antioxidant Cerium Oxide Nanoparticles in Biology and Medicine. *Antioxidants* **2016**, *5*, 15. [CrossRef] [PubMed]
24. Alpaslan, E.; Yazici, H.; Golshan, N.H.; Ziemer, K.S.; Webster, T.J. pH-Dependent Activity of Dextran-Coated Cerium Oxide Nanoparticles on Prohibiting Osteosarcoma Cell Proliferation. *ACS Biomat. Sci. Eng.* **2015**, *1*, 1096–1103. [CrossRef]
25. Thiagarajan, S.; Price, E.; Connors, L.; Dettman, A.; Koh, A.S. Study of n-Alkanethiol Self-Assembly Behavior on Iron Particles: Effect of Alkyl Chain Length and Adsorption Solvent on Resulting Iron-Based Magnetorheological Fluids. *Langmuir* **2022**, *38*, 13506–13521. [CrossRef]
26. Spiridonov, V.V.; Liu, X.Y.; Zezin, S.B.; Panova, I.G.; Sybachin, A.V.; Yaroslavov, A.A. Hybrid nanocomposites of carboxymethyl cellulose cross-linked by in-situ formed Cu_2O nanoparticles for photocatalytic applications. *J. Organomet. Chem.* **2020**, *914*, 121180. [CrossRef]
27. Sun, S.; Zeng, H. Size-Controlled Synthesis of Magnetite Nanoparticles. *J. Am. Chem. Soc.* **2002**, *124*, 8204–8205. [CrossRef]
28. Kim, D.K.; Mikhaylova, M.; Zhang, Y.; Muhammed, M. Protective Coating of Superparamagnetic Iron Oxide Nanoparticles. *Chem. Mater.* **2003**, *15*, 1617–1627. [CrossRef]
29. Mkrtchyan, K.V.; Pigareva, V.A.; Zezina, E.A.; Kuznetsova, O.A.; Semenova, A.A.; Yushina, Y.K.; Tolordava, E.R.; Grudistova, M.A.; Sybachin, A.V.; Klimov, D.I.; et al. Preparation of Biocidal Nanocomposites in X-ray Irradiated Interpolyelectrolyte Complexes of Polyacrylic Acid and Polyethylenimine with Ag-Ions. *Polymers* **2022**, *14*, 4417. [CrossRef]
30. Thivya, P.; Akalya, S.; Sinija, V.R. A comprehensive review on cellulose-based hydrogel and its potential application in the food industry. *Appl. Food Res.* **2022**, *2*, 100161. [CrossRef]
31. Xu, W.; Li, Z.; Shi, S.; Qi, J.; Cai, S.; Yu, Y.; O'Carroll, D.M.; He, F. Carboxymethyl cellulose stabilized and sulfidated nanoscale zero-valent iron: Characterization and trichloroethene dechlorination. *Appl. Cat. B* **2020**, *262*, 118303. [CrossRef]
32. Salem, S.S.; Hashem, A.H.; Sallam, A.-A.M.; Doghish, A.S.; Al-Askar, A.A.; Arishi, A.A.; Shehabeldine, A.M. Synthesis of Silver Nanocomposite Based on Carboxymethyl Cellulose: Antibacterial, Antifungal and Anticancer Activities. *Polymers* **2022**, *14*, 3352. [CrossRef]
33. Altam, A.A.; Zhu, L.; Huang, W.; Huang, H.; Yang, S. Polyelectrolyte complex beads of carboxymethylcellulose and chitosan: The controlled formation and improved properties. *Carbohyd. Polym. Technol. Appl.* **2021**, *2*, 100100. [CrossRef]
34. Spiridonov, V.V.; Panova, I.G.; Makarova, L.V.; Afanasov, M.I.; Zezin, S.B.; Sybachin, A.V.; Yaroslavov, A.A. The one-step synthesis of polymer-based magnetic γ-Fe2O3/carboxymethyl cellulose nanocomposites. *Carbohyd. Polym.* **2017**, *177*, 269–274. [CrossRef] [PubMed]
35. Klimov, D.I.; Zezina, E.A.; Lipik, V.C.; Abramchuk, S.S.; Yaroslavov, A.A.; Feldman, V.I.; Sybachin, A.V.; Spiridonov, V.V.; Zezin, A.A. Radiation-induced preparation of metal nanostructures in coatings of interpolyelectrolyte complexes. *Rad. Phys. Chem.* **2019**, *162*, 23–30. [CrossRef]
36. Glavee, G.N.; Klabunde, K.J.; Sorensen, C.M.; Hadjapanayis, G.C. Borohydride reductions of metal ions. A new understanding of the chemistry leading to nanoscale particles of metals, borides, and metal borates. *Langmuir* **1992**, *8*, 771–773. [CrossRef]

37. Calvache-Muñoz, J.; Prado, F.A.; Rodríguez-Páez, J.E. Cerium oxide nanoparticles: Synthesis, characterization and tentative mechanism of particle formation. *Colloids Surf. A* **2017**, *529*, 146–159. [CrossRef]
38. Cuba-Chiem, L.T.; Huynh, L.; Ralston, J.; Beattie, D.A. In Situ Particle Film ATR FTIR Spectroscopy of Carboxymethyl Cellulose Adsorption on Talc: Binding Mechanism, pH Effects, and Adsorption Kinetics. *Langmuir* **2008**, *24*, 8036–8044. [CrossRef] [PubMed]

Disclaimer/Publisher's Note: The statements, opinions and data contained in all publications are solely those of the individual author(s) and contributor(s) and not of MDPI and/or the editor(s). MDPI and/or the editor(s) disclaim responsibility for any injury to people or property resulting from any ideas, methods, instructions or products referred to in the content.

Article

Centrifugal Force-Spinning to Obtain Multifunctional Fibers of PLA Reinforced with Functionalized Silver Nanoparticles

María Dolores Martín-Alonso [1,2], Valentina Salaris [1], Adrián Leonés [1], Víctor Hevilla [1], Alexandra Muñoz-Bonilla [1], Coro Echeverría [1], Marta Fernández-García [1], Laura Peponi [1,*] and Daniel López [1,*]

[1] Instituto de Ciencia y Tecnología de Polímeros (ICTP-CSIC), Calle Juan de la Cierva 2, 28006 Madrid, Spain
[2] IMDEA Materials Institute, Calle Eric Kandel 2, 28906 Getafe, Spain
* Correspondence: lpeponi@ictp.csic.es (L.P.); daniel.l.g@csic.es (D.L.); Tel.: +34-915-622-900 (L.P. & D.L.)

Citation: Martín-Alonso, M.D.; Salaris, V.; Leonés, A.; Hevilla, V.; Muñoz-Bonilla, A.; Echeverría, C.; Fernández-García, M.; Peponi, L.; López, D. Centrifugal Force-Spinning to Obtain Multifunctional Fibers of PLA Reinforced with Functionalized Silver Nanoparticles. *Polymers* **2023**, *15*, 1240. https://doi.org/10.3390/polym15051240

Academic Editor: Sergio Torres-Giner

Received: 31 January 2023
Revised: 23 February 2023
Accepted: 24 February 2023
Published: 28 February 2023

Copyright: © 2023 by the authors. Licensee MDPI, Basel, Switzerland. This article is an open access article distributed under the terms and conditions of the Creative Commons Attribution (CC BY) license (https://creativecommons.org/licenses/by/4.0/).

Abstract: The design and development of multifunctional fibers awakened great interest in biomaterials and food packaging materials. One way to achieve these materials is by incorporating functionalized nanoparticles into matrices obtained by spinning techniques. Here, a procedure for obtaining functionalized silver nanoparticles through a green protocol, using chitosan as a reducing agent, was implemented. These nanoparticles were incorporated into PLA solutions to study the production of multifunctional polymeric fibers by centrifugal force-spinning. Multifunctional PLA-based microfibers were obtained with nanoparticle concentrations varying from 0 to 3.5 wt%. The effect of the incorporation of nanoparticles and the method of preparation of the fibers on the morphology, thermomechanical properties, biodisintegration, and antimicrobial behavior, was investigated. The best balance in terms of thermomechanical behavior was obtained for the lowest amount of nanoparticles, that is 1 wt%. Furthermore, functionalized silver nanoparticles confer antibacterial activity to the PLA fibers, with a percentage of killing bacteria between 65 and 90%. All the samples turned out to be disintegrable under composting conditions. Additionally, the suitability of the centrifugal force-spinning technique for producing shape-memory fiber mats was tested. Results demonstrate that with 2 wt% of nanoparticles a good thermally activated shape-memory effect, with high values of fixity and recovery ratios, is obtained. The results obtained show interesting properties of the nanocomposites to be applied as biomaterials.

Keywords: centrifugal force-spinning; PLA; nano/micro fibers; biopolymers; nanoparticles; nanocomposites

1. Introduction

Nowadays, the growing interest in multifunctional polymeric-based materials is a key factor to be focused on. Among them, shape-memory polymers (SMPs) play an important role. In particular, shape-memory polymers are a class of smart materials able to recover their original shape from a previously temporary programmed shape when exposed to external stimuli of different nature such as temperature, moisture, pH, light, etc. [1–4]. Therefore, the shape-memory effect (SME) results from a combination of polymer chemistry, morphology, and a specific processing condition. SMPs possess tremendous potential in many fields as minimally invasive medical devices, actuators, sensors, and smart textiles, among others [5–9]. Depending on the processing of the materials and of their final structures, different types of shape-memory blocks, shape-memory foams, shape-memory fibers, and shape-memory films can be obtained [10–12]. Recently, SMPs with fibrous structures are gaining interest in applications that imply contact with the human body such as smart textiles or scaffolds for regenerative medicine. In this sense, two main factors have to be considered, that is, on the one hand, the obtention of polymeric fibrous structures, and the use of biocompatible and or biodegradable polymers, on the other hand [13,14]. In this regard, the electrospinning process has been recognized as one of the most attractive technologies to produce fibrous structures, thanks to the advantageous high porosity and specific surface area obtained [14–17]. The electrospinning process is a simple, extremely versatile, and low-cost process for obtaining polymeric fibers recently scaled up at the industrial level, that can find several applications in biomedicine but also in food packaging and agriculture [18–24]. In fact, by solution electrospinning, it is possible to produce multifunctional thin polymeric materials in the form of woven non-woven fibers mats from polymeric solutions subjected to high electric voltage and at room temperature [25,26]. Electrospinning is currently the most popular method for producing polymer nano and microfibers. However, high voltage, which is sensitive to the dielectric constant, safety concerns, and low fiber yields limit the application of electrospinning [27]. The search for a method that would minimize, or even eliminate, many of the limitations encountered in electrospinning is focused on increasing material choice, improving production rate, and lowering fiber costs through an environmentally friendly process. In the centrifugal force-spinning method, the electric field, used in the electrospinning process, is replaced by centrifugal forces [28,29]. The combination of centrifugal forces with multiple configurations of easily interchangeable spinnerets makes centrifugal force-spinning a versatile method that overcomes many of the limitations of existing processes, namely, high electric fields and a solution that is typically dielectric. These changes significantly increase the selection of materials by allowing both non-conductive and conductive solutions to be spun into nanofibers [28,30,31]. If necessary, a high-temperature solvent can also be used by heating the spinneret holding the material of interest. Additionally, solid materials can be melted and spun without the need for chemical preparation and, therefore, for solvent recovery since no solvent is involved in the process [32].

Among polymers frequently used in solution electrospinning, poly(lactic acid), PLA, a non-petrochemical biodegradable polymer, is considered of high interest for food packaging applications as well as for the biomedicine field. PLA shows high tensile strength and elastic modulus, biocompatibility, and degradability under physiological conditions into non-toxic products, making of PLA an ideal material for practical use in contact with the human body. This versatile polymer is known to exhibit a thermo-responsive shape-memory effect, activated by its glass transition temperature (T_g) acting as transition temperature, around 60 °C [13,33,34]. However, its T_g is much higher than the human body temperature, together with its brittleness, poor toughness, and elongation at break < 10%, which limits its direct use in biomedical applications. Thermally-activated shape-memory effect, in general, and, in particular, in PLA is achieved by heating up to a switching temperature (T_{sw}) higher than its T_g in the case of PLA, at which polymer chains have enough mobility to recover the original form from a temporary shape previously fixed. Thus, to consider the thermally-activated shape-memory effect of PLA as a suitable way to change the temporary form of devices for biomedical applications, a T_{sw} closer to the human body has to be achieved [13,35].

Moreover, unique mechanical, electrical, chemical, and optical properties have been achieved by modifying polymers, fibers or not, with the addition of nanoparticles (NPs), obtaining polymeric nanocomposites with both enhanced and functional properties just compared with the neat matrix. Amounts, size, and proper distribution of NPs into the polymer matrix can effectively control the properties of these nanocomposites [5]. Recently, the utilization of polymers with embedded silver (Ag) NPs have attracted much attention mainly due to their antimicrobial activities [36]. Silver NPs show antibacterial activity toward germs on contact without the release of toxic biocides. Therefore, the antimicrobial properties of Ag NPs can be considered non-toxic and environmentally friendly materials in biomedical applications. The Ag-NPs-filled polymeric materials have to release the Ag ions to a pathogenic environment continuously in order to be efficacious. Therefore, silver nanoparticles gain attention due to their remarkable and unique optical, mechanical, and catalytic properties in addition to their conductivity, heat transfer, and many other excellent applications in biomedical fields. Moreover, Ag NPs have been approved for use in food-contact polymers in the USA and the European Union, after previous consideration of the use conditions [37,38]. Consequently, the development of stable NPs is presented in a wide range of applications in the field of biomedical science. Inorganic noble metal NPs are proven for their antimicrobial properties [37–39].

NPs can be synthesized by different physicochemical methods and they need to reach an optimal dispersion into the polymeric matrix, obtained by modifying them through physical or chemical functionalization. Recently, attention has been paid to the synthesis under eco-friendly conditions of silver and gold (Au) nanoparticles [37] by using chitosan instead of sodium borohydride, $NaBH_4$, as reducing agent and citrate, or ascorbate, which are associated with environmental toxicity or biological hazards. Chitosan (CH), is one of the well-known and most used biopolymers, derivative of chitin. In fact, thanks to their large amount of free amino and hydroxyl groups, under the proper thermal conditions, they can be used as reducing and stabilizing agents for the synthesis of Ag and Au nanoparticles, preventing also their aggregation [37,39–41]. This fact is possible, thus, considering that in proper thermal conditions, as previously reported in the literature, the hydroxyl groups can be converted into carboxyl groups by air oxidation, which reduces the silver ions [37].

The synthesis of nanoparticles from chitosan has been paid great attention to due to their biocompatibility, biodegradability, and hydrophilic properties, which endow them with opportunities for various applications [37]. The polymeric feature of chitosan interacts with negatively-charged molecules and polymers. Due to the interaction that takes place between the active amino groups in chitosan and metal nanoparticles, chitosan is chosen as a protecting agent in the synthesis of metal nanoparticles. Both active amino and hydroxyl functional groups in chitosan show many remarkable biological activities including antimicrobial activity for disease resistance toward a number of human cell types [41]. In this regard, chitosan can play the role of a reducing agent. The reduction of Ag+ ions is coupled to the oxidation of the hydroxyl groups in molecular CH and/or its hydrolyzates. The reducing ability of chitosan depends on its concentration and reaction temperature. Thus, by optimizing the experimental conditions such as temperature, Ag+, and CH concentration, monodisperse AgCH NPs with highly antibacterial and antiproliferative activities could be obtained. Moreover, it is interesting to note that, in this case, chitosan is able to act as both a reducing agent and stabilizer for the NPs. In fact, the excess of amine and hydroxyl groups present in the chitosan supports the nucleation as well as the stabilization of Ag NPs during the process. In other words, the Ag NPs adsorbed on the surface of the polymer are prevented from further aggregation [39–42].

With this background, in this study, eco-friendly synthesis conditioning to obtain functionalized silver nanoparticles with a green protocol, based on non-toxic biodegradable chitosan as a reducing agent, has been implemented to incorporate the nanoparticles into a PLA solution, with the final aim to study PLA-based reinforced fibers obtained by centrifugal force-spinning.

2. Materials and Methods

2.1. Materials

Polylactic acid (PLA3051D, 3% of D-lactic acid monomer, molecular weight 142×10^4 g/mol, density 1.24 g/cm^3) was supplied from NatureWorks® ((NatureWorks LLC, Min- netonka, MN, USA). Silver nitrate ($AgNO_3$) and chitosan from shrimp shells with a deacetylation degree > 75% were purchased from Sigma-Aldrich (St. Quentin Fallavier, France). Acetic acid and sodium hydroxide were purchased from Fluka (Seelze, Germany).

2.2. Synthesis of Based Chitosan Silver Nanoparticles

Chitosan-based silver nanoparticles (AgCH-NPs) were synthesized by a method described elsewhere [36]. Typically, 4.5 mL of a solution 52.0 mM of $AgNO_3$ and 10 mL of a solution of chitosan of concentration 6.92 mg/mL in 1% acetic acid were mixed and heated at 95 °C under stirring for 12 h in an argon-purified reactor. The dispersions obtained were dialyzed for six days with distilled water using dialysis membranes with a MWCO of 6000–8000 Da [36].

2.3. Preparation of Poly(Lactic Acid) Forced-Fibers Containing AgCH-NPs

Experimental conditions for centrifugal force-spinning were established after conducting an exploratory study in which the concentration of the PLA solution was varied from 2 to 20 wt% and the rotation speed was varied between 6000 and 10,000 rpm, with a fixed distance spinneret-collector of 10 cm. The criteria for fixing the spinning parameters were based on the best results obtained in terms of fiber production, mesh homogeneity, and absence of defects. In this way, appropriate amounts of PLA and AgCH-NPs were mixed in a two-mouth flask with a shovel shaker at 700 rpm. PLA pellets were previously dried overnight at 60 °C, in order to avoid the presence of moisture. The final volume (PLA solution + NPs solution) was kept constant at 25 mL. Briefly, PLA solution was first prepared by mixing the polymer with the appropriate amount of $CHCl_3$, in order to maintain a fixed concentration of PLA 12 $\%w/v$ in the final solution, by magnetic stirring. AgCH NPs dispersed in the appropriate amount of $CHCl_3$ were added dropwise while the PLA solution was stirred with a shovel shaker at 300–400 rpm. Once the amount of nanoparticles was added, the solution was mixed for 20 min at 700 rpm of stirring to ensure the homogenous dispersion of NPs. Finally, these mixtures were taken to the force-spinning equipment (see Scheme 1) to produce forced-fibers, f-fibers. A total amount of 25 mL of the PLA/AgCH-NPs solution was injected into the needle-based spinneret (needle gauge of 0.4 mm). The solution was spun at 6500 rpm and collected as mats, which were wrapped into an aluminum foil (see Scheme 1). The spinneret-collector distance was kept constant at 10 cm.

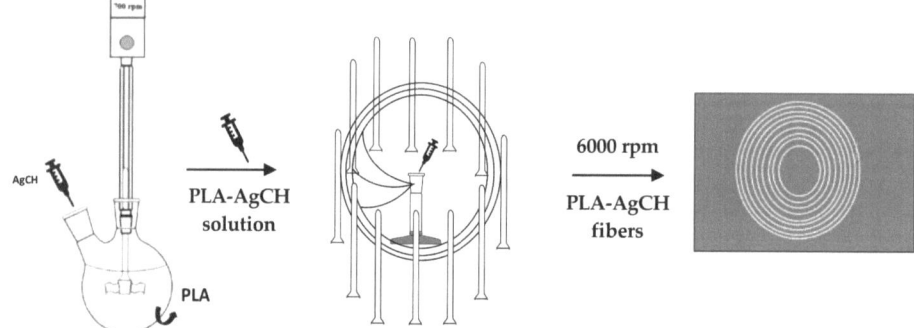

Scheme 1. Methodology followed in the experimental procedure to obtain the f-fibers.

2.4. Characterization Techniques

A Philips XL30 scanning electron microscope ((SEM, Phillips, Eindhoven, The Netherlands), with an accelerating voltage of 10 kV, work distance of 10–15 mm was used to record SEM micrographs of samples and observe changes produced by the different amount of nanofillers (AgCH-NPs). Each specimen was gold-coated (~5 nm thickness) in a Polaron SC7640 Auto/Manual Sputter (Quorum Technoligies, Newheaven, U.K.).

Images of nanoparticles were taken using a Field Emission Scanning electron microscopy (FE-SEM) Hitachi SU 8000 with an acceleration voltage of 30 kV.

UV-vis spectroscopy was performed on a Lamda 35 Perkin Elmer instrument (Perkin Elmer Spain, S.L., Madrid, Spain).

Fourier transform infrared (FTIR) spectra were recorded for all the samples using a Spectrum-One Perkin Elmer (Perkin Elmer Spain, S.L., Madrid, Spain) spectrometer between 650 and 4500 cm^{-1} spectral range with a 4 cm^{-1} resolution. A background spectrum was acquired before every sample and all samples were vacuum-dried prior to measurement.

Thermal transitions and stability of neat PLA as well as nanocomposites with different amounts of AgCH-NPs f-fibers, were studied by differential scanning calorimetry (DSC) and thermogravimetric analysis (TGA) in a DSC Q2000 and TA-Q500 apparatus both from TA Instruments (TA Instruments, New Castle, DE, USA), respectively. In the DSC, samples were heated from −60 °C up to 180 °C at 10 °C/min under a N_2 atmosphere (flow rate of 50 mL/min). Glass transition temperatures (T_g), calculated as the midpoint of the transition, melting temperatures (T_m), and both cold crystallization (ΔH_{cc}) and melting (ΔH_m) enthalpies were calculated by analyzing the thermograms in the TA Universal Analysis software. The degree of crystallinity (X_c%) was, therefore, obtained from Equation (1), using 93.6 J/g as the crystallization enthalpy value for pure crystalline PLA (ΔH_{0m}) [43]. W_f represents the weight fraction of pure PLA present in the sample ($W_f = 1 - m_f$, where mf is the weight fraction of the nanoparticles in the nanocomposite):

$$Xc\% = \frac{\Delta Hm - \Delta Hcc}{\Delta H0m} \times \frac{1}{Wf} \times 100 \tag{1}$$

Tensile tests were carried out in an Instron instrument equipped with a 100 N load cell, operated at room temperature and at a crosshead speed of 10 mm/min. The initial length between clamps was set at 10 mm, samples of 5 mm width and ~80 µm of average thickness were measured and results from five to ten specimens were averaged. The Young's modulus (slope of the curve from 0–2% deformation), maximum stress, ultimate tensile strength, and elongation at break were calculated.

Thermally-activated shape-memory characterization has been performed by dynamic mechanical thermal analysis (DMTA) in a DMA Q800 TA instruments (TA Instruments, New Castle, DE, USA), by using thermo-mechanical cyclic tests. The instrument was set in controlled force mode and four different stages were defined for each cycle:

- the sample was equilibrated at the chosen switching temperature (T_{sw}) for 5 min, in this case, the T_g of PLA matrix, 60 °C;
- ramp stress of 0.2 MPa min^{-1} was applied until the sample reached 50% of deformation;
- the sample was subsequently cooled at a fixing temperature (T_{fix}) of 10 °C under constant stress in order to fix the temporary shape;
- after releasing the stress at 0.50 MPa min^{-1}, the sample was heated at 3 °C min^{-1} up to T_{sw} and maintained for 30 min in order to recover the initial permanent shape.

Sample dimensions for the DMTA and shape-memory tests were the same as for the previously described mechanical tests. The quantification of the shape-memory behavior was carried out by calculating the strain fixity ratio (R_f) and the strain recovery ratio (R_r) given by Equations (2) and (3), respectively [13,44]:

$$R_f(N) = 100 \times \frac{\varepsilon_u(N)}{\varepsilon_m(N)} \tag{2}$$

$$R_r(N) = 100 \times \frac{\varepsilon_m(N) - \varepsilon_p(N)}{\varepsilon_m(N) - \varepsilon_p(N-1)} \tag{3}$$

where ε_m is the maximum strain after cooling to T_{fix} and before releasing the stress, ε_u is the fixed strain after releasing the stress at T_{fix} and ε_p is the residual strain after retaining the sample at T_{sw} for 30 min. In particular, R_f indicates the ability of the material to fix the temporary shape, and R_r, its capability to recover its original shape.

To study the crystalline structure of the plasticized PLA nanocomposite films, a Bruker D8 Advance X-ray diffractometer (Bruker, Madrid, Spain) equipped with a CuKα (λ = 0.154 nm) source was used. Samples were mounted on an appropriate holder and scanned between 2° and 60° (2θ) with a scanning step of 0.02°, a collection time of 10 s per step, and 40 kV of operating voltage.

Antimicrobial activity of the prepared nanocomposites was determined following the E2149-13a standard method of the American Society for Testing and Materials (ASTM) [45] against *Staphylococcus aureus* (*S. aureus*, ATCC 29213) bacteria. Each nanocomposite was placed in a sterile falcon tube containing bacterial suspension (ca. 10^4 colony forming units (CFU)/mL in phosphate-buffered saline, pH 7.4). Falcon tubes with only the inoculum and neat PLA were also prepared as control experiments. The samples were shaken at room temperature at 150 rpm for 24 h. Bacterial concentrations at time 0 and after 24 h were calculated by the plate count method. The measurements were made at least in triplicate.

The biodegradation test for all the samples was conducted under aerobic composting conditions in a laboratory-scale reactor following the ISO-20200 standard [46,47]. Briefly, samples were cut into square geometries of 15 mm × 15 mm and buried 4–6 cm in depth inside reactors containing solid biodegradation media 10 wt% of compost (Compo, Spain), 30 wt% of rabbit food, 10 wt% of starch, 5 wt% of sugar, 1 wt% of urea, 4 wt% of corn oil, 40 wt% of sawdust and water in a 45:55 wt% ratio) and incubated at 58 °C for 30 days. Samples were kept in textile meshes to allow easy removal from the composting medium, while when buried access to microorganisms and moisture was ensured. Water was periodically added to the reaction containers in order to maintain the relative humidity in the medium, and the aerobic conditions were guaranteed by regular mixing of the compost medium. Samples were recovered from the disintegration medium at different time intervals (7, 17, 21, 28, 36, and 44 days), cleaned carefully, and dried in an oven at 37 °C until constant weight. The mass loss weight % was calculated by normalizing the sample weight at different incubation times to the initial weight value. Photographs were taken from samples once extracted from the composting medium.

3. Results and Discussion

Firstly, the synthesis of AgCH-NPs was carried out and confirmed by FE-SEM, UV-Vis analysis as well as by crystallography analysis, as reported in Figure 1a–c, respectively. In particular, the morphology of AgCH-NPs was studied by FE-SEM. Figure 1a shows a representative microphotograph, which reveals a spherical morphology with an average diameter of ~8 ± 1 nm. This result agrees with previously reported values for samples obtained by the same procedure [48]. In addition, from the UV–Vis analysis, the characteristic surface plasmon resonance of silver, centered in 420 nm is obtained by studying the AgCH-NPs solution after lyophilization of the reduced solution of silver nitrate with chitosan [49].

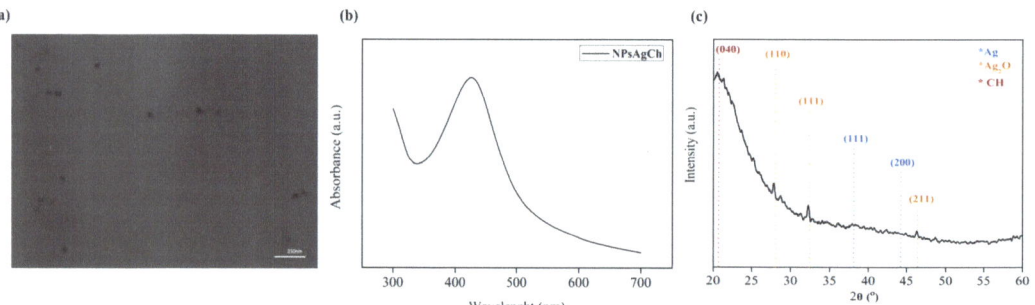

Figure 1. (**a**) FE-SEM micrograph at 100,000 magnification of chitosan-based silver nanoparticles (AgCH-NPs) prepared by reducing AgNO$_3$ in a chitosan solution. (**b**) UV–Vis absorption spectra of chitosan-based silver nanoparticles (AgCH-NPs). (**c**) XRD spectrum of synthesized AgCH-NPs.

Meanwhile, in Figure 1b, XRD has been used in order to identify the characteristic peaks at 38° and 44°, which indicates face-centered cubic Ag crystals, the presence of chitosan at about 20° and the peaks at about 28°, 32° and 46° which correspond to the plane (110), (111) and (211) of Ag$_2$O [50].

In fact, it is known that chitosan always contains bound water (~5%) even when it has been dried to the extreme. The incorporation of bound water into the crystal lattice, commonly termed hydrated crystals, generally gives rise to a more dominated polymorph which can be detected by a broad crystalline peak in the corresponding X-ray pattern. The peak registered near 20.2° is reported to be the indication of the relatively regular crystal lattice (040) of chitosan [51].

Therefore, once the functionalized nanoparticles are obtained, different concentrations of them have been used to be dispersed into a PLA solution, as indicated in Table 1, in order to obtain the reinforced forced-fibers, f-fibers, with different formulations.

Table 1. PLA-AgCH-NPs formulations.

Samples	PLA (wt%)	AgCH-NPs (wt%)
PLA	100	0
PLA-AgCH-1%	99	1
PLA-AgCH-1.5%	98.5	1.5
PLA-AgCH-2%	98	2
PLA-AgCH-3%	97	3
PLA-AgCH-3.5%	96.5	3.5

Once the f-fibers based on PLA reinforced with different concentrations of AgCH-NPs were obtained, their morphology was studied by scanning electron microscopy (SEM) as reported in Figures 2 and 3 for the different concentrations of NPs. In general, for each case, it is possible to observe that, by adding NPs, the fiber morphology changes, resulting in fibers with an increasing number of defects and lower homogeneity in size. In particular, the distribution of f-fibers diameters can be centered into different zones depending on the amount of NPs. For the neat PLA f-fibers, the diameter distribution is centered on a single point with an average value of about 10 µm, while for f-fibers reinforced with small amounts of NPs, that is 1, 1.5, and 2 wt%, the diameter distribution obtained presents a bimodal distribution (Figure 2), with one of them centered at low diameter values, less than 5 µm, and the other at larger ones between 10 and 20 µm, increased with an increasing amount of NPs.

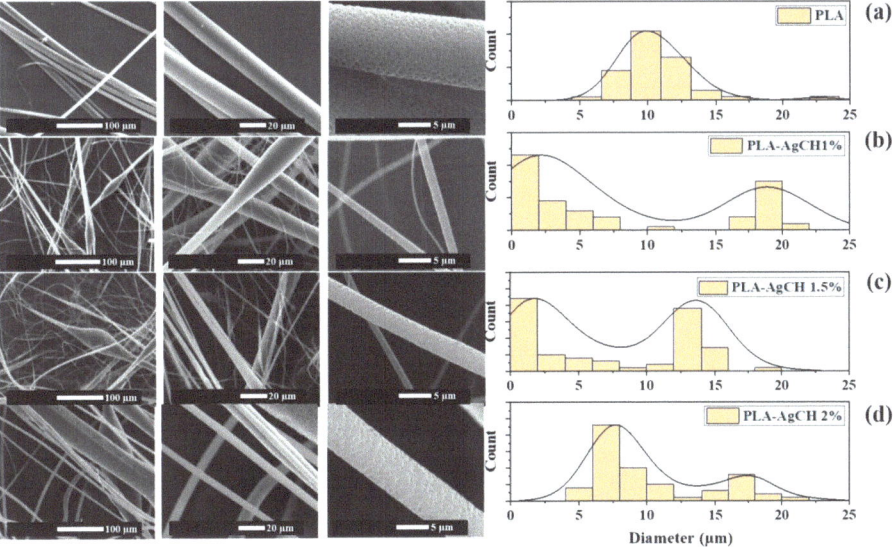

Figure 2. SEM images of PLA and PLA-AgCH-NPs reinforced f-fibers at different magnification, with their corresponding diameter distribution. (a) PLA, (b) 1 wt% AgCH-NPs, (c) 1.5 wt% AgCH-NPs, (d) 2 wt% AgCH-NPs.

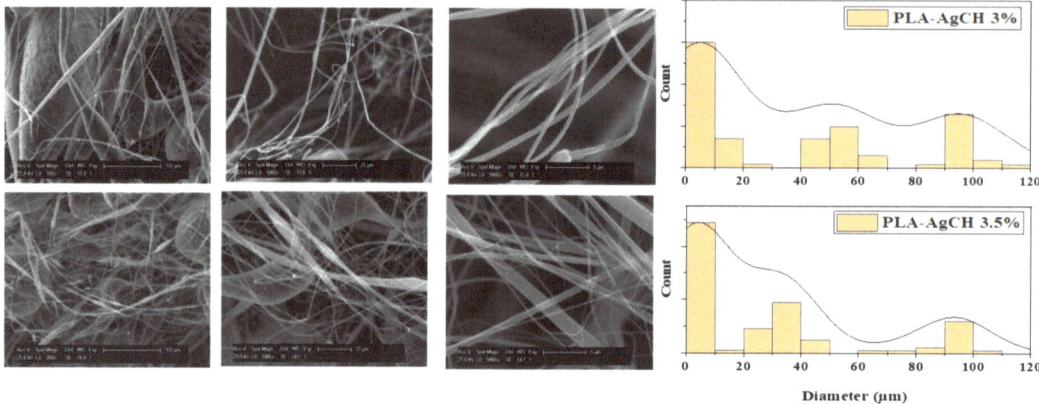

Figure 3. SEM images of PLA-AgCH-NPs reinforced f-fibers at different magnification, with their corresponding diameter distribution for high amount of NPS, that is 3 and 3.5 wt% AgCH-NPs.

Different behavior can be found when a high amount of NPs, such as 3 and 3.5 wt% have been added to the PLA f-fibers, as reported in Figure 3. In fact, the images reflect a material more heterogeneous, with a big variation on the diameter distribution, in respect to the others. In particular, in these f-fibers, a new distribution zone is obtained at higher diameters, centered at about 90 μm.

This phenomenon can be explained due to a critical reinforced effect in PLA when 3 wt% or more AgCH-NPs were added, where the solution cannot be spun correctly, producing fiber aggregates of ~90 μm in diameter. In fact, a significative change in the morphology of the reinforced f-fibers is evidenced between PLA reinforced with a low amount of NPs, PLA-AgCH 1/1.5/2 wt%, and PLA reinforced with a high amount of NPs PLA-AgCH 3/3.5%.

FTIR spectra of neat PLA, AgCH-NPs, and PLA-AgCH-NPs reinforced f-fibers are shown in Figure 4a. At 1748 cm^{-1} there is a carbonyl group stretching vibration band, and at 1454 cm^{-1}, 1383 cm^{-1}, and 1366 cm^{-1} appear -CH$_3$ groups, -CH deformation, and asymmetric bands, respectively. The -C-O- stretching bands appear at 1180 cm^{-1}, 1130 cm^{-1} and 1085 cm^{-1}. AgCH-NPs spectrum shows the characteristics absorption bands of chitosan at 1639.49 cm^{-1} for primary amine, N-H band, 1527.62 cm^{-1} for secondary amine, N-H band, and 1388.75 cm^{-1} for methylene = CH$_2$ CH$_3$ band [52]. A wide band at 1625 cm^{-1} can be observed in all the reinforced f-fibers spectra, which directly demonstrates the presence of nanoparticles in all the f-fibers. This wide band can be attributed to N-H bands of nanoparticles, slightly modified by the presence of PLA. Furthermore, the difference in height between the characteristic PLA peak of the carbonyl group vibration and the N-H band characteristic of nanoparticles is represented in Figure 4b for the different NPs concentrations. It is easy to observe that, for high amounts of NPs, this difference is reduced, confirming the increased presence of nanoparticles within the f-fibers. Moreover, it is remarkable the presence of the peak at ~2300 cm^{-1}, only in the reinforced f-fibers. This indicates an interaction between PLA and AgCH nanoparticles and, therefore, this peak may be due to C=O-N vibrations, which produces new symmetric and asymmetric vibrations of the COO$^-$ anion [53].

In Figure 5, XRD patterns were recorded for all the samples. Neat PLA exhibits a broad reflection indicative of its amorphous nature. With the addition of nanoparticles, a slight tendency to crystallize can be observed, by the appearance of a diffraction peak at 2θ = 16.7°, corresponding to (110/200) planes of PLA, the reflection of the α-form crystals [54]. In addition, new peaks at 2θ = 15.5° and 2θ = 19.2° corresponding to (010) and (203) plane reflections of PLA chains belonging to α and α′ type crystals were visible, which was also consistent with the literature values [54]. This is indicative of the nanoparticles nucleating effect, and the increase of the long-range ordering of crystal packing. Apparently, if no intermolecular interactions exist between the two components, each component will form its own crystalline domains [51] and, thus, the diffraction patterns of the reinforced f-fibers mats should be the simple superposition of those of each component. The obtained results signify that two components, chitosan and PLA, have interacted with each other in a certain manner so that the original crystalline structures of each component have been disturbed or partially damaged to a different extent, leading to various crystalline structures of the reinforced f-fibers.

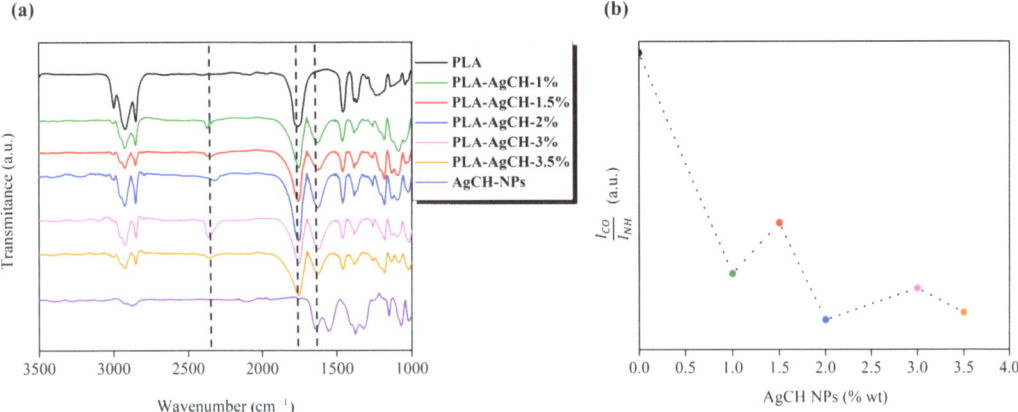

Figure 4. FTIR spectra of PLA, AgCH-NPs and PLA-AgCH-NPs reinforced f-fibers (a) and (b) the relation in height between the characteristic PLA peak of the carbonyl group vibration (1748 cm^{-1}) to the N-H band characteristic of nanoparticle (1625 cm^{-1}).

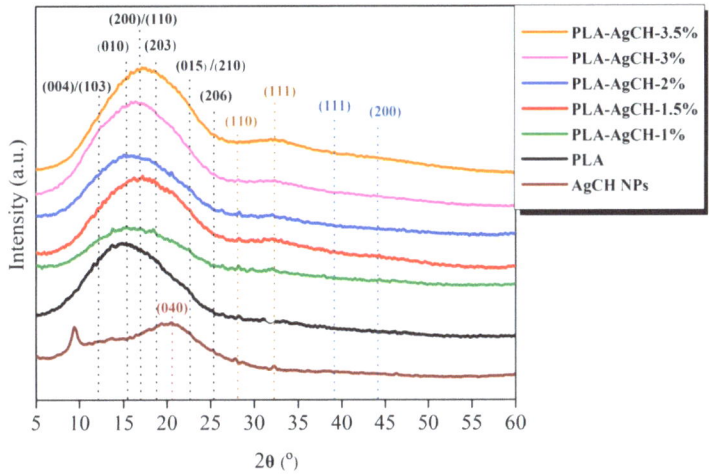

Figure 5. XRD spectra of PLA, AgCH-NPs and PLA-AgCH-NPs reinforced f-fibers.

Moreover, in order to study the effect of the addition of AgCH-NPs on the thermal stability of PLA as well as on the thermal degradation mechanism, thermogravimetric analysis was conducted under nitrogen atmosphere. Figure 6 shows weight loss vs. temperature (Figure 6a) as well as their corresponding derivative curves (Figure 6b), for neat PLA and PLA-AgCH-NPs reinforced f-fibers.

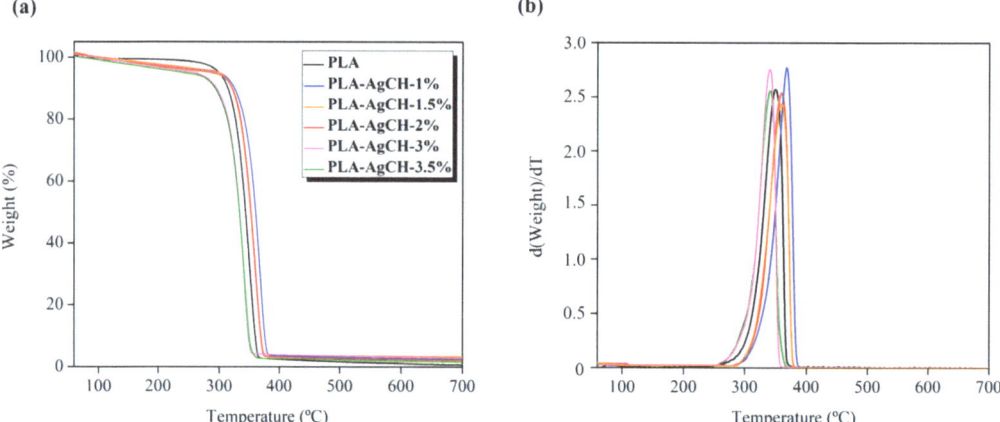

Figure 6. Thermogravimetric curves of neat PLA and its nanocomposites with different AgCH-NPs contents: (**a**) weight vs temperature curves, (**b**) derivative curves.

Table 2 collects temperatures at 5% of weight loss for each sample as well as the maximum degradation temperature and their residue obtained at 700 °C. It can be noted that the degradation mechanism of all the samples is carried out in one step, in Figure 6b. Thus, in general, TGA results show that the addition of the different amounts of AgCH-NPs did not significantly affect the thermal degradation of the PLA matrix. It is worth noticing that there are differences in the curves at high temperatures (700 °C), due to the presence of nanoparticles that decomposed at higher temperatures. This translates into a residue that fairly matches the real amount of NPs present in each fiber formulation.

Table 2. Temperatures at different weight losses of neat PLA and the different formulations containing AgCH-NPs.

Samples	T_{max}	T_5	T_{700}	Tg (°C)	Tm (°C)	Xc (%)
PLA	350	300	0.73	60	145	0.29
PLA AgCH1%	341	246	1.67	41	148	4.20
PLA AgCH1.5%	357	288	2.22	46	147	6.31
PLA AgCH2%	367	298	2.56	45	146	7.61
PLA AgCH3%	340	262	3.23	44	148	8.05
PLA AgCH3.5%	363	297	3.67	38	147	9.96

Differential scanning calorimetry (DSC) was used to study the Tg, melting temperature, and degree of crystallization of PLA and their reinforced f-fibers. Figure 7 shows the thermograms of neat PLA and its AgCH-NPs nanocomposites f-fibers. Glass transition, melting temperatures, and the degree of crystallinity values are summarized in Table 2. The addition of AgCH-NPs into PLA resulted in significant differences over Tg values with respect to the neat matrix.

The addition of AgCH-NPs produced a decrease in Tg values from 60 °C to 38–45 °C, as indicated also in Table 2. This plasticizing effect can be explained since AgCH-NPs may occupy intermolecular spaces between polymer chains, reducing the energy for molecular motion and the formation of hydrogen bonding between the polymer chains which, in turn, increases free volume and molecular mobility. Tm values did not show notable differences when nanoparticles were present.

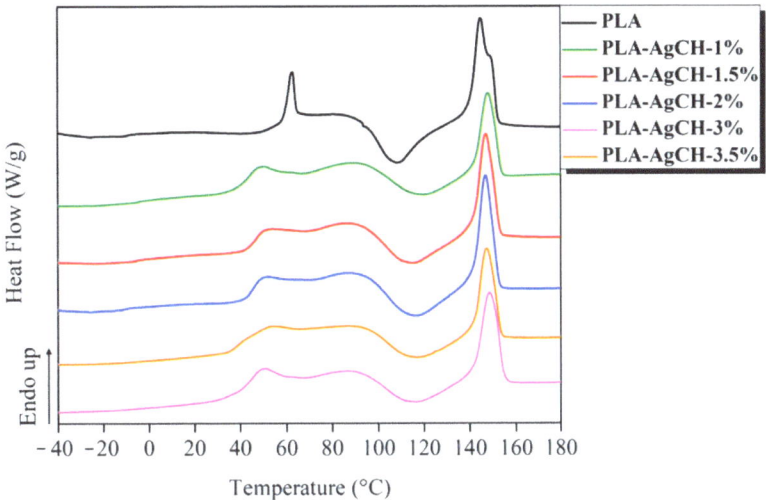

Figure 7. Differential scanning calorimetry (DSC) curves of PLA based f-fibers with different amount of AgCH-NPs contents.

The characteristic mechanical behavior of all the f-fibers formulations studied is summarized in Table 3. Neat PLA showed high elastic modulus, (E), (neat PLA E = 209 MPa; AgCH-NPs nanocomposites E~1–77 MPa) compared to nanocomposites. All the nanocomposites possessed a reduction in the elastic modulus and maximum tensile strength (σmax). Elongation at break, (ε), decreases with the addition of low amounts of nanoparticles (1–1.5 wt%), and increases when higher loads are presented in the nanocomposite. The best balance of mechanical behavior in the nanocomposites was obtained for the lowest amount of AgCH-NPs tested (1 wt%) having the highest Young's modulus and maximum tensile strength of the whole nanocomposites while retaining similar elongation at break. This sample (1 wt% of AgCH-NPs) is totally amorphous, therefore, the low amount of filler helps to retain mechanical properties closer to the neat PLA matrix that also possesses the lowest crystallinity. Thus, it seems that the nanoparticles did not disturb the chain mobility, allowing for an increased elongation at break, which is also enhanced.

Table 3. Mechanical properties for neat PLA and for formulations of PLA containing AgCH-NPs.

Sample	E (MPa)	σ (MPa)	ε Break (%)
PLA	209.07 ± 46.71 [a]	9.51 ± 1.79 [a]	15.79 ± 8.89 [a,b]
PLA-AgCH 1%	76.73 ± 18.14 [b]	3.32 ± 0.89 [b]	11.31 ± 1.69 [a,b]
PLA-AgCH 1.5%	43.98 ± 15.89 [b]	1.34 ± 0.47 [b]	5.92 ± 1.74 [a]
PLA-AgCH 2%	35.32 ± 6.79 [b]	1.67 ± 0.74 [b]	10.04 ± 3.55 [a]
F ratio	14.02	14.39	3.69
p-Value	0.0000 *	0.0000 *	0.0150 *

Different letters in the column indicate significant differences according to Tukey's test ($p < 0.05$). * Values are significant at $p < 0.05$.

Once the mechanical and thermal characterizations of the materials were carried out and in order to evaluate the thermally-activated shape-memory response of the material, three thermo-mechanical cycles were performed by DMA analysis for each formulation at 60 °C. In our previous work, the thermally-activated shape-memory response of woven non-woven PLA fibers obtained by a similar processing method based on the electrospinning technique was successfully demonstrated at 60 °C [13]. Thus, the shape-memory capability of fibers obtained by the force-spinning technique will be studied in this section in order to evaluate the suitability of the force-spinning processing technique for producing smart materials.

As a preliminary study, we verify that reinforced f-fibers are able to show thermally-activated shape-memory response, activated by Tg, which means that we studied how the addition of the higher amount of nanoparticles, that is 2 wt% AgCH NPs, affects the shape-memory behavior. With this aim, the thermo-mechanical cycles for the reinforced formulation are reported in Figure 8. In particular, we obtain an important result, that is PLA f-fibers reinforced with 2 wt% AgCH NPs show thermally-activated shape-memory response at 60 °C. In this case, adding 2 wt% AgCH NPs successfully enhances the ability to fix the temporary form by obtaining values of higher than 80% for R_f during all the thermo-mechanical cycles.

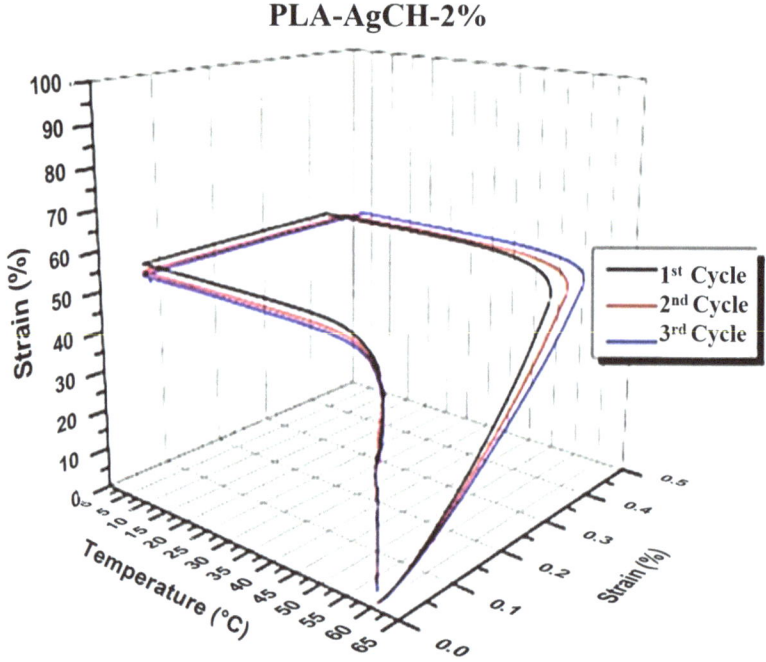

Figure 8. 3D thermo-mechanical cycles performed at 60 °C for PLA-AgCH-2%.

Additionally, good recovery ratios were reported for each thermo-mechanical cycle showing the excellent shape recovery of the original shape at 60 °C during all the tests performed with R_r values higher than 90% for all the thermo-mechanical cycles.

Looking at the results obtained by TGA analysis, the degradation temperature obtained for this formulation was 367 °C (the highest of all the formulation tested) which translates into greater compatibility between nanofillers and the polymeric matrix. This good interaction between AgCH-NPs and PLA matrix improves the fixing capacity of the fibers while the recoverability of the original shape is preserved, showing recovery ratios higher than 90% in all the thermo-mechanical cycles.

The effectiveness of these nanocomposites against gram-positive S. aureus bacteria was evaluated by the ASTM standard method [45]. Figure 9 displays the antibacterial activity of all nanocomposites represented as the percentage of bacteria kill and their confidence interval of 95%, which was calculated by the difference between the CFU after contact with control substrates (PLA and none) and CFUs after contact with the polymeric nanocomposites. When AgCH-NPs are introduced, this confers antibacterial activity to the matrix. The behavior against gram-positive bacteria is very effective in fibers since with

1 wt% of nanoparticles, the killing percentage of bacteria reaches 65% and from 1.5 wt%, this percentage becomes 90%.

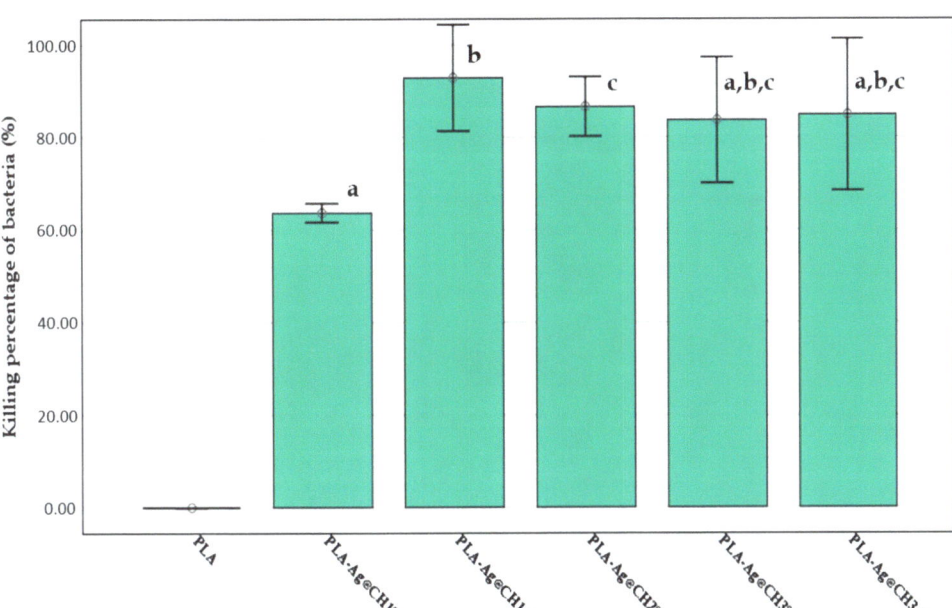

Figure 9. Percentage of killing bacteria for the different nanocomposites. Values having the same letter are not significantly different for the Tukey test (significance level of $p \leq 0.05$).

In order to evaluate the ability of PLA and PLA-AgCH-NPs f-fibers to undergo disintegration, firstly, a visual examination of samples at different times when subjected to composting conditions was performed and results are collected in Figure 10. It should be taken into account that the degradation experiments took place at 58 °C, which is around the T_g of PLA and higher than the T_g of the nanocomposite f-fibers. The degradation rate is much greater above the glass transition temperature as polymer chains become more flexible and water absorption increases, accelerating hydrolysis and microbial attachment. PLA is susceptible to hydrolysis due to the hydrolyzable functional groups in its backbone. On the seventh day, fragmentation and weight loss of the composites were already observed for all samples, especially in low-load fibers.

The environmental degradation process of PLA is affected by its material properties such as molecular first-order structural (molecular weight, optical purity) and higher order structures (crystallinity, T_g, and T_m), and by environmental factors such as humidity, temperature, and catalytic species (pH and the presence of enzymes or microorganisms) [55]. Crystalline regions hydrolyze much more slowly than the amorphous regions as water diffuses more readily into the less organized amorphous regions compared to the more ordered crystalline regions, causing greater rates of hydrolysis and increased susceptibility to biodegradation [46]. This explains the results obtained in Figure 10, the faster degradation of PLA and PLA AgCH 1–2%, which are less crystalline than PLA AgCH 3–3.5%, as we have seen in the DSC analysis.

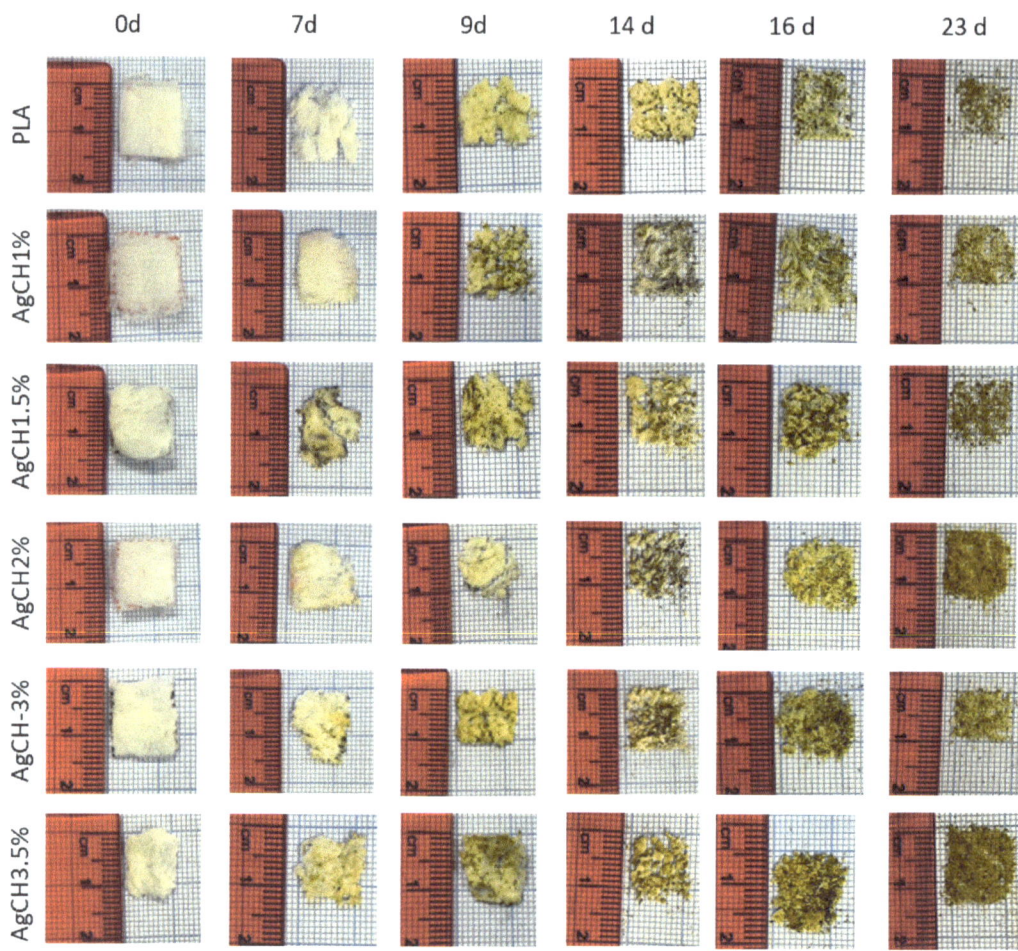

Figure 10. Disintegrated samples under composting conditions.

4. Conclusions

In this work, PLA-based fiber mats, containing between 0 and 3.5% by weight of chitosan-functionalized silver nanoparticles obtained through a green protocol, have been prepared by means of centrifugal force-spinning. Their morphological, thermal, and mechanical properties have been profusely characterized, as well as their functional properties in terms of biodegradation, antibacterial and smart properties. Homogeneous fiber mats with average fiber diameters in the micrometer scale have been prepared. We found that the best balance in terms of mechanical behavior was obtained for the lowest amount of AgCH-NPs, that is 1 wt%, which presented the highest Young's modulus. Moreover, with 2 wt% of AgCH-NPs, we obtain a very good thermally-activated shape-memory effect with very high values for Rf and Rr. Furthermore, AgCH-NPs confer antibacterial activity to the matrix. In particular, the behavior against gram-positive bacteria is very effective in fibers, since with 1 wt% of nanoparticles, the killing percentage of bacteria reaches 65%, and from 1.5 wt%, this percentage becomes 90%. Finally, all the samples are disintegrable under composting conditions, conferring promising properties for possible food packaging application and in the biomedical field.

Author Contributions: Conceptualization, D.L.; methodology, D.L., L.P. and M.F.-G.; validation, D.L., L.P., M.F.-G. and A.M.-B.; investigation, M.D.M.-A., A.L., V.H., D.L. and C.E.; data curation, M.D.M.-A., V.S., D.L. and L.P.; writing—original draft preparation, M.D.M.-A.; writing—review and editing, L.P., V.S., A.M.-B. and D.L.; supervision, L.P., M.F.-G., A.M.-B. and D.L.; funding acquisition, L.P., A.M.-B. and D.L. All authors have read and agreed to the published version of the manuscript.

Funding: This research was funded by the Agencia Estatal de Investigación (AEI, MICINN, Spain) and Fondo Europeo de Desarrollo Regional (FEDER, EU), PID2021-123753NB-C31, PID2019-104600RB-I00, and TED2021-129335B-C21.

Institutional Review Board Statement: Not applicable.

Data Availability Statement: The data presented in this study are available on request from the corresponding author.

Conflicts of Interest: The authors declare no conflict of interest.

References

1. Sessini, V.; Raquez, J.M.; Lo Re, G.; Mincheva, R.; Kenny, J.M.; Dubois, P.; Peponi, L. Multiresponsive Shape Memory Blends and Nanocomposites Based on Starch. *ACS Appl. Mater. Interfaces* **2016**, *8*, 19197–19201. [CrossRef] [PubMed]
2. Karger-Kocsis, J.; Kéki, S. Review of Progress in Shape Memory Epoxies and Their Composites. *Polymers* **2017**, *10*, 34. [CrossRef] [PubMed]
3. Chow, W.S.; Mohd Ishak, Z.A. Smart Polymer Nanocomposites: A Review. *Express Polym. Lett.* **2020**, *14*, 416–435. [CrossRef]
4. Peponi, L.; Navarro-Baena, I.; Kenny, J.M. *Shape Memory Polymers: Properties, Synthesis and Applications*; Woodhead Publishing: Sawston, UK, 2014; ISBN 9780857096951.
5. Peponi, L.; Puglia, D.; Torre, L.; Valentini, L.; Kenny, J.M. Processing of Nanostructured Polymers and Advanced Polymeric Based Nanocomposites. *Mater. Sci. Eng. R Rep.* **2014**, *85*, 1–46. [CrossRef]
6. Zhang, Z.; Lu, S.; Cai, R.; Tan, W. Sensors. *Nano Today* **2021**, *38*, 101202. [CrossRef]
7. Qiu, X.; Hu, S. "Smart" Materials Based on Cellulose: A Review of the Preparations, Properties, and Applications. *Materials* **2013**, *6*, 738–781. [CrossRef]
8. Mehrpouya, M.; Vahabi, H.; Janbaz, S.; Darafsheh, A.; Mazur, T.R.; Ramakrishna, S. 4D Printing of Shape Memory Polylactic Acid (PLA). *Polymer* **2021**, *230*, 124080. [CrossRef]
9. Nathal, M.V.; Stefko, G.L. Smart Materials and Active Structures. *J. Aerosp. Eng.* **2013**, *26*, 491–499. [CrossRef]
10. Sessini, V.; López Galisteo, A.J.; Leonés, A.; Ureña, A.; Peponi, L. Sandwich-Type Composites Based on Smart Ionomeric Polymer and Electrospun Microfibers. *Front. Mater.* **2019**, *6*, 301. [CrossRef]
11. Merlettini, A.; Gigli, M.; Ramella, M.; Gualandi, C.; Soccio, M.; Boccafoschi, F.; Munari, A.; Lotti, N.; Focarete, M.L. Thermal Annealing to Modulate the Shape Memory Behavior of a Biobased and Biocompatible Triblock Copolymer Scaffold in the Human Body Temperature Range. *Biomacromolecules* **2017**, *18*, 2490–2508. [CrossRef]
12. Arrieta, M.P.; Sessini, V.; Peponi, L. Biodegradable Poly(Ester-Urethane) Incorporated with Catechin with Shape Memory and Antioxidant Activity for Food Packaging. *Eur. Polym. J.* **2017**, *94*, 111–124. [CrossRef]
13. Leonés, A.; Sonseca, A.; López, D.; Fiori, S.; Peponi, L. Shape Memory Effect on Electrospun PLA-Based Fibers Tailoring Their Thermal Response. *Eur. Polym. J.* **2019**, *117*, 217–226. [CrossRef]
14. Leonés, A.; Lieblich, M.; Benavente, R.; Gonzalez, J.L.; Peponi, L. Potential Applications of Magnesium-Based Polymeric Nanocomposites Obtained by Electrospinning Technique. *Nanomaterials* **2020**, *10*, 1524. [CrossRef]
15. Mujica-Garcia, A.; Navarro-Baena, I.; Kenny, J.M.; Peponi, L. Influence of the Processing Parameters on the Electrospinning of Biopolymeric Fibers. *J. Renew. Mater.* **2014**, *2*, 23–34. [CrossRef]
16. El-hadi, A.; Al-Jabri, F. Influence of Electrospinning Parameters on Fiber Diameter and Mechanical Properties of Poly(3-Hydroxybutyrate) (PHB) and Polyanilines (PANI) Blends. *Polymers* **2016**, *8*, 97. [CrossRef]
17. Manea, L.R.; Bertea, A.-P.; Nechita, E.; Popescu, C.V. Mathematical Modeling of the Relation between Electrospun Nanofibers Characteristics and the Process Parameters. In *Electrospinning Method Used to Create Functional Nanocomposites Films*; IntechOpen: London, UK, 2018; pp. 91–111. [CrossRef]
18. Rodríguez-Sánchez, I.J.; Fuenmayor, C.A.; Clavijo-Grimaldo, D.; Zuluaga-Domínguez, C.M. Electrospinning of Ultra-Thin Membranes with Incorporation of Antimicrobial Agents for Applications in Active Packaging: A Review. *Int. J. Polym. Mater. Polym. Biomater.* **2020**, *70*, 1–24. [CrossRef]
19. Basar, A.O.; Castro, S.; Torres-Giner, S.; Lagaron, J.M.; Turkoglu Sasmazel, H. Novel Poly(ε-Caprolactone)/Gelatin Wound Dressings Prepared by Emulsion Electrospinning with Controlled Release Capacity of Ketoprofen Anti-Inflammatory Drug. *Mater. Sci. Eng. C* **2017**, *81*, 459–468. [CrossRef]
20. Rana, S.; Cho, J.W. Core-Sheath Polyurethane-Carbon Nanotube Nanofibers Prepared by Electrospinning. *Fibers Polym.* **2011**, *12*, 721–726. [CrossRef]
21. Arrieta, M.P.; Gil, A.L.; Yusef, M.; Kenny, J.M.; Peponi, L. Electrospinning of PCL-Based Blends: Processing Optimization for Their Scalable Production. *Materials* **2020**, *13*, 3853. [CrossRef] [PubMed]

22. Merlettini, A.; Pandini, S.; Agnelli, S.; Gualandi, C.; Paderni, K.; Messori, M.; Toselli, M.; Focarete, M.L. Facile Fabrication of Shape Memory Poly(ε-Caprolactone) Non-Woven Mat by Combining Electrospinning and Sol-Gel Reaction. *RSC Adv.* **2016**, *6*, 43964–43974. [CrossRef]
23. Dziemidowicz, K.; Sang, Q.; Wu, J.; Zhang, Z.; Zhou, F.; Lagaron, J.M.; Mo, X.; Parker, G.J.M.; Yu, D.-G.; Zhu, L.-M.; et al. Electrospinning for Healthcare: Recent Advancements. *J. Mater. Chem. B* **2021**, *9*, 939–951. [CrossRef] [PubMed]
24. Zhang, F.; Xia, Y.; Wang, L.; Liu, L.; Liu, Y.; Leng, J. Conductive Shape Memory Microfiber Membranes with Core-Shell Structures and Electroactive Performance. *ACS Appl. Mater. Interfaces* **2018**, *10*, 35526–35532. [CrossRef] [PubMed]
25. Li, M.; Qiu, W.; Wang, Q.; Li, N.; Liu, L.; Wang, X.; Yu, J.; Li, X.; Li, F.; Wu, D. Nitric Oxide-Releasing Tryptophan-Based Poly(Ester Urea)s Electrospun Composite Nanofiber Mats with Antibacterial and Antibiofilm Activities for Infected Wound Healing. *ACS Appl. Mater. Interfaces* **2022**, *14*, 15911–15926. [CrossRef]
26. Qi, Y.; Wang, C.; Wang, Q.; Zhou, F.; Li, T.; Wang, B.; Su, W.; Shang, D.; Wu, S. A Simple, Quick, and Cost-Effective Strategy to Fabricate Polycaprolactone/Silk Fibroin Nanofiber Yarns for Biotextile-Based Tissue Scaffold Application. *Eur. Polym. J.* **2023**, *186*, 111863. [CrossRef]
27. Lu, Y.; Li, Y.; Zhang, S.; Xu, G.; Fu, K.; Lee, H.; Zhang, X. Parameter Study and Characterization for Polyacrylonitrile Nanofibers Fabricated via Centrifugal Spinning Process. *Eur. Polym. J.* **2013**, *49*, 3834–3845. [CrossRef]
28. Zhang, X.; Lu, Y. Centrifugal Spinning: An Alternative Approach to Fabricate Nanofibers at High Speed and Low Cost Centrifugal Spinning: An Alternative Approach to Fabricate Nanofibers at High Speed and Low Cost. *Polym. Rev.* **2014**, *3724*, 677–701. [CrossRef]
29. Krifa, M.; Hammami, M.A.; Wu, H. Occurrence and Morphology of Bead-on-String Structures in Centrifugal Forcespun PA6 Fibers. *J. Text. Inst.* **2015**, *106*, 284–294. [CrossRef]
30. Zhang, Q.; Bao, N.; Wang, X.; Hu, X.; Miao, X.; Chaker, M.; Ma, D. Advanced Fabrication of Chemically Bonded Graphene/TiO2 Continuous Fibers with Enhanced Broadband Photocatalytic Properties and Involved Mechanisms Exploration. *Sci. Rep.* **2016**, *6*, 38066. [CrossRef]
31. Zannini Luz, H.; Loureiro dos Santos, L.A. Centrifugal Spinning for Biomedical Use: A Review. *Crit. Rev. Solid State Mater. Sci.* **2022**, 1–16. [CrossRef]
32. Sarkar, K.; Gomez, C.; Zambrano, S.; Ramirez, M.; de Hoyos, E.; Vasquez, H.K.L. Electrospinning to Forcespinning. *Mater. Today* **2010**, *13*, 42–44. [CrossRef]
33. Raquez, J.M.; Habibi, Y.; Murariu, M.; Dubois, P. Polylactide (PLA)-Based Nanocomposites. *Prog. Polym. Sci.* **2013**, *38*, 1504–1542. [CrossRef]
34. Hamley, I.W.; Parras, P.; Castelletto, V.; Castillo, R.V.; Müller, A.J.; Pollet, E.; Dubois, P.; Martin, C.M. Melt Structure and Its Transformation by Sequential Crystallization of the Two Blocks within Poly(L-Lactide)-Block-Poly(ε-Caprolactone) Double Crystalline Diblock Copolymers. *Macromol. Chem. Phys.* **2006**, *207*, 941–953. [CrossRef]
35. Navarro-Baena, I.; Kenny, J.M.; Peponi, L. Thermally-Activated Shape Memory Behaviour of Bionanocomposites Reinforced with Cellulose Nanocrystals. *Cellulose* **2014**, *21*, 4231–4246. [CrossRef]
36. Sonseca, A.; Madani, S.; Muñoz-Bonilla, A.; Fernández-García, M.; Peponi, L.; Leonés, A.; Rodríguez, G.; Echeverría, C.; López, D. Biodegradable and Antimicrobial Pla–Ola Blends Containing Chitosan-Mediated Silver Nanoparticles with Shape Memory Properties for Potential Medical Applications. *Nanomaterials* **2020**, *10*, 1065. [CrossRef]
37. Rizki, I.N.; Klaypradit, W. Patmawati Utilization of Marine Organisms for the Green Synthesis of Silver and Gold Nanoparticles and Their Applications: A Review. *Sustain. Chem. Pharm.* **2023**, *31*, 100888. [CrossRef]
38. Beyth, N.; Farah, S.; Domb, A.J.; Weiss, E.I. Antibacterial Dental Resin Composites. *React. Funct. Polym.* **2014**, *75*, 81–88. [CrossRef]
39. Vimala, K.; Mohan, Y.M.; Sivudu, K.S.; Varaprasad, K.; Ravindra, S.; Reddy, N.N.; Padma, Y.; Sreedhar, B.; MohanaRaju, K. Fabrication of Porous Chitosan Films Impregnated with Silver Nanoparticles: A Facile Approach for Superior Antibacterial Application. *Colloids Surf. B Biointerfaces* **2010**, *76*, 248–258. [CrossRef]
40. Huang, H.; Yang, X. Synthesis of Chitosan-Stabilized Gold Nanoparticles in the Absence/Presence of Tripolyphosphate. *Biomacromolecules* **2004**, *5*, 2340–2346. [CrossRef]
41. Antunes, J.C.; Domingues, J.M.; Miranda, C.S.; Silva, A.F.G.; Homem, N.C.; Amorim, M.T.P.; Felgueiras, H.P. Bioactivity of Chitosan-Based Particles Loaded with Plant-Derived Extracts for Biomedical Applications: Emphasis on Antimicrobial Fiber-Based Systems. *Mar. Drugs* **2021**, *19*, 359. [CrossRef] [PubMed]
42. Chandrasekaran, M.; Kim, K.D.; Chun, S.C. Antibacterial Activity of Chitosan Nanoparticles: A Review. *Processes* **2020**, *8*, 1–21. [CrossRef]
43. Peponi, L.; Navarro-Baena, I.; Báez, J.E.; Kenny, J.M.; Marcos-Fernández, A. Effect of the Molecular Weight on the Crystallinity of PCL-b-PLLA Di-Block Copolymers. *Polymer* **2012**, *53*, 4561–4568. [CrossRef]
44. Peponi, L.; Navarro-Baena, I.; Sonseca, A.; Gimenez, E.; Marcos-Fernandez, A.; Kenny, J.M. Synthesis and Characterization of PCL-PLLA Polyurethane with Shape Memory Behavior. *Eur. Polym. J.* **2013**, *49*, 893–903. [CrossRef]
45. ASTM E2149-13a; Standard Test Method for Determining the Antimicrobial Activity of Antimicrobial Agents under Dynamic Contact Conditions. American Society for Testing and Materials: West Conshohocken, PA, USA, 2013.
46. Arrieta, M.P.; Perdiguero, M.; Fiori, S.; Kenny, J.M.; Peponi, L. Biodegradable Electrospun PLA-PHB Fibers Plasticized with Oligomeric Lactic Acid. *Polym. Degrad. Stab.* **2020**, *179*, 109226. [CrossRef]

47. *UNE-EN ISO-20200*; Determination of the Degree of Disintegration of Plastic Materials under Simulated Composting Conditions in a Laboratory-Scale Test. AENOR: Madrid, Spain, 2015.
48. Sonseca, A.; Madani, S.; Rodríguez, G.; Hevilla, V.; Echeverría, C.; Fernández-García, M.; Muñoz-Bonilla, A.; Charef, N.; López, D. Multifunctional PLA Blends Containing Chitosan Mediated Silver Nanoparticles: Thermal, Mechanical, Antibacterial, and Degradation Properties. *Nanomaterials* **2020**, *10*, 22. [CrossRef]
49. Jia, X.; Ma, X.; Wei, D.; Dong, J.; Qian, W. Direct Formation of Silver Nanoparticles in Cuttlebone-Derived Organic Matrix for Catalytic Applications. *Colloids Surfaces A Physicochem. Eng. Asp.* **2008**, *330*, 234–240. [CrossRef]
50. Pawar, O.; Deshpande, N.; Dagade, S.; Waghmode, S.; Nigam Joshi, P. Green Synthesis of Silver Nanoparticles from Purple Acid Phosphatase Apoenzyme Isolated from a New Source Limonia Acidissima. *J. Exp. Nanosci.* **2016**, *11*, 28–37. [CrossRef]
51. Wan, Y.; Wu, H.; Yu, A.; Wen, D. Biodegradable Polylactide/Chitosan Blend Membranes. *Biomacromolecules* **2006**, *7*, 1362–1372. [CrossRef]
52. Senthilkumar, P.; Yaswant, G.; Kavitha, S.; Chandramohan, E.; Kowsalya, G.; Vijay, R.; Sudhagar, B.; Kumar, D.S.R.S. Preparation and Characterization of Hybrid Chitosan-Silver Nanoparticles (Chi-Ag NPs); A Potential Antibacterial Agent. *Int. J. Biol. Macromol.* **2019**, *141*, 290–297. [CrossRef]
53. Dhanapal, J.; Ravindrran, M.; Baskar, S. Toxic Effects of Aflatoxin B1 on Embryonic Development of Zebrafish (Danio Rerio): Potential Activity of Piceatannol Encapsulated Chitosan/Poly (Lactic Acid) Nanoparticles. *Anticancer. Agents Med. Chem.* **2015**, *15*, 248–257. [CrossRef]
54. Kawai, T.; Rahman, N.; Matsuba, G.; Nishida, K.; Kanaya, T.; Nakano, M.; Okamoto, H.; Kawada, J.; Usuki, A.; Honma, N.; et al. Crystallization and Melting Behavior of Poly (L-Lactic Acid). *Macromolecules* **2007**, *40*, 9463–9469. [CrossRef]
55. Rudnik, E.; Briassoulis, D. Degradation Behaviour of Poly(Lactic Acid) Films and Fibres in Soil under Mediterranean Field Conditions and Laboratory Simulations Testing. *Ind. Crops Prod.* **2011**, *33*, 648–658. [CrossRef]

Disclaimer/Publisher's Note: The statements, opinions and data contained in all publications are solely those of the individual author(s) and contributor(s) and not of MDPI and/or the editor(s). MDPI and/or the editor(s) disclaim responsibility for any injury to people or property resulting from any ideas, methods, instructions or products referred to in the content.

Review

A Review of Electrospun Nanofiber Interleaves for Interlaminar Toughening of Composite Laminates

Biltu Mahato *, Stepan V. Lomov, Aleksei Shiverskii, Mohammad Owais and Sergey G. Abaimov *

Center for Petroleum Science and Engineering, Skolkovo Institute of Science and Technology, Bolshoy Boulevard 30, bld. 1, Moscow 121205, Russia
* Correspondence: biltu.mahato@skoltech.ru (B.M.); s.abaimov@skoltech.ru (S.G.A.)

Abstract: Recently, polymeric nanofiber veils have gained lot of interest for various industrial and research applications. Embedding polymeric veils has proven to be one of the most effective ways to prevent delamination caused by the poor out-of-plane properties of composite laminates. The polymeric veils are introduced between plies of a composite laminate, and their targeted effects on delamination initiation and propagation have been widely studied. This paper presents an overview of the application of nanofiber polymeric veils as toughening interleaves in fiber-reinforced composite laminates. It presents a systematic comparative analysis and summary of attainable fracture toughness improvements based on electrospun veil materials. Both Mode I and Mode II tests are covered. Various popular veil materials and their modifications are considered. The toughening mechanisms introduced by polymeric veils are identified, listed, and analyzed. The numerical modeling of failure in Mode I and Mode II delamination is also discussed. This analytical review can be used as guidance for veil material selection, for estimation of the achievable toughening effect, for understanding the toughening mechanism introduced by veils, and for the numerical modeling of delamination.

Keywords: fracture toughness; electrospun veil/interleave; delamination; toughening mechanism; cohesive zone modeling

Citation: Mahato, B.; Lomov, S.V.; Shiverskii, A.; Owais, M.; Abaimov, S.G. A Review of Electrospun Nanofiber Interleaves for Interlaminar Toughening of Composite Laminates. *Polymers* **2023**, *15*, 1380. https://doi.org/10.3390/polym15061380

Academic Editors: Jesús-María García-Martínez and Emilia P. Collar

Received: 9 February 2023
Revised: 2 March 2023
Accepted: 8 March 2023
Published: 10 March 2023

Copyright: © 2023 by the authors. Licensee MDPI, Basel, Switzerland. This article is an open access article distributed under the terms and conditions of the Creative Commons Attribution (CC BY) license (https:// creativecommons.org/licenses/by/ 4.0/).

1. Introduction

Fiber-reinforced composite materials have excellent mechanical properties, corrosion resistance, and creep resistance compared with traditional materials [1–3], and consequently, are widely accepted and used for various structural applications in the aircraft, automobile, energy, ship, civil, sports, and offshore industries, to name a few. Such high-performance structural composite laminates are commonly produced either using autoclave technology or using liquid composite molding, based on preforms as layups of 2D plies with fibrous reinforcement. These laminates have high in-plane mechanical properties that are determined by the fibers, but suffer from low out-of-plane properties because interlaminar fracture toughness (FT) is only provided, besides the matrix, by partial fibrous involvement in the form of the fiber bridging effect. This makes a composite laminate highly susceptible to failure under through-the-thickness loads and out-of-plane low-velocity impacts. Events such as matrix and fiber cracking, as well as delaminations, are observed during such failure. The low interlaminar FT of a composite laminate remains one of the limiting factors during its service life. Poor interlaminar strength and interlaminar FT are thus major limitations of fiber-reinforced composite laminates.

The concentration of high interlaminar shear and transverse stress near the edges, possible pre-cracks or manufacturing defects, points of laminate curvature, and drilled holes are some of the possible causes of delamination initiation. In these cases, failure occurs via both modes (Mode I and Mode II) of failure, resulting in various in-service problems. Various methods are used to improve the interlaminar properties of structural composites:

matrix toughening [4,5], 3D reinforcement (3D weaving [6], stitching [6], Z-pinning [7]), nano-stitching [8], and fiber hybridization [9]. These methods improve the FT but come at the expense of either escalated complexity, increased cost/weight, or the loss of in-plane properties. For example, the involvement of 3D reinforcement, in the form of 3D weaving, stitching, or Z-pinning, solves the problem but cannot be implemented in structures with high load-carrying performance because of the fiber crimp and, thereby, loss in the targeted stiffness/mass ratio. In contrast to the 3D reinforcement method, nano-stitching improves FT and does not degrade the in-plane properties, but it is not a scalable method. Therefore, industry applications require measures to improve interlaminar FT in laminates (as layups of 2D plies), which would overcome the aforementioned drawbacks.

Recently, the use of polymeric non-woven nanofiber in the form of a thin mat has been a popular approach to toughening composite laminates. In this method, the thin mat is introduced as an additional layer between the laminae of a composite laminate. The thin mat is commonly known as a veil/interleave, and the method of introducing the veil/interleave is known as interleaving. The fiber diameter in the fibrous veil/interleave ranges from tens of nanometers to a few micrometers. The fine diameter and evenly distributed fibers of the veil ensure low areal density and low thickness [10]. Hence, the impact of introducing a veil/interleave on the laminate thickness and mass is negligible. The overall fiber volume fraction in the composite laminate is not affected much, guaranteeing the lowest level of compromise on the in-plane mechanical properties of the laminate. Similarly, the veil/interleaves are highly porous, and thus, do not disrupt the resin flow during impregnation or curing [11]. The effectiveness of veils/interleaves in toughening composite laminates has been demonstrated without doubt [12–14]. Melt blowing, solution blowing, electrospinning, etc., are some methods of producing such non-woven polymeric veils.

Four parameters were used to quantify and characterize the effectiveness of toughening, measured in terms of FT for the initiation and propagation of Mode I and Mode II delamination. These parameters are extensively reported for various fiber, matrix, and interleave systems of laminates. The toughening effect depends on the veil material, thickness, fiber diameter, areal density, form (melted vs. non-melted, solid vs. hollow), reinforcing fibers, and compatibility amongst the veil, fiber, and matrix. Most of the literature on this topic (see the publication statistics below) has been partially summarized in several reviews [15–17], with the most recent published in 2017.

This analytical review focuses on the toughness efficiency of the veils/interleaves produced via electrospinning. It offers a systematic comparative analysis of data from the literature up to 2022 from the viewpoint of the targeted increase of interlaminar FT using interleaving polymeric veils. Both modes of delamination failure observed in the baseline and interleaved composite laminates are covered for various veil materials. The toughening mechanisms added through the introduction of veils are listed, analyzed, and discussed. The numerical modeling of failure by both modes of delamination is also discussed. The present analytical review can be used as guidance for veil material selection, for estimation of the achievable toughening effect, for understanding the toughening mechanisms introduced by veils, and for the numerical modeling of delamination.

Section 2 presents the methodology followed for data gathering and analysis. It also shows the statistics and trends of publications on the topic of the review. Section 3 describes the electrospinning process used for the production of veils and briefly explains its history. Section 4 presents a comparative analysis of the toughening effects achieved by interleaving electrospun veils manufactured from different polymers. Section 5 analyzes the toughening mechanism noted for electrospun veil-interleaved laminates. Section 6 outlines a statistical analysis of the collected data on the basis of mode ratio and areal density. Section 7 introduces a general methodology for developing a numerical model of the toughening effect of electrospun veils using cohesive zone modeling. Finally, conclusions are drawn in Section 8.

2. Methodology

The factual data on toughening were structured based on the veil material (polyethylene terephthalate (PET), polyphenylene sulfide (PPS), polyamide (PA), polyacrylonitrile (PAN), and polycaprolactone (PCL)), including their hybrids and modifications, such as metal-, CNT-, or graphene-modified veils. The structured data contained information on reinforcement, the matrix and veil, the production methods, the FT of the baseline and interleaved laminates for Mode I and Mode II, improvements in the FT of interleaved laminates compared to the baseline laminate, and identification of their application area and toughening mechanics. The collected factual data were used for analysis and can be obtained from http://bit.ly/3TEDvzn (accessed on 7 March 2023). The analysis was carried out using MS Excel^{TM} 2010 and MATLAB^{TM} R2021.

Publications derived from keywords related to the topic, such as "veil OR electrosp*" (herein referred to as VE), "VE" AND "composite OR laminate" (herein referred to as CL), and "VE" AND "CL" AND "toughness", were searched in the Web of Science database (core collection, all years up to September 2021). This revealed the ongoing dynamics in the field's development. The search was further narrowed using the filter "fiber OR fibre". The results are shown in Figure 1; note the logarithmic scale on the vertical axis in Figure 1a.

"VE" as a keyword for the search provided extensive results. However, the narrowing to fibrous materials helped to focus the search from 117,862 to 15,621 documents, including articles, conference papers, reviews, book chapters, and editorial materials. Similarly, there were 5272 and 240 documents on "VE AND CL" and "VE AND CL AND toughness", respectively, when narrowed using the filter "fiber OR fibre", as presented in Figure 1a. The data show escalating interest in the proposed topic, which grew fast in the last decade, with the number of publications on "VE AND CL AND toughness" increasing at a rate of 10x. This increase reflects general growth in publication activity. Therefore, the numbers were compared with the total number of publications on the toughness of fiber-reinforced composites ("Composite AND Toughness" refined using "fiber OR fibre") to calculate the percentage contribution of VE as a toughening technology. The total contribution increased from below 1% in 2001 to ~6% in 2022, as presented in Figure 1b. A linear fit to the data shows a steadily increasing trend of publications in the field. So, the topic of the present research is in the growth phase of the technology evolution curve.

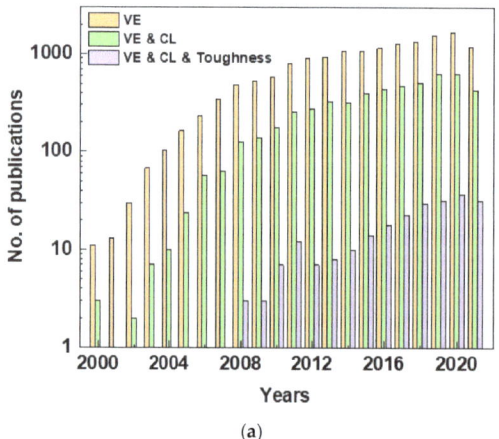
(a)

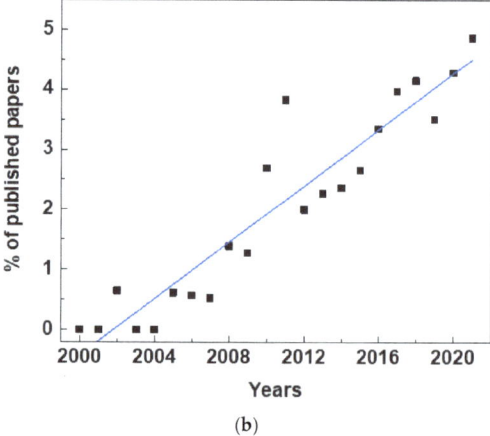
(b)

Figure 1. Statistics of publications on electrospun toughening veil interleaves: (**a**) various keywords (log scale on vertical axis); (**b**) growing trend of using VE for toughening composite laminate: articles on use of VE as a proportion of all papers on toughening FRP composite laminates.

3. Production of Veil

Electrospinning is a technology that is more than a century-old [18]. However, applications for laminate interleaving appeared much later [17], and the US patent, generally accepted as the pioneering invention, was awarded in 1999 to Y.A. Dzenis and D.H. Reneker [19]. The patent focused on interleaving an "electrospun sheet" as a veil between two plies of a composite laminate to improve its mechanical performance.

Electrospinning is a flexible, simple, and cost-effective technology that is used to produce extremely fine fibers for a wide range of materials, with diameters ranging from tens of nanometers to a few micrometers [18]. It is a top-down technique for manufacturing, [20] where the millimeter-sized polymer pellets are dissolved in an organic solvent, and then, electrospun. Electrospun nanofibers are long, continuous, easily aligned, and inexpensive. These nanofibers have unique properties, such as high surface area-to-volume ratios, high aspect ratios (length/diameter), and high mechanical properties (stiffness and strength) because of the high molecular orientation along the fiber axis. Electrospinning can easily be scaled for mass production in industrial applications [18,20,21].

Figure 2a shows a schematic representation of electrospinning in the manufacture of an electrospun veil. The major parts of the setup include a syringe with a nozzle at its tip, a conducting collector plate, and a high-voltage source that connects the collector and the nozzle. The syringe with the nozzle tip holds the polymer solution. When high voltage is applied, the polymeric solution in the syringe is pulled out of the syringe. Liquid droplets are formed at the tip of the nozzle, which are further converted into a jet of polymeric liquid, finally being collected on the conducting collector plate; this results in the formation of a continuous polymeric fiber. After collecting layers of this continuous fiber, one over the other, it forms a non-woven, porous nano-fiber veil (Figure 2b). The diameter of the fiber depends on the applied voltage and polymeric solution. Similarly, the veil's areal density and thickness depend on the duration of manufacturing. Once the desired thickness is achieved, the electrospun veil can be separated from the collector and transferred onto a substrate.

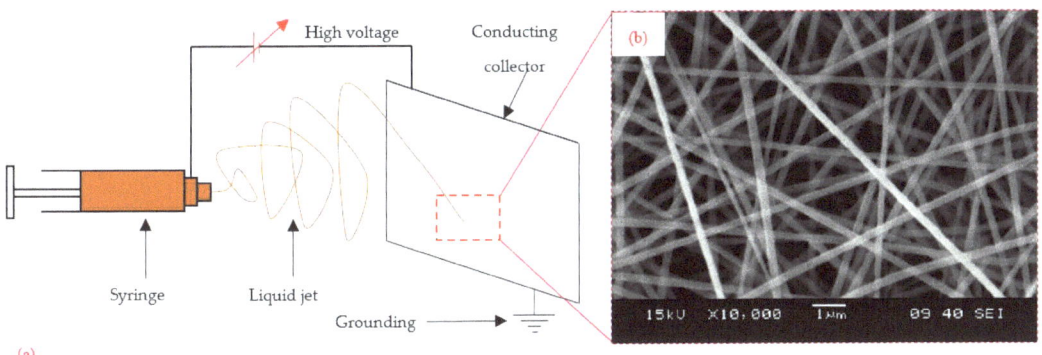

Figure 2. Electrospinning: (**a**) schematic representation; (**b**) a typical PAN veil (courtesy of H. He, K. Molnar, Budapest University of Technology and Economics).

The electrospun veils are placed between two plies of fibrous reinforcement at desired locations in the laminate layup. Then, the fiber-reinforced polymer (FRP) laminate is manufactured as per the standard manufacturing procedure. Manufacturing methods include vacuum infusion, compression molding or press-clave, autoclave, and hand wetlayup followed by vacuum bagging, to name a few. The veil thickness under the processing pressure is sufficiently small. Hence, the laminate's fiber volume fraction is not affected much, and the laminate's in-plane mechanical properties are preserved [10].

Depending on the test procedure, the samples used to measure toughness are manufactured with the interleaves placed either at the laminate mid-plane or at every ply/ply interface. In particular, to measure the Mode I initiation and propagation energy, a DCB test is conducted as per the ASTM D5528 standard [22]. Similarly, the ENF test measures the Mode II initiation and propagation FT per the ASTM D7905 standard [23].

4. Comparative Analysis of the Toughening Effect

The FT values reported in the literature were analyzed based on the veil materials and their modifications, and a comparative analysis of attainable FT with the veil was conducted. The results are discussed below.

4.1. Polyethylene Terephthalate

Tzetzis et al. [24,25], Kuwata et al. [26,27], Quan et al. [14,28,29], Fitzmaurice et al. [30], and Del Saz-Orozco et al. [31] studied the effect of interleaving a PET veil in Mode I and Mode II on the toughening of glass fiber and carbon fiber laminate composites. A general trend of improvement in all the FT parameters was observed, with an exception noted in Mode I properties by Del Saz-Orozco et al. [31].

4.1.1. Neat PET Veil

Tzetzis et al. [24,25] explored the possibility of using a veil as an interleave for repair purposes (to attach patches) in GFRP laminates [24,25]. Significant improvements in G_{1i} and G_{1p} were reported with increasing veil areal density, reaching up to 740% and 770%, respectively, at a veil areal density of around 25–40 g/m^2. However, it should be noted that these significant improvements were relative to the values reported for the "as-received" in-service GFRP surfaces. When these "as received" in-service laminates were treated using hand abrasion and grit blasting, the toughness increased significantly, even in the absence of interleaves. The improvement due to veil interleaves can be re-calculated relative to "treated" surfaces, after which it becomes only moderate of 16–49%.

In contrast, Kuwata et al. [26,27] explored the effect of a PET veil as an interleave in a newly manufactured CFRP laminate. The study investigated the veil's effect on various epoxy matrices (in civil and aerospace applications) and CF reinforcement architectures (UD, satin weave, and plain weave). For the UD-reinforced composite, a stable but moderate increase of 15–56% was observed, which occurred almost independently of resin type. The low value of this increment is probably explained by the interleave replacing the intrinsic UD fiber's bridging effect with the tough thermoplastic bridging of nanofibers, thus replacing the strong UD plie bridging. For woven composites, the effect of PET interleaves was stronger, as expected, due to the weak fiber bridging of the original reinforcement. For the satin weave composites, the PET veils had a moderately positive effect (up to 94% for epoxy; up to 33% for vinyl ester) on Mode I toughness and a very pronounced effect on Mode II toughness, increasing the values to 3100–3630 J/m^2 for initiation and 3750–4760 J/m^2 for propagation, with the effect being stronger for epoxy comparing to vinyl ester. For the plain weave composites, the veils' introduction generated a significant positive effect only for epoxy resin (Mode I: up to 175%, Mode II: up to 88%), while for vinyl ester, the effect, although always positive, was less significant. A similar effect of moderate improvement in the UD CF reinforcement but stronger improvement in the weave CF reinforcement (5H and NCF weave) was also noted by Quan et al. [14,29]. The values of FT noted by Kuwata et al. [26,27] and Quan et al. [14,29] are in a comparable range despite slight differences in their weave architectures (satin and plain weaves vs. 5H and NCF weaves).

Similarly, Fitzmaurice et al. [30] studied the effects of different numbers of veil layers in a laminate (one vs. two veil layers between each pair of plies) in weave GF reinforcement, and noted a crack deviation from the veil region to the fiber–matrix interface region. Similar crack deviation was also noted by Del Saz-Orozco et al. [31] in UD GF reinforcement. The phenomenon of crack deviation was attributed to a strong interaction between the PET veil and the matrix. In spite of crack deviation, Fitzmaurice et al. [30] noted moderate improvement in FT, but Del Saz-Orozco et al. [31] noted a decrease in FT because of the difference in the GF reinforcement architectures (weave vs. UD).

4.1.2. PET-CF Hybrid Veil

Tzetzis et al. [25] repaired an in-service FRP structure by grit blasting the surface before the CF or PET-CF hybrid veil was inserted as an interleave. It was inserted between the repaired structure and the patch, followed by infusion. Grit blasting of the in-service FRP structure induced many new sites for stronger bonding between the structure and the patch material. The implemented veil did not impede the repair and enhanced the fiber bridging effects. Great improvements (up to 600% for initiation and 1100% for propagation) were reported for Mode I toughness, but these were mainly attributed to grit blasting.

The use of CF or PET-CF hybrid veils in newly manufactured laminates had a mixed effect [26,27]. For satin weave composites, the detrimental effect of CF implementation was masked by the improving toughness induced by PET. Still, the tendency was clear: the higher the share of CF implemented, the poorer the results were. A pure CF interleave led to a decrease in all values of FT. For the plain weave laminate, the tendency was not pronounced but still present. CF interleaves improved Mode II toughness for the UD laminates, but their use cannot be recommended because of the degradation of the Mode I toughness properties.

4.1.3. Nano-Modified PET Veil

PET veils have demonstrated their toughening capabilities; however, introducing a PET veil decreases a laminate's electric conductivity [28]. To address this problem, two different concentrations of MWCNTs were airbrushed on a PET veil surface. It was found that airbrushing as little as 0.4 g/m^2 of MWCNTs on the veil improved the laminate's FT (up to 65% for Mode I and 100% for Mode II) and overall electrical conductivity (up to 65%), which is a significant improvement in FT and electrical conductivity compared to the baseline laminate properties, whereas the neat PET veil improved FT only. Airbrushing a higher concentration of MWCNTs increased the electrical conductivity further but decreases the FT.

In summary, PET veils have a positive impact on the toughening mechanism in both modes for both epoxy and vinyl ester resins, for both UD and weave CF reinforcement, and for weave GF reinforcement architectures. The effects with epoxy are typically much more pronounced than with vinyl ester. PET-CF hybrid veils show mixed results. Airbrushing a lower concentration of nano-reinforcement on the veil surface is enough to improve the FT and electrical conductivity. Based on these results, a PET veil and its nano-modified counterpart are recommended for structural delamination control applications.

4.2. Polyphenylene Sulfide

Quan et al. [14,29,32], Ramirez et al. [33], and Ramji et al. [34] reported the application of a PPS veil to improve the fracture performance of a CFRP composite. No studies have been reported for GFRP. Quan et al. studied the effect of an interleaving neat PPS veil [14,29] and a nano-modified PPS veil [32] on FT. The neat PPS veil was interleaved for different CF reinforcement architectures (UD vs. weave), whereas the nano-modified PPS veil was interleaved for UD CF only. Ramirez et al. [33] explored the impact of the manufacturing directionality (compared to reinforcing UD CF) of the veil on FT. Ramji et al. [34] focused on the combined effect of interleaving the PPS veil and the interfacial orientation of CF on delamination migration and FT.

4.2.1. Neat PPS Veil

Interleaving the neat PPS veil in the CF laminate showed a clear trend of high FT for weave CF reinforcement in comparison to UD CF reinforcement [14,29]. A similar trend was also observed for the neat PET veil. This trend was observed to be moderate for Mode

I (up to a ~75% increase) and profound for Mode II loading, increasing the value of FT to 2200–2600 J/m² for initiation and 3000–3200 J/m² for propagation.

Ramirez et al. [33] manufactured CF laminates with PPS veils oriented in the manufacturing direction (MD) and in the cross-direction (CD), perpendicular to the MD. A significant change (up to ~2x) in Mode I FT was noted for a thick veil of 38 g/m² irrespective of veil orientation. It is necessary to note that the mechanical anisotropy of the PPS veil showed no impact on the achievable FT because a PPS veil is a non-woven structure with randomly oriented fibers.

Ramji et al. [34] observed delamination migration when an element of 90° or 45° plies was present in the midplane. Delamination migration added additional crack propagation sites, leading to high FT. It can be noted that the highest FT value was reported for 90°/90° compared to all other interfacial orientation combinations of 0°, 45°, and 90°.

4.2.2. Nano-Modified PPS Veil

Higher improvement in the FT was noted when the neat PPS veil was modified through doping with a lower concentration (up to 0.6 g/m²) of MWCNTs in both Mode I and Mode II. However, when MWCNT concentration was increased to 1.45 g/m², the Mode I FT decreased, becoming worse than for baseline laminates. The Mode II FT also decreased only moderately, but was still better than the baseline. Irrespective of concentration, doping GNP on the neat PPS veil showed the detrimental effects on Mode I FT. A lower concentration of GNP showed good improvement in Mode II FT, which decreased slightly at a higher concentration.

Such behaviors of MWCNT and GNP can be attributed to their 1D and 2D shapes, respectively. The addition of a small wt.% of MWCNTs introduced additional interactions, such as MWCNT pull-out and breakage between the PPS nanofibers and the epoxy matrix. This resulted in improved PPS fiber/epoxy adhesion, and subsequently, led to additional PPS fiber breakage and an improved nanofiber bridging mechanism as a toughening mechanism during the fracture process. For these reasons, the fracture energy was further increased by doping a small amount of MWCNTs on the PPS veils. However, at a high wt.% of MWCNTs, the PPS nanofiber/epoxy adhesion increased to a sufficient level to prevent the PPS nanofiber pull-out and fiber bridging (the primary toughening mechanism). This caused a considerable drop in fracture energy. In contrast, the 2D-structured GNPs significantly agglomerated and attached to the PPS nanofibers. This resulted in a decline in PPS nanofiber/epoxy adhesion [32].

To sum up, interleaving a neat veil improves laminate fracture performance irrespective of the fiber architecture; however, higher improvement was noted in the weave CF compared to the UD CF. The airbrushing of a lower concentration of MWCNTs on the veil surface improves the fracture performance, as well as the electrical conductivity, of a laminate. Airbrushing GNPs increases electrical conductivity but decreases the FT. It is noteworthy that a higher concentration of MWCNTs and all concentrations of GNPs also cause drastic drop in FT. Based on these results, neat and low-concentration MWCNT doping of PPS interleaves is recommended for delamination control in structural applications. Doping nano-reinforcement using materials such as MWCNTs and GNPs improves electrical conductivity; hence, it can be recommended for lightning strike protection, electro-magnetic shielding, and damage detection applications.

4.3. Polyamide

Beckermann et al. [11], Meireman et al. [12], Garcia-Rodriguez et al. [13], Quan et al. [14], Kuwata et al. [26,27], Del Saz-Orozco et al. [31], Saghafi et al. [35,36], Nash et al. [37,38], Guo et al. [39], Chen et al. [40], Ognibene et al. [41], Pozegic et al. [42], Beylergil et al. [43–45], Alessi et al. [46], Barjasteh et al. [47], Monteserin et al. [48,49], Daelemans et al. [50–52], De Schoenmaker et al. [53], O'Donovan et al. [54], and Hamer et al. [55] reported that PA veils as interleaves for FRP composite laminates are applicable in either carbon or glass fiber reinforcement. PAs of different classes were used in these studies, such as PA 6 (sometimes, the brand name Nylon is used), PA 11, PA 12, PA 66, and PA 69. PA is the most popular material for veils amongst the presented materials, as demonstrated by the higher number of studies on PA veils. It has been extensively considered as an interleave for composites manufactured via not only VI and VARTM, but also compression molding and autoclave manufacturing methods.

4.3.1. Neat PA Veil

A general trend of improvement in the overall fracture performance of an FRP laminate has been observed for both CF and GF, with the exception of decreases in Mode I FT [23,24,34,39,43,47,50]. These exceptions are caused by various factors, such as compatibility [26,27,37], mesoscale inhomogeneity due to either a thick veil [53] or a thick veil fiber [46], and a weak interface [42]. PA veils in CF reinforcement have been identified as more compatible with vinyl ester-based matrixes than epoxy-based matrixes [26,27]. Amongst the epoxies, the PA veil is more compatible with BZ9120 epoxy than BZ9130 epoxy reinforced with CF [37].

Quan et al. [14], Kuwata et al. [26,27], and Daelemans et al. [47] observed high FT for weave CF reinforcement and low FT for UD reinforcement interleaved with a PA veil. Similar trends are also noted in PET- and PPS-interleaved veils. Woven reinforcement showed promising improvements due to the fabric architecture. This resulted in plastically failed PA nanofibers zones, indicating good load transfer to the nanofibers. Comparatively, in UD reinforcement, the nanofibers blocked the formation of the carbon fiber bridging zone and delamination propagation between the nanofibers. This resulted in relatively low improvements in FT.

4.3.2. Metal-Modified PA Veil

PA veils are modified with different kinds of metal-based chemicals for various purposes, which include altering the stiffness and hardness of nanofibers [11], adding antibacterial effects [48], and developing electrically conductive laminates [39,56]. Modification techniques include precipitating $AgNO_3$ throughout the nanofiber veil [11], dispersing TiO_2 nanoparticles on the nanofiber veil [48], painting AgNW solution on veil surface [39], and coating Ag (silver-based salt) via electroless plating of the veil [56].

Interleaving a modified PA veil with a $AgNO_3$ coating [11], AgNW painting [39], a pure Ag paste coating [56], and TiO_2 nanoparticle dispersion [48] demonstrated some improvement in fracture performance. These studies show that improvement in the modified conditions was comparatively lower than for the neat PA-interleaved laminate; however, it was still higher than for the baseline laminate. Such slight deterioration in FT improvement for the modified veil was caused by a weak interface formed by the metal surface. However, it should be noted that AgNWs have the strongest interface amongst them.

An Ag-modified veil adds the multifunctionality of improved conductivity to a laminate. For instance, AgNW-painted veils improve laminate conductivity by up to $100\times$ in-plane and by $10\times$ in the thickness direction [39]. Similarly, Ag-plated veils improves conductivity by $1500\times$ in-plane and by $25\times$ in the thickness direction [56]. TiO_2 adds antibacterial functionality, making the laminate suitable for marine applications [48]. There is strong, growing interest in the development of multifunctional interleaves. Metal-modified veils can be used as multifunctional interleaves, but slight deterioration in FT must be considered. The multifunctionality of upgraded conductivity and improved antibacterial properties can be successfully achieved; however, altering the stiffness and hardness of a PA nanofiber with a $AgNO_3$ coating has not been achieved [11].

To sum up, the interleaving of a PA veil always increases Mode II FT, irrespective of the reinforcement type or the binding matrix. However, for Mode I FT, improved fracture performance depends on various factors, such as the compatibility of the PA veil with the binding matrix, crack path, reinforcement type, and areal density, or the thickness of the veil. Based on these results, the PA veil is recommended for delamination control in FRP laminates for structural applications. Similarly, metal-modified PA veils are also recommended for the multifunctionality of laminates.

4.4. Polyacrylonitrile

VanderVennet et al. [57], Zhang et al. [58], Chiu et al. [59], Razavi et al. [60], Molnar et al. [61], and Eskizeybek et al. [62] studied a PAN interleave's effect on CFRP laminates and its effect on FT. Zhang et al. compared a PAN veil to other veils, whereas others have studied the PAN veil and its nano modifications.

4.4.1. Neat PAN Veil

The neat PAN-interleaved laminates showed a significant depreciating effect of up to -70% in G1i compared to the baseline laminate [57–59,62]. They were outperformed by veils of other material, including doped PAN [62]. This reduction in toughness was mainly attributed to problematic impregnation in the presence of a dense veil. A minor positive effect in G1i, with an increase of around 18%, was observed only in two cases: for VARTM [60] with very low veil areal density of 1 g/m^2, and the autoclaving [57] of prepregs, where, for both cases, impregnation does not present a problem. The effect of neat PAN interleaves on G1p was explored, and a mixed result of 15% deprecation [58] and 22% improvement [62] was found.

4.4.2. Nano-Modified PAN Veil

Some examples of nano-modifications include nanoparticles of Al_2O_3 [60] and CNT [61,62], mixed in PAN's electrospinning solution, i.e., incorporated into the veil fiber. Compared to the baseline laminate, nano-modified PAN veils showed better overall performance depending on the wt.% of the nano-reinforcement. Al_2O_3 nanoparticles improved FT by up to 47%, and CNT improved FT by 6–27% in G1i and 45–77% in G1p [62]. Similarly, introducing CNT to an electrospun PAN fiber also improved the electrical conductivity (up to 50%) and thermal conductivity (~3x) of the laminate [61].

To summarize, neat PAN interleaves generally show minor improvement but can be detrimental if impregnated with viscous resin. However, doping nano-reinforcements, such as Al_2O_3 or CNT, demonstrates significant improvement, as well as providing multifunctionality. This merit can probably be attributed to the nano-reinforcement delivered to the ply/ply interface via veil placement rather than via PAN's direct involvement.

4.5. Polycaprolactone

Beckermann et al. [11], Saghafi et al. [32], Cohades et al. [63], and Heijden et al. [64] reported the effect of PCL veils on G/CFRP laminates. A similar trend of higher FT for woven reinforcement has also been noted for PCL interleaves.

Beckermann et al. studied several veil materials and found a correlation between Mode I FT and the ultimate elongation of the bulk polymer used to make the veil. PES, PAI, PA66, and PVB followed this trend, but PCL did not, despite it having the highest ultimate elongation (679–948%). Upon further investigation, it was found that the PCL veils melted during oven curing of the laminate due to their low melting temperature. Thus, the primary toughening mechanism (plastic deformation-identified) was replaced by a phase-separated microstructure, leading to a slight improvement in toughness (up to 14%). Similarly, a phase-separated microstructure where epoxy particles are surrounded by a PCL matrix was identified as a toughening mechanism by Saghafi et al. and Cohades et al. Similarly, Beckermann et al. also noted a correlation between the Mode II FT and the tensile strength of the bulk polymer used to manufacture a veil. This correlation was followed by all the veils investigated. The tensile strength of PCL was low compared to that of PA66 (9 MPa vs. 85 MPa); hence, a small improvement of 7% in Mode II toughness was noted in PCL veil-interleaved laminate (compared to 29–69% for PA66 veil). Similarly, a 24% improvement in PCL (vs. 68% in PA66) was also noted in [35]. It is noteworthy that the authors did not describe the effect of PCL melting in Mode II toughness.

Such atypical behavior of the PCL veil was further investigated, and the authors obtained up to 94% improvement in Mode I fracture toughening [64]. It was noted that a room-temperature pre-cure (before oven curing at 80 °C) of laminate eliminates such atypical behavior and acts as a crucial step when the PCL veil is interleaved. The impact of such a step has not been explored in Mode II experiments. It is worth mentioning that Heijden et al. used single- and double-layered PCL as interleaves, increasing the net areal density of the veil.

Cohades et al. assessed the possibility of using a PCL veil for toughening and healing cracks. Thermal treatment at 150 °C was applied for 30 min to the cracked specimens to assess PCL's capability to bleed, flow, and bridge the cracked faces, thus healing the cracked laminate. However, such healing was not observed because of the high viscosity of the high-molecular-weight PCL, which was used in this study, with a required healing time of around 100 h for the fine nanostructure of the pores (compared to the 30 min applied in tests). The authors concluded that self-healing properties could be achieved either by increasing the diameters of the nanofibers, which may compromise the FT, or by decreasing the viscosity with the application of the low-molecular-weight PCL. However, in this case, the electrospinning becomes unstable due to low molecular entanglement.

These analyses show that PCL interleaves provide only mild improvements compared to other polymers, so they are not recommended for structural applications. Although a room-temperature pre-cure removes atypical PCL behavior, such effects are not studied to Mode II.

To summarize the comparative analysis, it is clear that the introduction of a polymeric nanofiber veil improves the overall FT, with some exceptions. A comparison of attainable FT upon interleaving these veils is plotted in Figure 3, which shows cloud point plots of four FT parameters, plotted as the Mode I vs. Mode II initiation FTs in Figure 3a, and the Mode I vs. Mode II propagation FTs in Figure 3b. Mode lines are also added, which is discussed separately in Section 6. This plot can be used as a reference plot for material selection as per the FT requirement in the design of composite structures in engineering applications. For instance, for composite design anticipating a higher G1i and a moderate G2i, a PA veil is recommended. Similarly, for a higher G2i and a moderate G1i, a PET veil is recommended.

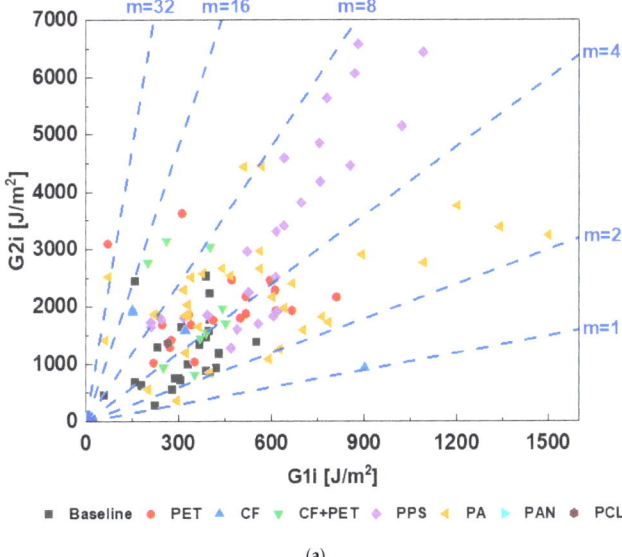

(a)

Figure 3. *Cont.*

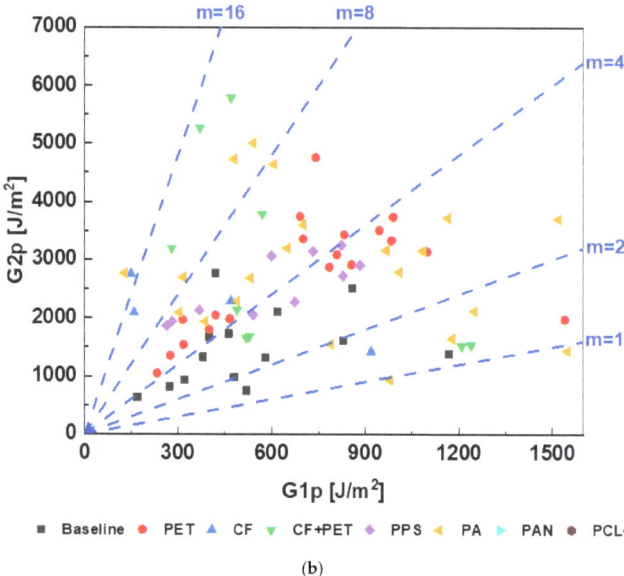

(b)

Figure 3. Mode ratio for (**a**) initiation and (**b**) propagation FTs for baseline vs. interleaved laminate. Interleaves of various polymers are considered.

5. Toughening Mechanism

Cracks propagate in FRP laminates through the epoxy-rich areas, such as the interlaminar region, resin pockets, etc. The brittle nature of the epoxy aids in such crack propagation. However, interleaving the laminates with veils results in non-linear fracture patterns. This change is brought about by several toughening mechanisms added by the veil to the resulting laminates. Nanofiber bridging or crossings, nanofiber pull-out and debonding, nanofiber plastic deformation and breakage, crack pinning, crack deflection, strong adhesion bonds between the resin and veil due to compatibility or melting of the veil, and fusion-bonded dots are the toughening mechanisms that have been identified. (Fusion-bonded dots are semi-spherical dots formed by melting the veil at regular locations within the veil, which reduces crack propagation and improves the FT.) Some of these toughening mechanisms are shown in Figure 4. Doping veils with nano-reinforcements complicates the aforementioned behavior, mostly by changing the veil/resin adhesion, delivering nano-reinforcement to the fracture zone, and modifying the functional properties of the laminate, such as conductivity.

The nanofiber bridging or crossing formed by the nanofibers that compose the veil has been identified as the most common toughening mechanism. The nanofibers bridge the two laminae on the opposite sides of the veil, preventing crack initiation and delaying propagation. Upon further loading, the nanofiber is pulled out of the epoxy, inducing debonding. This is followed by nanofiber plastic deformation and nanofiber breakage. These toughening mechanisms consume much energy, resulting in an increase in the overall FT of the interleaved laminates. The effectiveness of the nanofiber bridging toughening mechanism depends on proper load transfer to the nanofibers. Crack propagation under Mode II loading results in much higher improvements than Mode I loading due to the alignment of the loading with the nanofiber direction in the veil plane. In Mode I crack propagation, the loading of the nanofibers is less optimal and is shown to be dependent on both the primary reinforcement fabric architecture and the presence of a reinforcement fiber bridging zone.

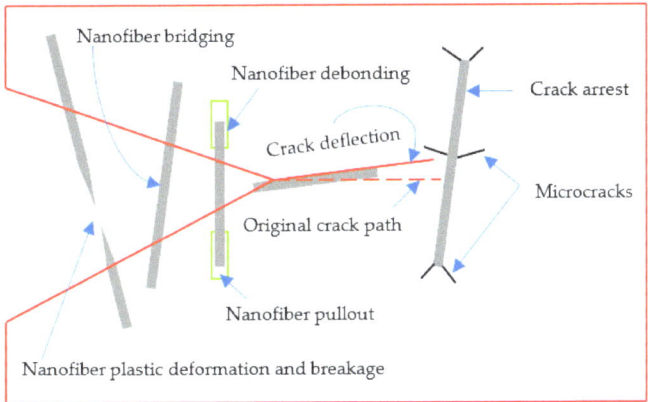

Figure 4. Schematic diagram of the toughening mechanism identified in interleaved laminates.

Crack pinning and crack deflection are also observed as toughening mechanisms that consume energy to improve toughness. Crack deflection and delamination migration occur when a strong adhesion bond is formed between the lamina and the veil due to material compatibility or ply orientation. Material compatibility can be observed in its neat form or through phase changes noted during curing for meltable veils.

6. Statistical Analysis
6.1. Mode Ratio

The mode ratio $\left(m = \frac{G2}{G1}\right)$ inherits the nature of the resin used in the composite. The introduction of an electrospun veil changes the overall behavior of the resin in the interlaminar region. For a baseline laminate, a low mode ratio close to unity is expected for ductile resins. In contrast, a much higher ratio is observed for brittle resins. The change in the ratio due to interleaving depends on the changes in the individual properties, which further depends on the type of veil, its areal density, and its production method, as noted in Section 3.

Figure 3 shows the initiation and propagation FT of the baseline and interleaved composite laminates. The scatter of points for baseline vs. interleaved laminates indicates a change in the nature of the resin at the interlaminar region due to the addition of the electrospun veil in the interlaminar region of the laminate. The FT of baseline laminates is concentrated near the bottom left corner of the plot, in contrast to the FT of interleaved laminates, which is distributed throughout. The highest Mode I initiation and propagation FTs are noted for laminate interleaved with the PA veil, whereas the highest Mode II initiation and propagation FTs are noted for laminate interleaved with the PPS-and-PET-CF hybrid veil, respectively.

Similarly, adding mode ratio (m = 1 ... 32) lines shows some other characteristic behaviors. For instance, the initiation FT of baseline laminates lies between mode ratios of 1 to 16, and most of the values are concentrated around a mode ratio of 4, whereas the propagation FT of baseline laminates lies between mode ratios of 1 to 8, and most of the values are concentrated between mode ratios of 2 and 4. Similarly, in the case of interleaved laminates, the mode ratio reaches up to more than 32 for the initiation FT and more than 16 for the propagation FT.

6.2. Areal Density

Further, an areal density of zero indicates the baseline laminates, and a non-zero areal density indicates that the laminate is interleaved with a veil. The minimal, mean, and maximal areal density values are 0, 10, and 40 g/m², respectively. The veil's areal density and FT, as presented in Table 1, show a positive correlation between the four FT parameters and areal density. The mode II FT is more strongly correlated with the areal density ($r > 0.5$) than the Mode I FT ($r < 0.5$). Figure 5 plots a cloud of points of the Mode I and Mode II initiation and propagation FTs for UD CF laminates when interleaved with an electrospun veil produced with different materials and areal densities. The denser cloud of points in Figure 5a,c shows that extensive research has been carried to study the interleaving effect on initiation energy. However, only limited research has focused on measuring propagation FT (applicable to both modes). The maximal attainable value of G1i is ~1600 J/m^2, of G1p is ~1600 J/m^2, of G2i is ~7000 J/m^2, and of G2p is ~6000 J/m^2.

Table 1. Correlation between areal density and FT.

	G1i	G1p	G2i	G2p
Correlation coefficient (r)	0.44	0.34	0.59	0.58

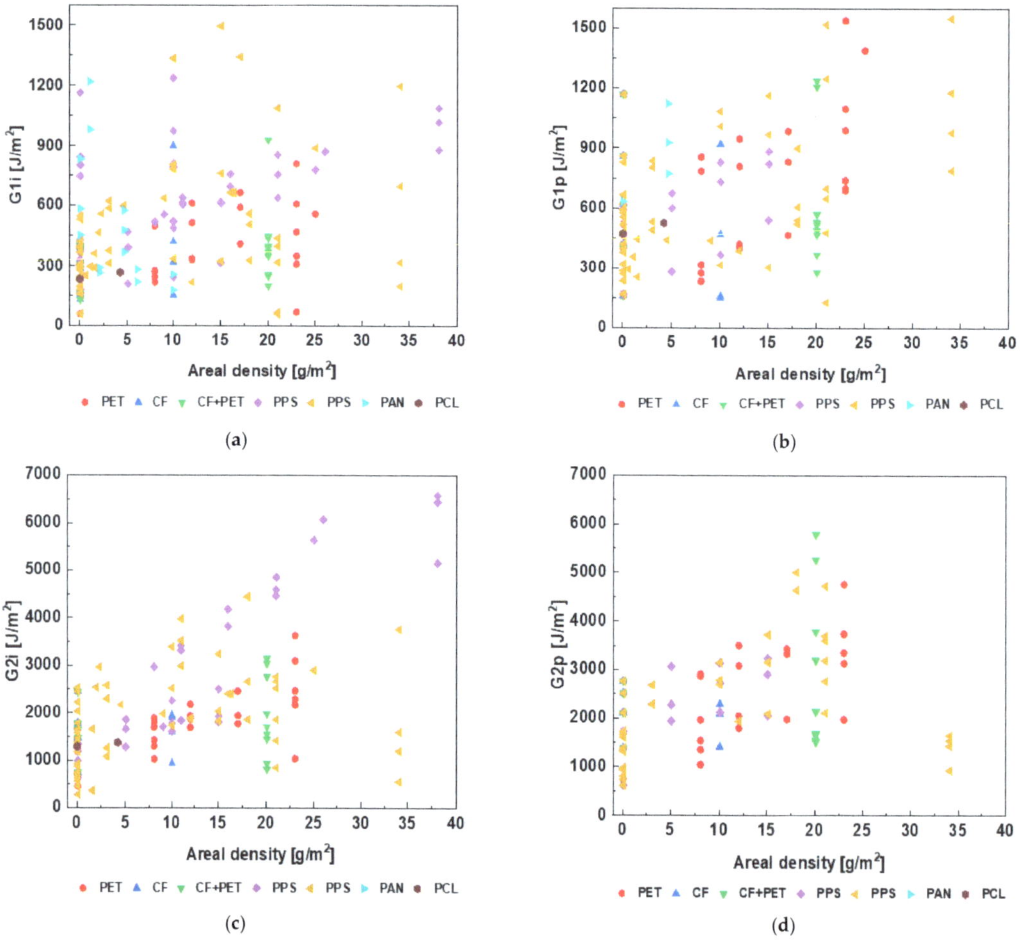

Figure 5. Dependence of (**a**) G1i, (**b**) G1p, (**c**) G2i, and (**d**) G2p on areal density of various electrospun veils. The areal density of 0 means no electrospun veil was placed in the laminate.

PA has been identified as the most common choice of polymeric material for interleaving based on the reported number of data points, as shown in Figure 5. Various modifications of these materials are also used. These modifications include coating/doping with nano-reinforcement, such as CNT, GNP, metal modification/plating, or mixing one polymeric material or CF with another. Such modifications of the veil add a new feature, making the veil multifunctional. A broad range of areal densities has been investigated for PA interleaves, showing a stable interleaving effect on improving FT. However, such a wide range of studies is missing for other veil materials.

7. Numerical Modeling of Delamination

Numerical modeling is typically carried out using a 2D strain finite element model to simulate the DCB and ENF beam while cohesive elements are used to model the delamination behavior. "Traction" and "Separation" at the interface with the possibility of delamination by crack propagation exist, and they govern cohesive element modeling. The cohesive elements initially behave linearly until a threshold level, usually known as the damage initiation point, is reached. The stiffness and strength decrease progressively upon reaching the initiation point, which is known as damage evolution or softening. This continues until the surfaces reach a total separation point. There are various forms of traction–separation law. The similarity amongst them is that each of them describes linear behavior until the damage initiation point. The difference is the damage evolution or softening phase, which may be linear, exponential, polynomial, or bi- or tri-linear, to name a few.

Bi-linear cohesive law (Figure 6a) reasonably models predominantly brittle fracture and matrix cracking in baseline laminates. Significant fiber bridging is observed in veil-interleaved laminates. Such a bridging effect is modeled by modifying the bi-linear to tri-linear law (Figure 6b). The three parameters required to model bi-linear law are cohesive strength (t_m), penalty stiffness (K), and cohesive energy (Γ_m). An additional three parameters are needed to model tri-linear law, which include bridging strength (t_b), bridging energy (Γ_b), and the size of the tougher zone (L_a). These parameters are different and should be determined for each fracture mode. Appendix A shows a collection of parameters used in the literature to model delamination of FRP laminates with bi- and tri-linear cohesive laws.

Bi-linear and tri-linear laws model the delamination of composite laminates [65,66]. Though a significant fiber bridging effect was observed, only bi-linear law was considered to model delamination of the baseline and interleaved laminates. However, it should be noted that bi-linear law only models the first linear part of the traction–separation curve until damage initiation [67–69]. The damage evolution part of interleaved laminates cannot be modeled using bi-linear law only; hence the model presented does not reflect the actual behavior.

The FT, experimentally obtained using a DCB [65–67] or ENF [65,66,68,69] test, is the initial cohesive energy value for numerical modeling. The cohesive strength and penalty stiffness are calculated using a trial-and-error process. The values of these two parameters are changed and tuned until the numerical model's force–displacement curve coincides with the experimental curve.

The tuning of parameters starts from penalty stiffness, a numerical parameter chosen to compromise the computational need for the value to be low and the physical requirement for it to be as high as possible. The cohesive strength is tuned until the sharpness of the peak in the experimental force–displacement curve matches with the thus obtained numerical curve. Once these two parameters are adjusted, the FT value is also changed and tuned to match the experimental and numerical softening curve.

A similar trial-and-error principle is followed for tuning bridging strength and bridging energy [65,66] after the bi-linear parameters are obtained until the experimental and numerical softening curves are matched. The size of the tougher zone (L_a) is only required to model the ENF test. The method to determine L_a is not clearly defined.

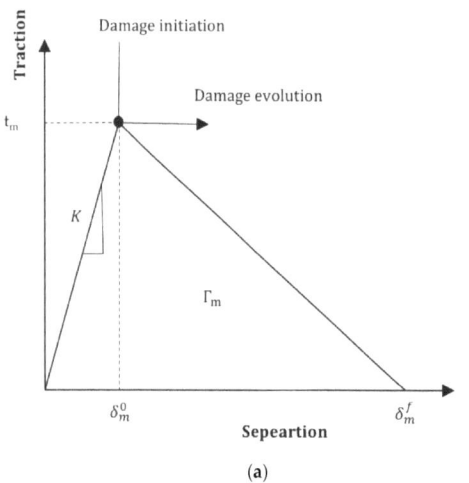 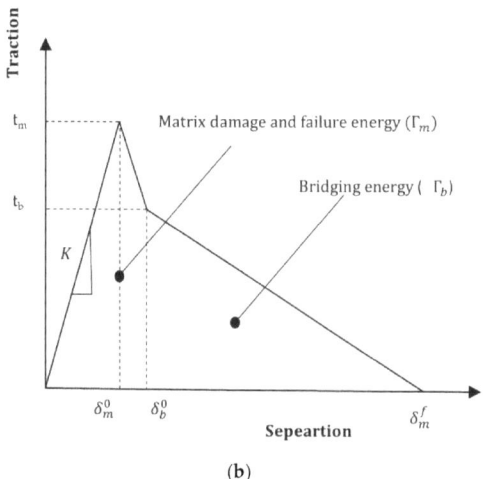

Figure 6. Traction–separation law for cohesive zone modeling: (**a**) bi-linear (**b**) tri-linear cohesive law.

Numerical modeling was carried using a 2D plane-strain finite element model [65–69]. CPE4R elements were used to model the composite laminate, and COH2D4 elements modeled the cohesive surface in front of the pre-existing crack front. As the size of cohesive elements was sensitive, a size of 0.1–0.2 mm (DCB sample length = 125 mm, ENF sample length = 160 mm) was used to model the cohesive layer. Thus, the obtained cohesive parameters can further be utilized to model the behavior of composite components and parts.

8. Conclusions

In this paper, a general trend of improvement in all four parameters used to characterize the FT was observed. Interleaving was proven to be one of the most effective methods for improving the FT of composite laminate. The growing number of publications on this topic indicates that it is in the growth phase of the technology growth curve. The method of production of interleaves, electrospinning, is stable, scalable and well understood, with lots of choices for material selection. Some of the materials that are widely used as interleaves were analyzed and reviewed. The highest attainable FT for various interleave materials was identified and discussed (see Section 6 for comparative analysis). Figure 3 was plotted as a reference for veil material selection as per the FT requirement in the composite design. Various toughening mechanisms introduced by these polymeric interleaves were examined, listed, and discussed. The functionalization of interleaves was discussed, and the key parameters and methodology for the numerical modeling of delamination using cohesive zone modeling was reviewed and outlined.

Author Contributions: Conceptualization, B.M., S.V.L. and S.G.A.; methodology, B.M., S.V.L. and S.G.A.; software, B.M.; validation, B.M., S.V.L. and S.G.A.; formal analysis, B.M.; investigation, B.M.; resources, B.M.; data curation, B.M.; writing—original draft preparation, B.M.; writing—review and editing, B.M., S.V.L., A.S., M.O. and S.G.A.; visualization, B.M.; supervision, S.V.L. and S.G.A.; project administration, S.G.A.; funding acquisition, S.G.A. All authors have read and agreed to the published version of the manuscript.

Funding: No funding was received for this research.

Institutional Review Board Statement: Not applicable.

Informed Consent Statement: Not applicable.

Data Availability Statement: The data presented in this study are available on request from the corresponding authors.

Conflicts of Interest: The authors declare no conflict of interest.

Abbreviations

2D	Two-dimensional
3D	Three-dimensional
5H	5-harness
$AgNO_3$	Silver nitrate
AgNW	Silver nanowire
ASTM	American Society for Testing and Materials
CF	Carbon fiber
CFRP	Carbon fiber-reinforced polymer
CNT	Carbon nanotubes
COH2D4	Four-node two-dimensional cohesive
CPE4R	Plane-strain quadrilateral with reduced integration
DCB	Double-cantilever beam
ENF	End notch flexure
FRP	Fiber-reinforced polymer
FT	Fracture toughness
G/CFRP	Glass/carbon fiber-reinforced polymer
G1i	Mode I initiation [J/m^2]
G1p	Mode I propagation [J/m^2]
G2i	Mode II initiation [J/m^2]
G2p	Mode II propagation [J/m^2]
GF	Glass fiber
GNP	Graphene nanoparticles
K	Penalty stiffness
L_a	Size of tougher zone
m	Mode ratio
MWCNT	Multi-walled carbon nanotube
NCF	Non-crimp fabric
PA	Polyamide
PAN	Polyacrylonitrile
PAI	Polyamide-imide
PCL	Polycaprolactone
PES	Polyethersulfone
PET	Polyethylene terephthalate
PPS	Polyphenylene sulfide
PVB	Polyvinyl butyral
PVDF	Polyvinylidene fluoride
r	Correlation coefficient
t_b	Bridging strength
t_m	Cohesive strength
TiO_2	Titanium dioxide
UD	Unidirectional
VARTM	Vacuum-assisted resin transfer molding
VI	Vacuum infusion
Γ_m	Cohesive energy

Appendix A. Collection of Parameters to Model Delamination of FRP Laminates

Materials	Test	Cohesive Parameters						Ref.
		K [MPa/mm]	tm [MPa]	Γ_m [N/mm]	tb [MPa]	Γ_b [N/mm]	La [mm]	
-/CF/Epoxy prepreg	DCB	80,000	60	0.8	-	-	-	Moroni 2013, [65]
Nylon 66/"		64,000	48	1	0.8	0.6	-	
-/CF/Epoxy prepreg	ENF	40,000	30	3	30	2.2	10	
Nylon 66/"		"	35	3.5	35	3	15	
-/Woven CF/Epoxy	DCB	55,000	60	0.75	-	-	-	Giuliese 2015, [66]
150 nm/25 μm thick/random Nylon 66/"		"	48	1	1	0.45	-	
150 nm/50 μm thick/random Nylon 66/"		"	38	0.3	-	-	-	
500 nm/25 μm thick/random Nylon 66/"		"	25	0.33	-	-	-	
500 nm/50 μm thick/random Nylon 66/"		"	40	0.17	-	-	-	
150 nm/25 μm thick/oriented Nylon 66/"		"	23	0.28	1	0.22	-	
150 nm/50 μm thick/oriented Nylon 66/"		"	20	0.25	1	0.22	-	
500 nm/25 μm thick/oriented Nylon 66/"		"	30	0.22	-	-	-	
500 nm/50 μm thick/oriented Nylon 66/"		"	8	0.2	-	-	-	
-/Woven CF/Epoxy	ENF	55,000	30	3	30	2.2	10	
150 nm/25 μm thick/random Nylon 66/"		"	40	3.3	51	4.6	20	
150 nm/50 μm thick/random Nylon 66/"		"	17.5	1	35	2	25	
500 nm/25 μm thick/random Nylon 66/"		"	30	1.1	30	1.45	25	
500 nm/50 μm thick/random Nylon 66/"		"	30	1.65	-	-	-	
150 nm/25 μm thick/oriented Nylon 66/"		"	35	1.8	48	3	25	
150 nm/50 μm thick/oriented Nylon 66/"		"	45	3.1	48	3.45	10	
500 nm/25 μm thick/oriented Nylon 66/"		"	14	0.7	7	0.28	10	
500 nm/50 μm thick/oriented Nylon 66/"		"	17	1.25	8	0.35	20	
-/2 twill CF/Epoxy prepreg	DCB	80,000	50	0.25	-	-	-	Saghafi 2016, [67]
Thin PVDF/"		"	50	0.36	-	-	-	
Thick PVDF/"		"	50	0.47	-	-	-	
-/2 twill CF/Epoxy prepreg	ENF	50,000	70	0.18	-	-	-	Saghafi 2019, [68]
Thin PVDF/"		"	73	0.29	-	-	-	
Thick PVDF/"		"	85	0.38	-	-	-	
-/E-GF/Phenolic resin	ENF	40,000	15	0.581	-	-	-	Barzoki 2019, [69]
100 nm-PVB/"		"	50	0.143	-	-	-	
165 nm-PVB/"		"	12.5	0.749	-	-	-	
314 nm-PVB/"		"	10	0.44	-	-	-	
500 nm-PVB/"		100,000	9	0.423	-	-	-	
165 nm (aligned)-PVB/"		"	8	0.393	-	-	-	

": Same as in the row above; -: not available.

References

1. Lu, Z.; Li, Y.; Xie, J. Durability of BFRP Bars Wrapped in Seawater Sea Sand Concrete. *Compos. Struct.* **2021**, *255*, 112935. [CrossRef]
2. Guo, R.; Li, C.; Xian, G. Water Absorption and Long-Term Thermal and Mechanical Properties of Carbon/Glass Hybrid Rod for Bridge Cable. *Eng. Struct.* **2023**, *274*, 115176. [CrossRef]
3. Xian, G.; Guo, R.; Li, C.; Wang, Y. Mechanical Performance Evolution and Life Prediction of Prestressed CFRP Plate Exposed to Hygrothermal and Freeze-Thaw Environments. *Compos. Struct.* **2022**, *293*, 115719. [CrossRef]
4. Quaresimin, M.; Schulte, K.; Zappalorto, M.; Chandrasekaran, S. Toughening Mechanisms in Polymer Nanocomposites: From Experiments to Modelling. *Compos. Sci. Technol.* **2016**, *123*, 187–204. [CrossRef]
5. Islam, A.B.M.I.; Kelkar, A.D. Prospects and Challenges of Nanomaterial Engineered Prepregs for Improving Interlaminar Properties of Laminated Composites—a Review. *MRS Commun.* **2017**, *7*, 102–108. [CrossRef]
6. Mouritz, A.P.; Cox, B.N. A Mechanistic Interpretation of the Comparative In-Plane Mechanical Properties of 3D Woven, Stitched and Pinned Composites. *Compos. Part A* **2010**, *41*, 709–728. [CrossRef]
7. Mouritz, A.P. Review of Z-Pinned Composite Laminates. *Compos. Part A Appl. Sci. Manuf.* **2007**, *38*, 2383–2397. [CrossRef]
8. Guzman de Villoria, R.; Hallander, P.; Ydrefors, L.; Nordin, P.; Wardle, B.L. In-Plane Strength Enhancement of Laminated Composites via Aligned Carbon Nanotube Interlaminar Reinforcement. *Compos. Sci. Technol.* **2016**, *133*, 33–39. [CrossRef]
9. Swolfs, Y.; Gorbatikh, L.; Verpoest, I. Fibre Hybridisation in Polymer Composites: A Review. *Compos. Part A Appl. Sci. Manuf.* **2014**, *67*, 181–200. [CrossRef]
10. Lomov, S.V.; Molnár, K. Compressibility of Carbon Fabrics with Needleless Electrospun PAN Nanofibrous Interleaves. *Express Polym. Lett.* **2016**, *10*, 25–35. [CrossRef]
11. Beckermann, G.W.; Pickering, K.L. Mode I and Mode II Interlaminar Fracture Toughness of Composite Laminates Interleaved with Electrospun Nanofibre Veils. *Compos. Part A Appl. Sci. Manuf.* **2015**, *72*, 11–21. [CrossRef]
12. Meireman, T.; Daelemans, L.; Rijckaert, S.; Rahier, H.; Van Paepegem, W.; De Clerck, K. Delamination Resistant Composites by Interleaving Bio-Based Long-Chain Polyamide Nanofibers through Optimal Control of Fiber Diameter and Fiber Morphology. *Compos. Sci. Technol.* **2020**, *193*, 108126. [CrossRef]
13. García-Rodríguez, S.M.; Costa, J.; Rankin, K.E.; Boardman, R.P.; Singery, V.; Mayugo, J.A. Interleaving Light Veils to Minimise the Trade-off between Mode-I Interlaminar Fracture Toughness and in-Plane Properties. *Compos. Part A Appl. Sci. Manuf.* **2020**, *128*, 105659. [CrossRef]
14. Quan, D.; Bologna, F.; Scarselli, G.; Ivankovic, A.; Murphy, N. Interlaminar Fracture Toughness of Aerospace-Grade Carbon Fibre Reinforced Plastics Interleaved with Thermoplastic Veils. *Compos. Part A Appl. Sci. Manuf.* **2020**, *128*, 105642. [CrossRef]
15. Sela, N.; Ishai, O. Interlaminar Fracture Toughness and Toughening of Laminated Composite Materials: A Review. *Composites* **1989**, *20*, 423–435. [CrossRef]
16. Wang, G.; Yu, D.; Kelkar, A.D.; Zhang, L. Electrospun Nanofiber: Emerging Reinforcing Filler in Polymer Matrix Composite Materials. *Prog. Polym. Sci.* **2017**, *75*, 73–107. [CrossRef]
17. Palazzetti, R.; Zucchelli, A. Electrospun Nanofibers as Reinforcement for Composite Laminates Materials—A Review. *Compos. Struct.* **2017**, *182*, 711–727. [CrossRef]
18. Mitchell, G.R. *Electrospinning: Principles, Practice and Possibilities*; Edited; Tang, B.Z., Abd-El-Aziz, A.S., Craig, S., Dong, J., Masuda, T., Weder, C., Eds.; The Royal Society of Chemistry: Cambridge, UK, 2015; ISBN 978-1849735568.
19. Yuris, A.; Dzenis, D.H.R. Delamination Resistant Composites Prepared by Small Diameter Fiber Reinforcement at Ply Interfaces. US Patent No. 6,265,333 B1, July 2001.
20. Molnar, K.; Nagy, Z.K. Corona-Electrospinning: Needleless Method for High-Throughput Continuous Nanofiber Production. *Eur. Polym. J.* **2016**, *74*, 279–286. [CrossRef]
21. Molnár, K.; Vas, L.M. Electrospun Composite Nanofibers and Polymer Composites. In *Synthetic Polymer-Polymer Composites*; Carl Hanser Verlag GmbH & Co. KG: München, Germany, 2012; pp. 301–349.
22. *ASTM D5528-01*; Standard Test Method for Mode I Interlaminar Fracture Toughness of Unidirectional Fiber-Reinforced Polymer Matrix Composites. ASTM: West Conshohocken, PA, USA, 2014.
23. *ASTM D7905*; Standard Test Method for Determination of the Mode II Interlaminar Fracture Toughness of Unidirectional Fiber-Reinforced Polymer Matrix Composites. ASTM: West Conshohocken, PA, USA, 2014.
24. Tzetzis, D.; Hogg, P.J.; Jogia, M. Double Cantilever Beam Mode-I Testing for Vacuum Infused Repairs of GFRP. *J. Adhes. Sci. Technol.* **2003**, *17*, 309–328. [CrossRef]
25. Tzetzis, D.; Hogg, P.J. Bondline Toughening of Vacuum Infused Composite Repairs. *Compos. Part A Appl. Sci. Manuf.* **2006**, *37*, 1239–1251. [CrossRef]
26. Kuwata, M.; Hogg, P.J. Interlaminar Toughness of Interleaved CFRP Using Non-Woven Veils: Part 2. Mode-II Testing. *Compos. Part A Appl. Sci. Manuf.* **2011**, *42*, 1560–1570. [CrossRef]
27. Kuwata, M.; Hogg, P.J. Interlaminar Toughness of Interleaved CFRP Using Non-Woven Veils: Part 1. Mode-I Testing. *Compos. Part A Appl. Sci. Manuf.* **2011**, *42*, 1551–1559. [CrossRef]
28. Quan, D.; Mischo, C.; Li, X.; Scarselli, G.; Ivanković, A.; Murphy, N. Improving the Electrical Conductivity and Fracture Toughness of Carbon Fibre/Epoxy Composites by Interleaving MWCNT-Doped Thermoplastic Veils. *Compos. Sci. Technol.* **2019**, *182*, 107775. [CrossRef]

29. Quan, D.; Bologna, F.; Scarselli, G.; Ivanković, A.; Murphy, N. Mode-II Fracture Behaviour of Aerospace-Grade Carbon Fibre/Epoxy Composites Interleaved with Thermoplastic Veils. *Compos. Sci. Technol.* **2020**, *191*, 108065. [CrossRef]
30. Fitzmaurice, K.; Ray, D.; McCarthy, M.A. PET Interleaving Veils for Improved Fracture Toughness of Glass Fibre/Low-Styrene-Emission Unsaturated Polyester Resin Composites. *J. Appl. Polym. Sci.* **2016**, *133*, 42877. [CrossRef]
31. Del Saz-Orozco, B.; Ray, D.; Stanley, W.F. Effect of Thermoplastic Veils on Interlaminar Fracture Toughness of a Glass Fiber/Vinyl Ester Composite. *Polym. Compos.* **2017**, *38*, 2501–2508. [CrossRef]
32. Quan, D.; Mischo, C.; Binsfeld, L.; Ivankovic, A.; Murphy, N. Fracture Behaviour of Carbon Fibre/Epoxy Composites Interleaved by MWCNT- and Graphene Nanoplatelet-Doped Thermoplastic Veils. *Compos. Struct.* **2020**, *235*, 111767. [CrossRef]
33. Ramirez, V.A.; Hogg, P.J.; Sampson, W.W. The Influence of the Nonwoven Veil Architectures on Interlaminar Fracture Toughness of Interleaved Composites. *Compos. Sci. Technol.* **2015**, *110*, 103–110. [CrossRef]
34. Ramji, A.; Xu, Y.; Yasaee, M.; Grasso, M.; Webb, P. Delamination Migration in CFRP Laminates under Mode I Loading. *Compos. Sci. Technol.* **2020**, *190*, 108067. [CrossRef]
35. Saghafi, H.; Zucchelli, A.; Palazzetti, R.; Minak, G. The Effect of Interleaved Composite Nanofibrous Mats on Delamination Behavior of Polymeric Composite Materials. *Compos. Struct.* **2014**, *109*, 41–47. [CrossRef]
36. Saghafi, H.; Palazzetti, R.; Zucchelli, A.; Minak, G. Influence of Electrospun Nanofibers on the Interlaminar Properties of Unidirectional Epoxy Resin/Glass Fiber Composite Laminates. *J. Reinf. Plast. Compos.* **2015**, *34*, 907–914. [CrossRef]
37. Nash, N.H.; Ray, D.; Young, T.M.; Stanley, W.F. The Influence of Hydrothermal Conditioning on the Mode-I, Thermal and Flexural Properties of Carbon/Benzoxazine Composites with a Thermoplastic Toughening Interlayer. *Compos. Part A Appl. Sci. Manuf.* **2015**, *76*, 135–144. [CrossRef]
38. Nash, N.H.; Young, T.M.; Stanley, W.F. The Influence of a Thermoplastic Toughening Interlayer and Hydrothermal Conditioning on the Mode-II Interlaminar Fracture Toughness of Carbon/Benzoxazine Composites. *Compos. Part A Appl. Sci. Manuf.* **2016**, *81*, 111–120. [CrossRef]
39. Guo, M.; Yi, X.; Liu, G.; Liu, L. Simultaneously Increasing the Electrical Conductivity and Fracture Toughness of Carbon-Fiber Composites by Using Silver Nanowires-Loaded Interleaves. *Compos. Sci. Technol.* **2014**, *97*, 27–33. [CrossRef]
40. Chen, G.; Zhang, J.; Liu, G.; Chen, P.; Guo, M. Controlling the Crack Propagation Path of the Veil Interleaved Composite by Fusion-Bonded Dots. *Polymers* **2019**, *11*, 1260. [CrossRef]
41. Ognibene, G.; Latteri, A.; Mannino, S.; Saitta, L.; Recca, G.; Scarpa, F.; Cicala, G. Interlaminar Toughening of Epoxy Carbon Fiber Reinforced Laminates: Soluble Versus Non-Soluble Veils. *Polymers* **2019**, *11*, 1029. [CrossRef]
42. Pozegic, T.R.; King, S.G.; Fotouhi, M.; Stolojan, V.; Silva, S.R.P.; Hamerton, I. Delivering Interlaminar Reinforcement in Composites through Electrospun Nanofibres. *Adv. Manuf. Polym. Compos. Sci.* **2019**, *5*, 155–171. [CrossRef]
43. Beylergil, B.; Tanoğlu, M.; Aktaş, E. Experimental and Statistical Analysis of Carbon Fiber/Epoxy Composites Interleaved with Nylon 6,6 Nonwoven Fabric Interlayers. *J. Compos. Mater.* **2020**, *54*, 4173–4184. [CrossRef]
44. Beylergil, B.; Tanoğlu, M.; Aktaş, E. Effect of Polyamide-6,6 (PA 66) Nonwoven Veils on the Mechanical Performance of Carbon Fiber/Epoxy Composites. *Compos. Struct.* **2018**, *194*, 21–35. [CrossRef]
45. Beylergil, B.; Tanoğlu, M.; Aktaş, E. Enhancement of Interlaminar Fracture Toughness of Carbon Fiber-Epoxy Composites Using Polyamide-6,6 Electrospun Nanofibers. *J. Appl. Polym. Sci.* **2017**, *134*, 45244. [CrossRef]
46. Alessi, S.; Di Filippo, M.; Dispenza, C.; Focarete, M.L.; Gualandi, C.; Palazzetti, R.; Pitarresi, G.; Zucchelli, A. Effects of Nylon 6,6 Nanofibrous Mats on Thermal Properties and Delamination Behavior of High Performance CFRP Laminates. *Polym. Compos.* **2015**, *36*, 1303–1313. [CrossRef]
47. Barjasteh, E.; Sutanto, C.; Reddy, T.; Vinh, J. A Graphene/Graphite-Based Conductive Polyamide 12 Interlayer for Increasing the Fracture Toughness and Conductivity of Carbon-Fiber Composites. *J. Compos. Mater.* **2017**, *51*, 2879–2887. [CrossRef]
48. Monteserín, C.; Blanco, M.; Murillo, N.; Pérez-Márquez, A.; Maudes, J.; Gayoso, J.; Laza, J.M.; Hernáez, E.; Aranzabe, E.; Vilas, J.L. Novel Antibacterial and Toughened Carbon-Fibre/Epoxy Composites by the Incorporation of TiO2 Nanoparticles Modified Electrospun Polyamide Nanofibre Veils. *Polymers* **2019**, *11*, 1524. [CrossRef]
49. Monteserín, C.; Blanco, M.; Murillo, N.; Pérez-Márquez, A.; Maudes, J.; Gayoso, J.; Laza, J.; Aranzabe, E.; Vilas, J. Effect of Different Types of Electrospun Polyamide 6 Nanofibres on the Mechanical Properties of Carbon Fibre/Epoxy Composites. *Polymers* **2018**, *10*, 1190. [CrossRef]
50. Daelemans, L.; van der Heijden, S.; De Baere, I.; Rahier, H.; Van Paepegem, W.; De Clerck, K. Nanofibre Bridging as a Toughening Mechanism in Carbon/Epoxy Composite Laminates Interleaved with Electrospun Polyamide Nanofibrous Veils. *Compos. Sci. Technol.* **2015**, *117*, 244–256. [CrossRef]
51. Daelemans, L.; van der Heijden, S.; De Baere, I.; Rahier, H.; Van Paepegem, W.; De Clerck, K. Using Aligned Nanofibres for Identifying the Toughening Micromechanisms in Nanofibre Interleaved Laminates. *Compos. Sci. Technol.* **2016**, *124*, 17–26. [CrossRef]
52. Daelemans, L.; Van Der Heijden, S.; De Baere, I.; Rahier, H.; Van Paepegem, W.; De Clerck, K. Damage-Resistant Composites Using Electrospun Nanofibers: A Multiscale Analysis of the Toughening Mechanisms. *ACS Appl. Mater. Interfaces* **2016**, *8*, 11806–11818. [CrossRef]
53. De Schoenmaker, B.; Van der Heijden, S.; De Baere, I.; Van Paepegem, W.; De Clerck, K. Effect of Electrospun Polyamide 6 Nanofibres on the Mechanical Properties of a Glass Fibre/Epoxy Composite. *Polym. Test.* **2013**, *32*, 1495–1501. [CrossRef]

54. O'Donovan, K.; Ray, D.; McCarthy, M.A. Toughening Effects of Interleaved Nylon Veils on Glass Fabric/Low-Styrene-Emission Unsaturated Polyester Resin Composites. *J. Appl. Polym. Sci.* **2015**, *132*, 41462. [CrossRef]
55. Hamer, S.; Leibovich, H.; Green, A.; Intrater, R.; Avrahami, R.; Zussman, E.; Siegmann, A.; Sherman, D. Mode I Interlaminar Fracture Toughness of Nylon 66 Nanofibrilmat Interleaved Carbon/Epoxy Laminates. *Polym. Compos.* **2011**, *32*, 1781–1789. [CrossRef]
56. Guo, M.; Yi, X.; Rudd, C.; Liu, X. Preparation of Highly Electrically Conductive Carbon-Fiber Composites with High Interlaminar Fracture Toughness by Using Silver-Plated Interleaves. *Compos. Sci. Technol.* **2019**, *176*, 29–36. [CrossRef]
57. VanderVennet, J.A.; Duenas, T.; Dzenis, Y.; Peterson, C.T.; Bakis, C.E.; Carter, D.; Roberts, J.K. Fracture Toughness Characterization of Nanoreinforced Carbon-Fiber Composite Materials for Damage Mitigation. In Proceedings of the Behavior and Mechanics of Multifunctional Materials and Composites 2011, San Diego, CA, USA, 7–9 March 2011; Ounaies, Z., Seelecke, S.S., Eds.; SPIE: Bellingham, WA, USA, 2011; Volume 7978, p. 797823.
58. Zhang, J.; Yang, T.; Lin, T.; Wang, C.H. Phase Morphology of Nanofibre Interlayers: Critical Factor for Toughening Carbon/Epoxy Composites. *Compos. Sci. Technol.* **2012**, *72*, 256–262. [CrossRef]
59. Chiu, K.R.; Duenas, T.; Dzenis, Y.; Kaser, J.; Bakis, C.E.; Roberts, J.K.; Carter, D. Comparative Study of Nanomaterials for Interlaminar Reinforcement of Fiber-Composite Panels. In Proceedings of the Behavior and Mechanics of Multifunctional Materials and Composites 2013, San Diego, CA, USA, 10–14 March 2013; Goulbourne, N.C., Naguib, H.E., Eds.; 2013; Volume 8689, p. 86891D.
60. Razavi, S.M.J.; Neisiany, R.E.; Khorasani, S.N.; Ramakrishna, S.; Berto, F. Effect of Neat and Reinforced Polyacrylonitrile Nanofibers Incorporation on Interlaminar Fracture Toughness of Carbon/Epoxy Composite. *Theor. Appl. Mech. Lett.* **2018**, *8*, 126–131. [CrossRef]
61. Molnár, K.; Szebényi, G.; Szolnoki, B.; Marosi, G.; Vas, L.M.; Toldy, A. Enhanced Conductivity Composites for Aircraft Applications: Carbon Nanotube Inclusion Both in Epoxy Matrix and in Carbonized Electrospun Nanofibers. *Polym. Adv. Technol.* **2014**, *25*, 981–988. [CrossRef]
62. Eskizeybek, V.; Yar, A.; Avcı, A. CNT-PAN Hybrid Nanofibrous Mat Interleaved Carbon/Epoxy Laminates with Improved Mode I Interlaminar Fracture Toughness. *Compos. Sci. Technol.* **2018**, *157*, 30–39. [CrossRef]
63. Cohades, A.; Daelemans, L.; Ward, C.; Meireman, T.; Van Paepegem, W.; De Clerck, K.; Michaud, V. Size Limitations on Achieving Tough and Healable Fibre Reinforced Composites through the Use of Thermoplastic Nanofibres. *Compos. Part A Appl. Sci. Manuf.* **2018**, *112*, 485–495. [CrossRef]
64. Van der Heijden, S.; Daelemans, L.; De Schoenmaker, B.; De Baere, I.; Rahier, H.; Van Paepegem, W.; De Clerck, K. Interlaminar Toughening of Resin Transfer Moulded Glass Fibre Epoxy Laminates by Polycaprolactone Electrospun Nanofibres. *Compos. Sci. Technol.* **2014**, *104*, 66–73. [CrossRef]
65. Moroni, F.; Palazzetti, R.; Zucchelli, A.; Pirondi, A. A Numerical Investigation on the Interlaminar Strength of Nanomodified Composite Interfaces. *Compos. Part B Eng.* **2013**, *55*, 635–641. [CrossRef]
66. Giuliese, G.; Palazzetti, R.; Moroni, F.; Zucchelli, A.; Pirondi, A. Cohesive Zone Modelling of Delamination Response of a Composite Laminate with Interleaved Nylon 6,6 Nanofibres. *Compos. Part B Eng.* **2015**, *78*, 384–392. [CrossRef]
67. Saghafi, H.; Ghaffarian, S.R.; Brugo, T.M.; Minak, G.; Zucchelli, A.; Saghafi, H.A. The Effect of Nanofibrous Membrane Thickness on Fracture Behaviour of Modified Composite Laminates—A Numerical and Experimental Study. *Compos. Part B Eng.* **2016**, *101*, 116–123. [CrossRef]
68. Saghafi, H.; Moallemzadeh, A.R.; Zucchelli, A.; Brugo, T.M.; Minak, G. Shear Mode of Fracture in Composite Laminates Toughened by Polyvinylidene Fluoride Nanofibers. *Compos. Struct.* **2019**, *227*, 111327. [CrossRef]
69. Barzoki, P.K.; Rezadoust, A.M.; Latifi, M.; Saghafi, H.; Minak, G. Effect of Nanofiber Diameter and Arrangement on Fracture Toughness of out of Autoclave Glass/Phenolic Composites—Experimental and Numerical Study. *Thin Walled Struct.* **2019**, *143*, 106251. [CrossRef]

Disclaimer/Publisher's Note: The statements, opinions and data contained in all publications are solely those of the individual author(s) and contributor(s) and not of MDPI and/or the editor(s). MDPI and/or the editor(s) disclaim responsibility for any injury to people or property resulting from any ideas, methods, instructions or products referred to in the content.

Article

Thermal Properties and In Vitro Biodegradation of PLA-Mg Filaments for Fused Deposition Modeling

Adrián Leonés [1], Valentina Salaris [1], Ignacio Ramos Aranda [1], Marcela Lieblich [2], Daniel López [1] and Laura Peponi [1,*]

[1] Instituto de Ciencia y Tecnología de Polímeros (ICTP-CSIC), Calle Juan de la Cierva 3, 28006 Madrid, Spain; aleones@ictp.csic.es (A.L.); v.salaris@ictp.csic.es (V.S.); daniel.l.g@csic.es (D.L.)
[2] Centro Nacional de Investigaciones Metalúrgicas (CENIM-CSIC), 28040 Madrid, Spain; marcela@cenim.csic.es
* Correspondence: lpeponi@ictp.csic.es; Tel.: +34-915622900

Abstract: Additive manufacturing, in particular the fused deposition method, is a quite new interesting technique used to obtain specific 3D objects by depositing layer after layer of material. Generally, commercial filaments can be used in 3D printing. However, the obtention of functional filaments is not so easy to reach. In this work, we obtain filaments based on poly(lactic acid), PLA, reinforced with different amounts of magnesium, Mg, microparticles, using a two-step extrusion process, in order to study how processing can affect the thermal degradation of the filaments; we additionally study their in vitro degradation, with a complete release of Mg microparticles after 84 days in phosphate buffer saline media. Therefore, considering that we want to obtain a functional filament for further 3D printing, the simpler the processing, the better the result in terms of a scalable approach. In our case, we obtain micro-composites via the double-extrusion process without degrading the materials, with good dispersion of the microparticles into the PLA matrix without any chemical or physical modification of the microparticles.

Keywords: PLA; magnesium microparticles; in vitro degradation; 3D filament; 3D printing; additive manufacturing; fused deposition modeling

Citation: Leonés, A.; Salaris, V.; Ramos Aranda, I.; Lieblich, M.; López, D.; Peponi, L. Thermal Properties and In Vitro Biodegradation of PLA-Mg Filaments for Fused Deposition Modeling. *Polymers* **2023**, *15*, 1907. https://doi.org/10.3390/polym15081907

Academic Editor: Chin-San Wu

Received: 14 March 2023
Revised: 10 April 2023
Accepted: 13 April 2023
Published: 16 April 2023

Copyright: © 2023 by the authors. Licensee MDPI, Basel, Switzerland. This article is an open access article distributed under the terms and conditions of the Creative Commons Attribution (CC BY) license (https://creativecommons.org/licenses/by/4.0/).

1. Introduction

Additive manufacturing is a quite new processing method to create a three-dimensional object by adding materials layer by layer [1]. Among the additive manufacturing methods, fused deposition modeling, FDM, is one of the most studied because of its low cost and the use of traditional thermoplastic polymers such as thermoplastic polyurethane (TPU) [2] poly(lactic acid) and PLA [3]. For instance, PLA is a biodegradable polymer widely used in 3D printing applications such as bone tissue repair [4] or biodegradable scaffolds [5]. Additionally, PLA chains are susceptible to being attacked by hydrolytic reactions, turning into non-toxic molecules with a high enough degradation rate for human body usage [6]. However, PLA shows some technical limitations in terms of poor hardness [7] and low degradation rate [8], which have to be improved in order for it to be used in biomedical applications. Moreover, when working with a 3D printer, a homogenous filament is necessary when it comes to the dimension diameter of the 3D printer extruder, which can be considered the greatest challenge when the use of non-commercial filament is desired. Therefore, the main strategy to obtain PLA-based filaments is mixing PLA matrix with other polymers or with nano/micro fillers, which not only allows for the obtention of PLA-based filaments with enhanced mechanical properties, but also the obtention of additional functional properties such as antioxidant or antimicrobial activities, etc. [9–11]. Composites of PLA with other metals have been produced in order to improve PLA properties; for example, the addition of iron powder leads to an increase in tensile strength [12], and the introduction of silver nanoparticles enhances the antimicrobial properties of PLA [13].

However, considering biomedical applications, among the inorganic fillers, magnesium metal, Mg, is an interesting reinforcement since Mg^{2+} ions are essential in metabolism at a biochemical level, such as protein synthesis, muscle function, and the nervous and immune system [14,15]. Additionally, the released Mg^{2+} ions present an extraordinary biological response, promoting the formation of collagen in skin wounds [16]. These advantages offer a promising opportunity for the applicability of Mg as a biomaterial. Moreover, PLA-Mg composites have been already fabricated before through FDM, exhibiting an improvement in the main properties of the polymeric matrix due to the incorporation of Mg particles [17].

However, as the physiological medium is water-based, the reaction of Mg with water has to be taken into account. In particular, Mg reacts with H_2O molecules, leading to the formation of hydrogen gas, H_2 (g), and hydroxyl groups, OH^-, which create two challenges to be overcome for biomedical applications [14,18]. The accumulation of H_2 (g) bubbles in the human body and variation in pH values can compromise the viability of the material for biomedical applications [14]. As can be found in the literature, some authors have reported the enhancement of PLA mechanical properties by adding Mg in terms of creep strength [19] and compression modulus [20]. In addition, in vitro degradation of these compounds has been studied in phosphate buffer saline, PBS [21], in simulated body fluid, SBF [22], and in keratinocyte basal media, KBM [23], demonstrating that Mg particles in PLA matrix prevent pH changes and improve the degradation rate of PLA [24]. However, no works studying the relation between the thermal properties and the later in vitro degradation properties of PLA-based filaments reinforced with Mg for 3D printing have been published yet. The main objective of research related to new PLA-based filaments for FDM is focused mainly on the correlations between the processing of filaments and their final mechanical properties in terms of elastic modulus, tensile strength, or elongation at break [25–27].

Furthermore, to the best of our knowledge, we studied for the first time the in vitro degradation of PLA-Mg filaments previous to the FDM processing step. In particular, in our work, we carried out thermal characterization via DSC and TGA, obtaining the thermal properties of PLA-Mg filaments, thus considering how the two-step extrusion process can affect the thermal degradation of the filaments. Moreover, considering that we want to obtain a functional filament for further 3D printing, the simpler the processing, the better the result in terms of a scalable approach. In our case, we obtained micro-composites using a double-extrusion process without degrading the materials, with good dispersion of the microparticles into the PLA matrix without any chemical or physical modification of the microparticles. Finally, an 84-day study in phosphate buffer saline, PBS, media was also performed in terms of H_2 release and pH changes to check for potential uses in biomedical applications.

2. Materials and Methods

Polylactic acid (PLA3051D), 3% D-lactic acid monomer, molecular weight 14.2×10^4 g·mol^{-1}, density 1.24 g·cm^{-3}), was supplied by NatureWorks®. Magnesium microparticles were supplied by Nitroparis (average size < 100 µm, purity > 99.90%).

Each PLA-Mg formulation was prepared according to the following process. Firstly, the PLA pellets were powdered in an A10 Basic IKA miller using liquid nitrogen to prevent thermal degradation at 25,000 rpm. Then, the corresponding amount of Mg microparticles was mixed with the PLA powder for 45 min at 240 rpm and, finally, the extrusion process was carried out in a Rondol co-rotating twin-screw extruder with a screw diameter of 10 mm and length/diameter ratio of 20, working at 40 rpm, with a residence time of 3 min and temperatures of 180, 190, 190, 165, and 120 °C. The PLA-Mg formulations obtained were pelletized and used as feedstock in the single-screw filament extruder.

Therefore, the filaments were obtained from a 3DEVO filament extruder working at 5.0 rpm, heater 1 at 165 °C, heater 2 at 190 °C, heater 3 at 185 °C, and heater 4 at 170 °C; the fan speed was set at 30%. The PLA-Mg filaments were stored in a dryer to avoid humidity spoilage of the PLA.

Once the PLA-Mg filaments were obtained, morphological characterization was carried out via scanning electron microscopy, SEM, (PHILIPS XL30 Scanning Electron Microscope, Phillips, Eindhoven, The Netherlands). All the samples were previously gold-coated (~5 nm thickness) in a Polaron SC7640 Auto/Manual Sputter (Polaron, Newhaven, East Sussex, UK).

Thermal transitions were studied using differential scanning calorimetry in a DSC Q2000 TA instrument under a nitrogen atmosphere (50 mL·min^{-1}). The thermal analysis was programmed as follows: First, heating was performed at 10 °C·min^{-1} from 0 °C up to 180 °C, obtaining the glass transition temperature (T_g) that was calculated as the midpoint of the transition, the cold crystallization enthalpy (ΔH_{cc}), and the melting enthalpy (ΔH_m).

Thermogravimetric analysis, TGA analysis, was performed to study the thermal degradation of the PLA-Mg filaments in a TA-TGA Q500 thermal analyzer. Dynamic TGA experiments were performed under a nitrogen atmosphere (flow rate of 50 mL·min^{-1}). Samples were heated from room temperature to 600 °C at 10 °C·min^{-1}. In this case, the maximum degradation temperature (T_{max}) was calculated as the peak from the first derivative of the TGA curves, and the onset degradation temperature, $T_{5\%}$, was taken at 5% of mass loss.

The degree of crystallinity ($X_c\%$) was calculated taking the value of crystallization enthalpy of pure crystalline PLA ($\Delta H_m°$) as 93.6 J·g^{-1} and W_f as the weight fraction of PLA in the sample [28].

$$X_c\% = 100 \times \left(\frac{\Delta H_m - \Delta H_{cc}}{\Delta H_m^{0\,\circ}} \right) \times \frac{1}{Wf} \qquad (1)$$

XRD measurements were performed using a Bruker D8 Advance instrument with a CuK as source (0.154 nm) and a Detector Vantec1. The scanning range was 5° to 60°, and the step size and count time per step were 0.023851° and 0.5 s, respectively.

Fourier transform infrared spectroscopy, FTIR, measurements were conducted using a Spectrum One FTIR spectrometer (Perkin Elmer instruments). Spectra were obtained in the 4000–400 cm^{-1} region at room temperature in transmission mode with a resolution of 4 cm^{-1}.

The in vitro degradation process was studied by immersing samples of 2 mm length of each PLA-Mg filament in the corresponding volume of phosphate-buffered saline solution, PBS, maintaining the ratio 20 mL PBS: 1 cm^2 surface area. The degradation was studied at different times, taking samples after 7, 14, 21, and 28 days to characterize the first month and after 84 days, corresponding to 3 months. The extraction days are named T_x, where x indicates the number of the corresponding day. The as-obtained PLA-Mg filaments are considered as time 0, T0, and are used as references. Each week, the PBS solution was renovated. The in vitro degradation process was run in an oven at a constant temperature of 37 ± 1 °C and the amount of H_2 released was recorded daily for up to 28 days.

The pH evolution was measured every 7 days for 130 days with a pH METER-02 (Homtiky) with an error of ±0.01. The mass of the PLA-Mg filament samples was measured before beginning the immersion in PBS and after each extraction day. Samples were removed from the solution, and their surfaces were dried with a paper towel, after which they were weighed, obtaining the "wet samples weight". Then, the samples were dried for 2 weeks under vacuum; afterward, their weights were measured, obtaining the "dry samples weight". Water accumulation was obtained from the difference between the mass of the wet and dried samples. The mass loss was calculated from the difference between the mass of the dried samples and the mass of the initial samples. A precision balance was used to weigh all samples within an error of 0.05 mg. The results are given in mass % with respect to the initial mass for each sample.

$$\text{Water uptake \%} = 100 \times \left(\frac{\text{Weight wet} - \text{Weight dry}}{\text{Weight dry}} \right) \qquad (2)$$

$$\text{Mass variation \%} = 100 \times \left(\frac{\text{Weight dry} - \text{Weight initial}}{\text{Weight initial}}\right) \quad (3)$$

3. Results and Discussion

Previously, to obtain PLA-Mg filaments, a cryogenic mill was used to perform the grinding of Mg powder to Mg microparticles. In Figure 1, SEM images at different magnifications of Mg microparticles are shown to study their average size and morphology. As can be seen, round morphology is observed in Mg microparticles with an average particle size of 12.5 ± 7.4 µm.

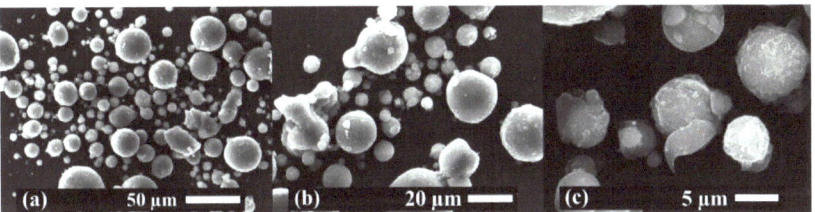

Figure 1. SEM images of Mg microparticles at (**a**) ×500, (**b**) ×1000 and (**c**) ×4000 magnifications.

Once the Mg microparticles were characterized, five different Mg-PLA filaments were obtained in a two-extrusion-step process. After the extrusion process, the extruded materials were cut and used as feedstock for a single-screw filament extruder to obtain filament with a diameter of 1.75 mm. In particular, Mg microparticles were added at 1.2, 5, 10, and 15 wt% with respect to PLA. Neat PLA was also extruded and then a filament of PLA was obtained. In Figure 2, a digital photograph of examples of filaments and their corresponding 3D printing pieces is shown.

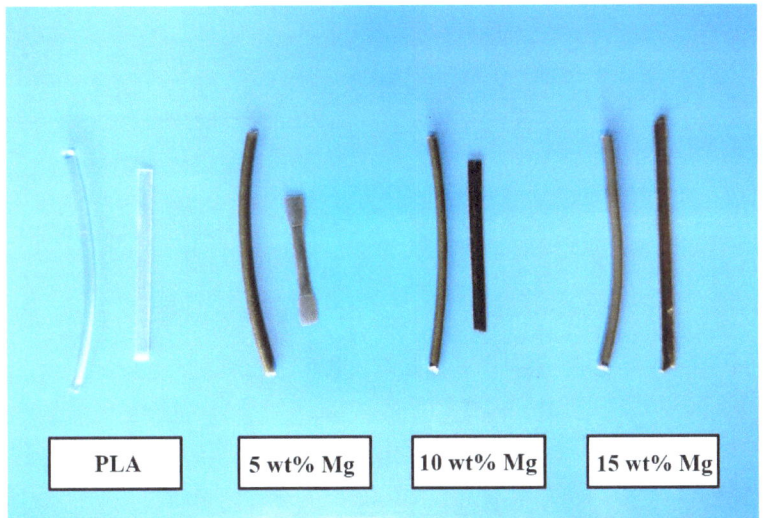

Figure 2. Digital photo of the different filaments and the corresponding 3D printing pieces.

To analyze the dispersion of Mg microparticles through the PLA filaments, SEM images were taken both on the filament surface and the fracture surface and are reported in Figure 3. Moreover, the average diameter values of the PLA-Mg filaments are also reported in Figure 3. As can be observed, the neat PLA filament shows smooth morphology with an average diameter of 1.79 ± 0.01 mm, which is in good agreement for its further use in 3D printers. Moreover, as the amount of Mg increases, more irregularities can be observed on the surface of the PLA-Mg filaments. In addition, the average diameters measured for 1.2, 5, 10, and 15 wt% Mg were 1.74 ± 0.01, 1.74 ± 0.07, 1.64 ± 0.03, and 1.71 ± 0.02 mm, respectively. The presence of Mg powder tends to slightly decrease the average diameter of PLA-Mg filaments; however, all values are in the range of diameter values possessed by other commercial filaments. On the other hand, the fracture surface was also studied using SEM analysis. In Figure 3, a fracture pattern characteristic of polymers such as PLA can be observed for the neat PLA filament. According to SEM analysis, in the fracture area, homogeneous distribution of Mg microparticles can be observed through the PLA matrix with good adhesion between the fillers and the matrix.

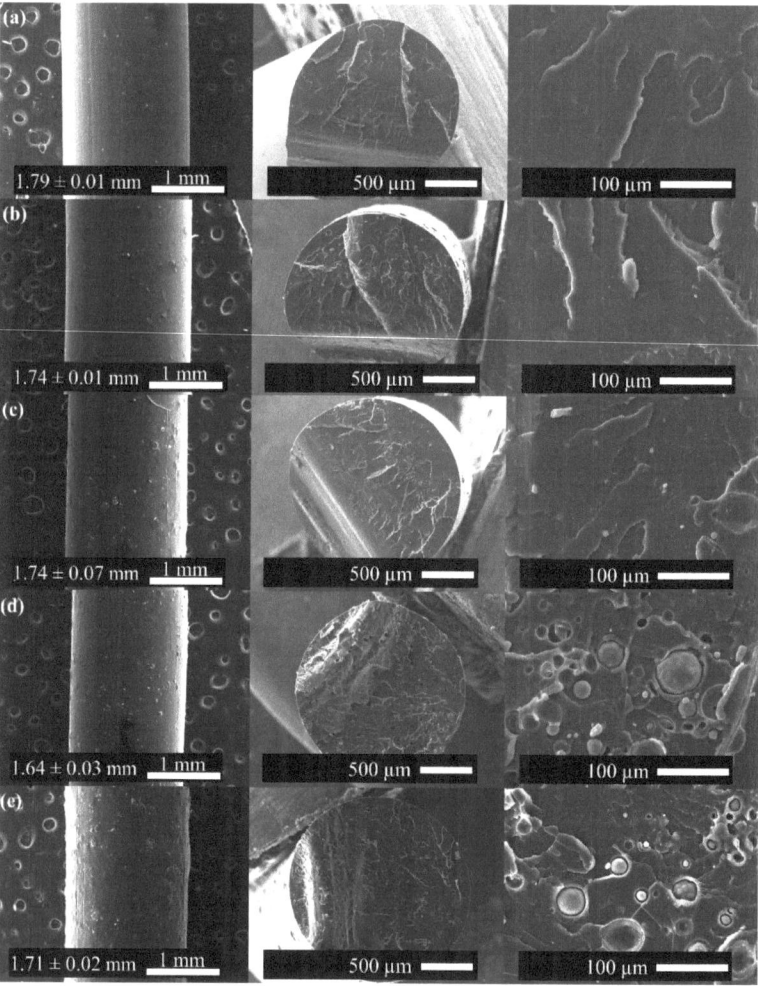

Figure 3. SEM images of (**a**) neat PLA, (**b**) 1.2 wt% Mg, (**c**) 5 wt% Mg, (**d**) 10 wt% Mg, and (**e**) 15 wt% Mg filaments and their fracture surfaces.

The thermal properties of extruded PLA-Mg filaments were properly obtained from DSC thermograms and are summarized in Figure 4 and Table 1. In particular, the DSC curves show that the glass transition temperature (T_g) appears at around 60 °C, and is always accompanied by a small endothermic peak due to physical aging, commonly observed in PLA matrices. In addition, the degree of crystallinity remained essentially unchanged, revealing the amorphous structure of PLA-Mg filaments. In particular, X_c values of 1.0, 1.0, 0.8, 0.9, and 1.1% were calculated for neat PLA, 1.2, 5, 10, and 15 wt% Mg, respectively. No significant changes in the melting temperatures, T_m, were observed for any PLA-Mg filaments, with all values being around 150 °C.

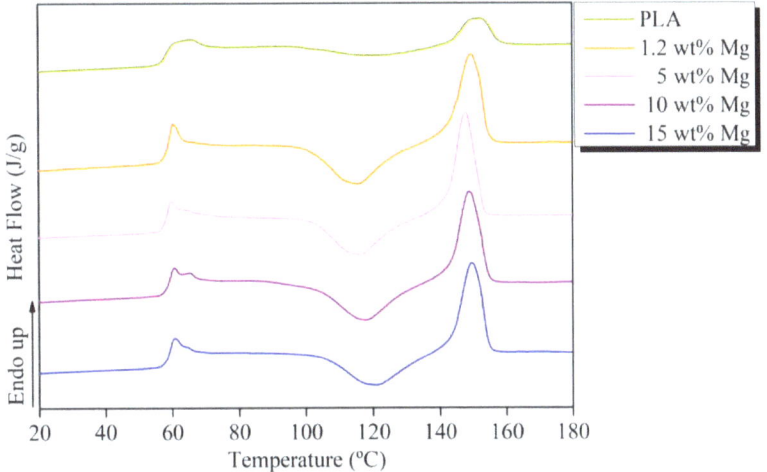

Figure 4. First heating DSC curves for each PLA-Mg filament.

Table 1. Thermal properties for each PLA-Mg filament obtained from DSC curves.

Sample	T_g (°C)	ΔH_{cc} (J/g)	T_{cc} (°C)	ΔH_m (J/g)	T_m (°C)	X_c (%)	$T_{5\%}$ (°C)	T_{max} (°C)	R. Mg wt%*
PLA	60	7.43	118	9.43	153	2.0	320	353	0.0
1.2 wt% Mg	59	22.29	114	25.16	150	3.1	320	339	1.1
5 wt% Mg	59	23.69	116	23.80	148	0.8	283	306	4.8
10 wt% Mg	59	23.34	118	23.48	149	0.9	269	295	9.9
15 wt% Mg	59	20.72	121	21.72	150	1.1	263	291	14.6

R. Mg wt%*: Residual Mg wt% measured at 600 °C via TGA.

To evaluate the effect of Mg microparticle addition on the thermal stability and degradation temperatures of the PLA matrix, TGA analysis was carried out on neat PLA and PLA-Mg filaments. As shown in Figure 5 and Table 1, the weight loss curves of Mg-PLA filaments shift towards lower temperatures with respect to neat PLA filament. In particular, the thermal degradation starts at 320 °C for neat PLA and 1.2 Mg wt% filaments, which are the highest temperatures measured, and falls to 263 °C for the highest amount of Mg, 15 wt%. These results show that Mg induces the degradation of PLA. This behavior has been reported in the literature, associated with MgO (present on the surface of Mg particles), which acts as a catalyst for the PLA depolymerization reaction when the polymer is subjected to elevated temperatures [29]. In addition, a second degradation at 425 °C can be observed related to the dehydration of $Mg(OH)_2$ on the surface of the Mg microparticles [30]. As can be seen, the residual inorganic amount of each PLA-Mg filament reported in Table 1 corresponds to the Mg wt% calculated, indicating the success of the extrusion protocol. Moreover, considering that the temperature usually used in the FDM process for PLA filaments is around 200 °C [31], our PLA-Mg filaments show suitable thermal stability across the range of Mg concentrations studied for FDM processing.

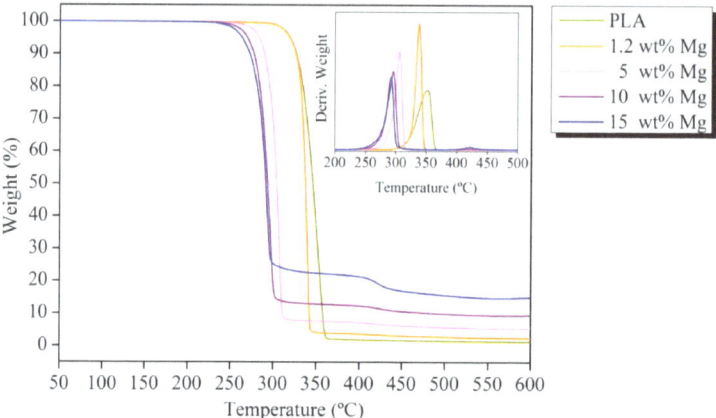

Figure 5. Thermograms and TGA derivate for each PLA-Mg filament.

Once the thermal characterization of PLA-Mg filaments was carried out, the in vitro degradation test in PBS was performed over the course of 84 days, T84, and compared with the original materials, named T0. To study how the degradation process could affect the morphology, average diameter, and surface of PLA-Mg filaments, fracture surface images at T0 and T84 are shown in Figure 6.

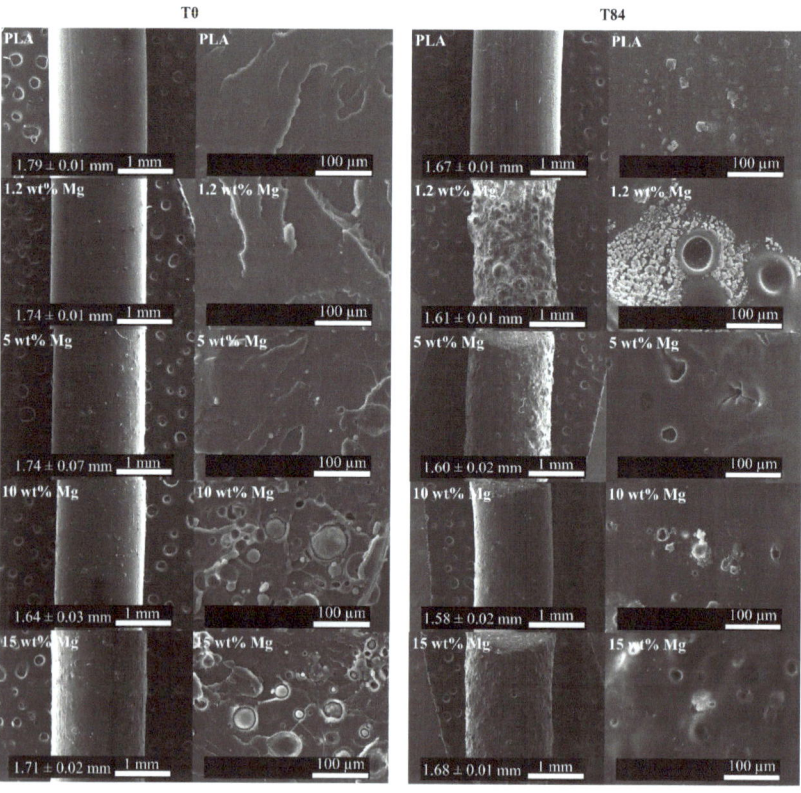

Figure 6. SEM images of PLA-Mg filaments and their fracture surface both at T0 and T84.

As can be appreciated, the PLA-Mg filaments after 84 days of immersion in PBS showed an irregular surface with a high number of holes and no presence of Mg microparticles. Remarkably, this phenomenon is observed in the 1.2 wt% Mg filament, which showed the highest eroded surface of all samples. Moreover, the immersion in PBS provoked the precipitation of salts deposited on the surface of the PLA-Mg filaments after 84 days. Parallel to the SEM study, a visual appearance study was carried out to visualize the color changes of the PLA-Mg filaments every seven days, as shown in Figure 7.

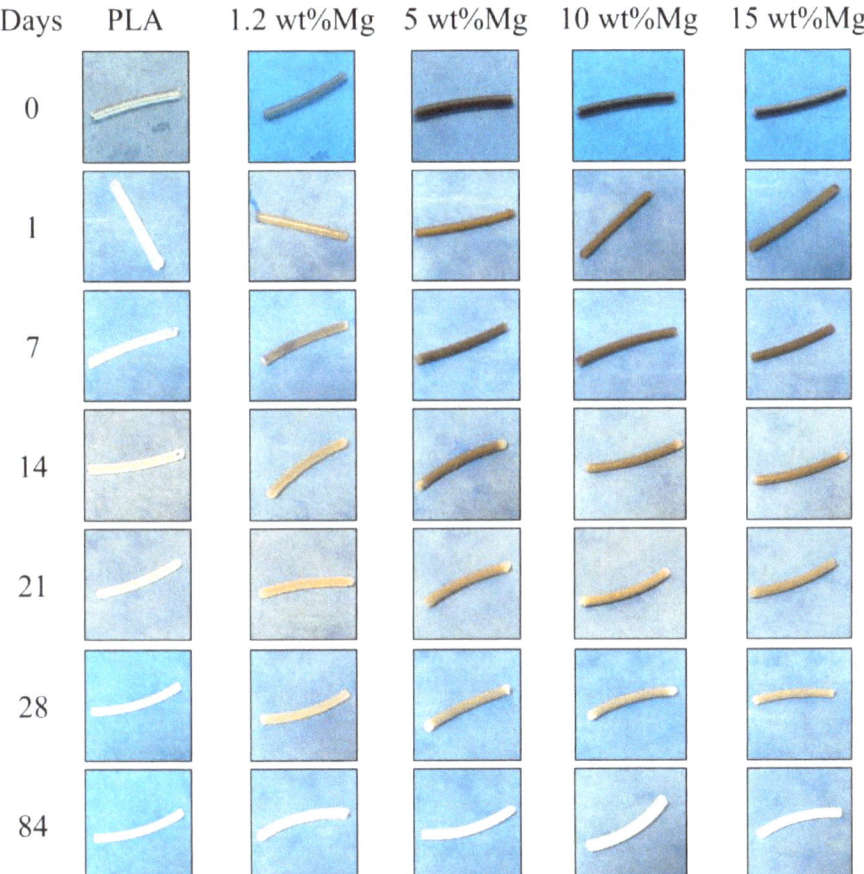

Figure 7. Visual appearance evolution of each PLA-Mg filament at different times.

Initially at T0, the filaments' color changed as the amount of Mg increased. In particular, from a white transparent color for neat PLA to soft gray for 1.2 wt% Mg and dark gray for 5, 10, and 15 wt% Mg. As the degradation time increased, a loss of gray color was observed attributed to the release of Mg microparticles, which is in agreement with the previous fracture surface observed via SEM. After 84 days, all PLA-Mg filaments turned white, as was the neat PLA sample.

To corroborate the release of Mg and also study the chemical nature of deposited salts, XRD analysis of PLA-Mg filaments was carried out after 84 days of immersion in PBS and compared with the XRD patterns at T0, shown in Figure 8.

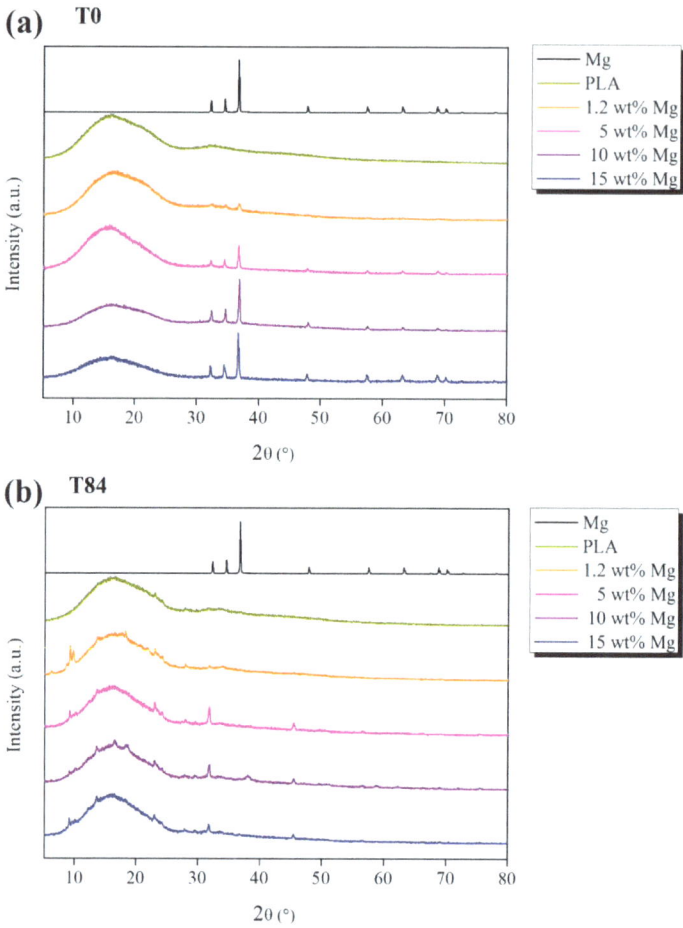

Figure 8. XRD patterns of Mg microparticles as well as PLA-Mg filaments (**a**) at T0 and (**b**) after 84 days, T84, in PBS media.

According to XRD patterns, the Mg microparticles show crystallographic peaks located at 2θ = 32.19°, 34.39°, 36.62°, 47.82°, 57.37°, 63.05°, 67.31°, 68.63°, 69.99°, 72.49°, and 77.82° attributed to the [100], [002], [101], [102], [110], [103], [200], [112], [201], [004], and [202] crystallographic planes, respectively [32]. In particular, the presence of Mg microparticles in the Mg-PLA filaments is confirmed by the fact that these peaks can be observed in all XRD patterns shown in Figure 8. However, after 84 days of immersion in PBS, no crystallographic peaks related to Mg were observed, confirming the release of Mg previously discussed. Additionally, the presence of NaCl is confirmed by the peaks observed at 2θ = 32.0°, 45.5°, and 56.5° attributed to the [200], [220], and [222] crystallographic planes of NaCl [32,33].

On the other hand, before starting the degradation test, neat PLA and PLA-Mg filaments show amorphous structure with no crystallographic peaks attributed to PLA, in good agreement with previously Xc (%) values measured using DSC. After 84 days of degradation, the crystallographic peaks located at 16.5°, 19.1°, and 22.5° related to the alpha crystalline phase of PLA [33] were observed for PLA-Mg filaments. As reported in the literature, water molecules showed selective hydrolytic attack towards long PLA chains in the amorphous phase, which highly affects the degree of crystallinity. In fact, the ester groups of aliphatic PLA chains are susceptible to being broken by water molecules, yielding short PLA chains with carboxylate and hydroxyl end groups [33].

Furthermore, the hydrolysis of ester groups can be monitored using FTIR analysis following the characteristic peaks related to the stretching of the amorphous carbonyl group at 1750 cm^{-1} [33], the stretching of carbonyl in the carboxylate group at 1600 cm^{-1} [33], and the stretching peaks related to O-H groups at 3690 cm^{-1} [33]. With this aim, FTIR analysis was performed on neat PLA and PLA-Mg filaments at T0 and T84 and reported in Figure 9.

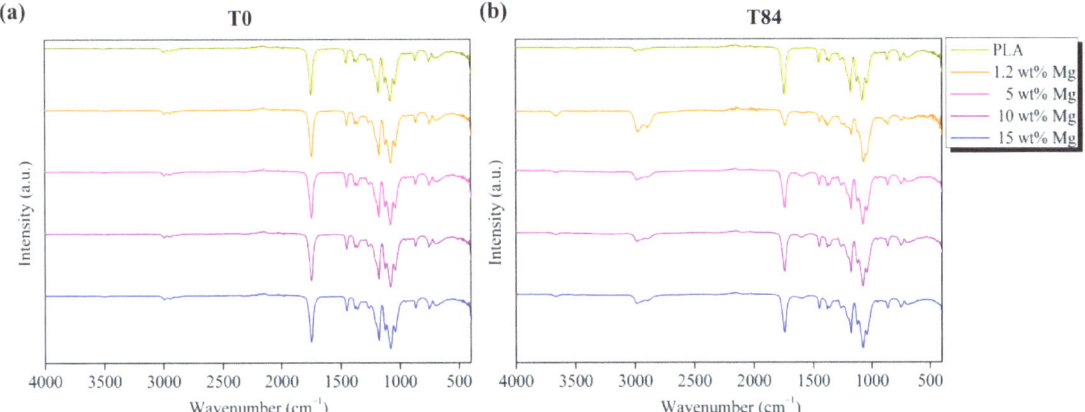

Figure 9. FTIR spectra of PLA-Mg filaments (**a**) at T0 and (**b**) after 84 days in PBS media.

It is important to remark that characteristic bands attributed to PLA chemical group vibrations can be observed in all spectra at T0, in particular between 2800 and 3200 cm^{-1}, the bands attributed to the symmetrical and asymmetrical stretching vibration of CH groups, and between 1000 and 1500 cm^{-1}, the bands attributed to the symmetrical and asymmetrical stretching vibration of C-O-C groups. Moreover, at 830–960 cm^{-1} and 1430–1520 cm^{-1}, the bands assigned to the stretching vibration and bending vibrations of CH$_3$, respectively, can be observed [33].

It is worth noting that the hydrolytic degradation of the PLA chains was accelerated by the Mg microparticles. Neat PLA filament showed the same characteristic bands at T0 and T84. However, for PLA-Mg filaments, the hydrolytic degradation after 84 days in PBS media was confirmed using FTIR following the intensity of the 1750 cm^{-1} characteristic band attributed to the amorphous carbonyl group at T84, which is slightly lower than at T0. Additionally, the presence of two new bands at 3690 and 1600 cm^{-1}, attributed to the stretching of O-H groups and the stretching of carbonyl in the carboxylate group, respectively, can be observed in Figure 9. This acceleration can be explained since Mg is highly hydrophilic, which increases the presence of water molecules that degrade the PLA matrix.

As the degradation rate of PLA is deeply affected by the amount of water, the percentage of water uptake of PLA-Mg filaments during the in vitro degradation test was calculated every 7 days up to 84 days and reported in Figure 10.

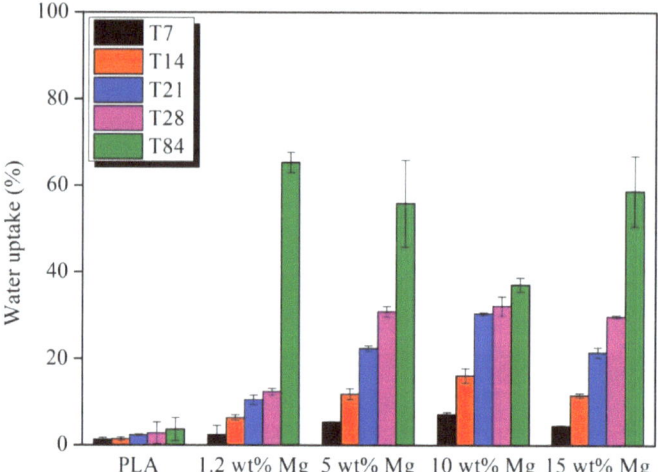

Figure 10. Percentage of water uptake for each PLA-Mg filament after 7, 14, 21, 28, and 84 days in PBS media.

The hydrolysis of PLA is controlled by the diffusion of water molecules in the amorphous phase [34]. As previously proposed, the presence of Mg highly increases the water uptake, which implies an increase in the water diffusion through the PLA matrix as seen in Figure 10. It is important to remark that after 84 days, the amount of water is almost constant in all PLA-Mg filaments, no matter the amount of Mg. As previously discussed, after 84 days of immersion in PBS, the Mg released provoked a high number of holes and erased surfaced sites where water molecules can degrade the PLA matrix.

The main concern in the application of Mg in the biomedical field is its reaction with water, as shown in Equation (4), in particular the generation of hydrogen, H_2 [14].

$$Mg + 2H_2O \rightarrow Mg^{2+} + 2(OH^-) + H_2 \tag{4}$$

The problem of H_2 released in the human body cannot be explained in terms of how much hydrogen is produced but instead has to be explained in terms of time. In Figure 11a, the H_2 release as a function of immersion time can be observed.

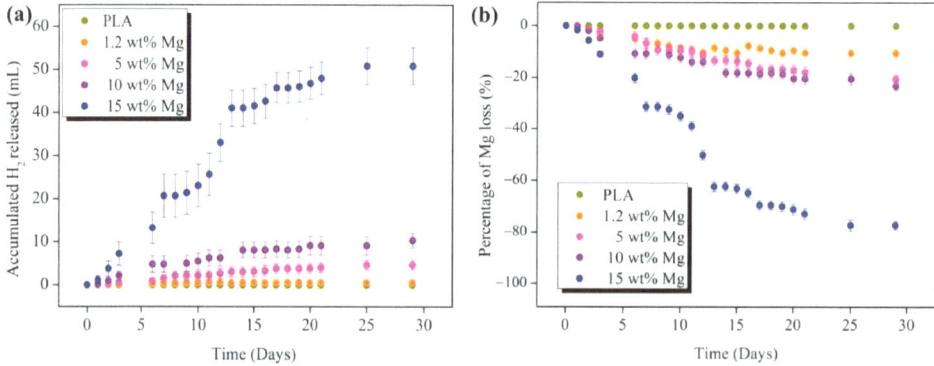

Figure 11. (a) Accumulated amount of hydrogen released as a function of immersion time in PBS and (b) Mg loss in terms of percentage of mass loss with respect to the Mg initial weight as a function of immersion time in PBS.

The accumulated H_2 released depends on the Mg content of the PLA-Mg filaments. As expected, with higher Mg content, the hydrogen release rate increases up to a volume of 50.9 mL of H_2 for the highest amount of 15 wt% Mg after 28 days in PBS. It is important to remark that in any case, the maximum amount of H_2 tolerated by the human body, (2.25 mL/cm^2/day) [14], is not achieved, which reveals that our materials are suitable for biomedical uses. Moreover, the H_2 release kinetics start to stabilize after the first two weeks, where a plateau in the accumulated H_2 released is observed in all PLA-Mg filaments and kept constant during the 28 days. In addition, the amount of Mg that reacted to produce the measured hydrogen was stoichiometrically calculated and is shown in Figure 11b as a percentage of Mg mass loss with respect to the initial Mg weight in each filament. It is important to point out that after 28 days, the 15 wt% Mg filament lost more than 70% of the Mg, which is in accordance with the large volume of H_2 measured for this sample. Moreover, as the amount of initial Mg increases, the percentage of Mg loss increases by increasing the time in PBS media. Considering that these percentages are stoichiometrically calculated by hydrogen measurements, we can approximate that, after 84 days, 100% of Mg will be released at least for the highest amount of 15 wt% Mg, which is in good agreement with the XRD analysis previously discussed.

Another concern regarding the Mg reaction with physiological environments is the alkalization of the surrounding area due to the OH$^-$ produced. To verify if the buffer capacity of the physiological environment can be surpassed by the alkalization produced in the reaction of Mg with water, the variation in pH in the PBS media was measured every seven days and is reported in Figure 12.

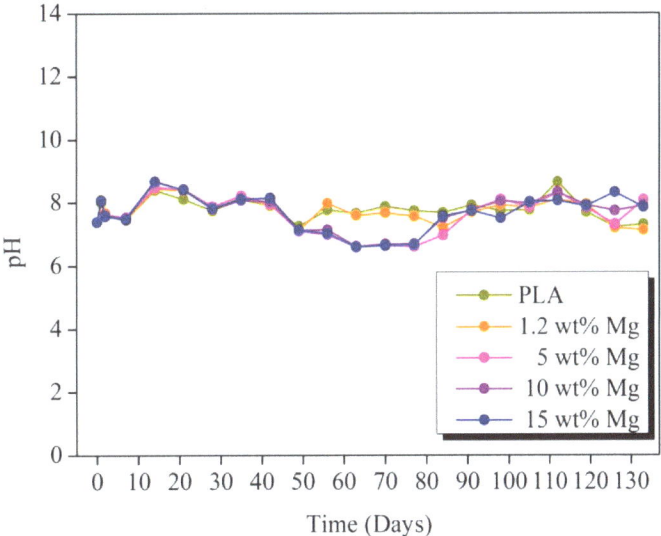

Figure 12. Evolution of pH values for each PLA-Mg filament during degradation in PBS media.

As can be observed, the pH values fluctuated around 7.6 for 5 months for all PLA-Mg filaments. It is important to point out that the pH evolution is quite similar during the first 50 days in PBS media, after which slightly different pH values were observed, in particular from 7.9 for neat PLA to 6.9 for 15 wt% Mg. This is a contradictory result, taking into account that 15 wt% Mg is the material that releases the highest amount of hydrogen, which implies that it must induce the largest increment in pH (alkaline) as well. This behavior may be understood considering that the polymeric matrix can degrade and release lactic acid to the medium, which induces a pH decrement that can compensate for the increment in pH caused by the reaction of Mg.

Finally, Figure 13 shows the mass variation of PLA-Mg filaments after 7, 14, 21, 28, and 84 days in PBS media. The PLA-Mg filaments gain mass during the first 28 days, probably due to NaCl precipitation on the surface of the filaments. However, as discussed above, the Mg released induces the formation of pores and holes through the surface of the filaments. Thus, the irregular morphology increases the surface where water molecules can attack the PLA chains. As can be seen in Figure 13, after 84 days, a high loss of mass was measured for all PLA-Mg filaments; remarkably, at this time, T84, the water uptake reported in Figure 10 was the highest. Furthermore, it seems that Mg release occurs after the 28 first days and then the water uptake increases up to the maximum values where the highest PLA degradation rate is observed, losing 3, 10, 15, and 10% of mass for 1.2, 5, 10, and 15 wt% Mg, respectively. On the other hand, the neat PLA filament started to lose mass since the first week in PBS but the PLA degradation was not enough to observe the FTIR bands related to carboxylate and hydroxyl end groups.

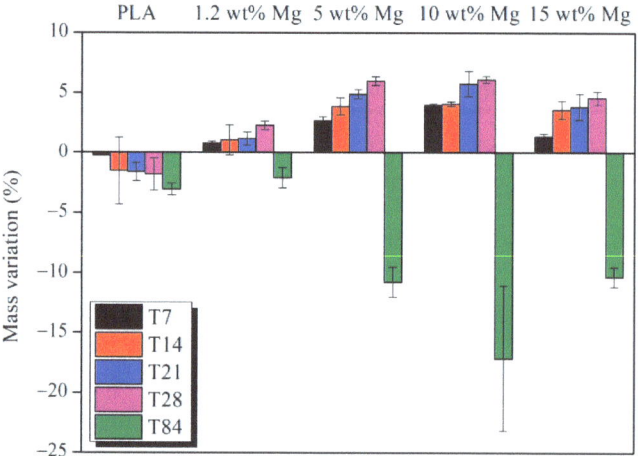

Figure 13. Percentage of mass variation in PLA-Mg filaments after 7, 14, 21, 28, and 84 days in PBS media.

4. Conclusions

In this work, PLA-Mg filaments for further 3D printing with different Mg microparticle concentrations were successfully obtained. In particular, Mg microparticles were added at 1.2, 5, 10, and 15 wt% with respect to the polymeric matrix, obtaining an average diameter of the filaments in the range of diameters used in 1.75 mm 3D printers. All the extruded PLA-Mg filaments showed a T_g around 60 °C and a T_m around 150 °C; however, the presence of Mg microparticles does not affect the small degree of crystallinity of pure PLA. Furthermore, PLA-Mg filaments showed high thermal stability, with PLA and 1.2 MgO wt% presenting the highest T_{max} values of 353 and 339 °C, respectively, this value further decreasing down to 291 °C at 15 wt% of Mg. This phenomenon can be related to the presence of Mg and in particular the MgO present on the surface of the microparticles inducing PLA degradation. However, these degradation temperatures are higher than the temperature used in the extrusion process as well as in 3D printing, which is generally 200 °C for PLA-based materials. After the fabrication and characterization of the materials, in vitro degradation in PBS was studied. From a morphological point of view, PLA-Mg filaments present an irregular surface and holes after 84 days due to the degradation process and the release of Mg microparticles. In fact, no presence of Mg microparticles was observed after 84 days through XRD analysis, confirming the release of Mg. Furthermore, all the samples with Mg microparticles presented a high water uptake capacity, which increases water diffusion through the material, improving degradation. The evaluation of

H$_2$ release at different immersion times revealed that, as expected, 15 wt% of Mg showed the highest volume of H$_2$ released, without, however, reaching the maximum amount of H$_2$ tolerated by the human body. Thus, taking into account the obtained results, we can conclude that our materials are suitable for potential 3D printed biomedical applications.

Author Contributions: Conceptualization, L.P. and A.L.; investigation, A.L., V.S., and I.R.A.; resources, L.P; writing—original draft preparation, A.L. and V.S.; writing—review and editing, M.L., D.L. and L.P.; funding acquisition, L.P. and D.L. All authors have read and agreed to the published version of the manuscript.

Funding: This research was funded by the Agencia Estatal de Investigación (AEI, MICINN, Spain) and Fondo Europeo de Desarrollo Regional (FEDER, EU), PID2021-123753NB-C31 and TED2021-129335B-C21.

Institutional Review Board Statement: Not applicable.

Data Availability Statement: Not applicable.

Acknowledgments: Financial support of the Spanish Ministry of Science and Innovation (AEI, MICINN) through PID2021-123753NB-C31 and TED2021-129335B-C21 is greatly acknowledged.

Conflicts of Interest: The authors declare no conflict of interest.

References

1. Liu, Z.; Wang, Y.; Wu, B.; Cui, C.; Guo, Y.; Yan, C. A critical review of fused deposition modeling 3D printing technology in manufacturing polylactic acid parts. *Int. J. Adv. Manuf. Technol.* **2019**, *102*, 2877–2889. [CrossRef]
2. de Castro Motta, J.; Qaderi, S.; Farina, I.; Singh, N.; Amendola, A.; Fraternali, F. Experimental characterization and mechanical modeling of additively manufactured TPU components of innovative seismic isolators. *Acta Mech.* **2022**, 1–12. [CrossRef]
3. Bhagia, S.; Bornani, K.; Agarwal, R.; Satlewal, A.; Ďurkovič, J.; Lagaňa, R.; Bhagia, M.; Yoo, C.G.; Zhao, X.; Kunc, V.; et al. Critical review of FDM 3D printing of PLA biocomposites filled with biomass resources, characterization, biodegradability, upcycling and opportunities for biorefineries. *Appl. Mater. Today* **2021**, *24*, 101078. [CrossRef]
4. Chen, X.; Chen, G.; Wang, G.; Zhu, P.; Gao, C. Recent Progress on 3D-Printed Polylactic Acid and Its Applications in Bone Repair. *Adv. Eng. Mater.* **2020**, *22*, 1901065. [CrossRef]
5. Serra, T.; Mateos-Timoneda, M.A.; Planell, J.A.; Navarro, M. 3D printed PLA-based scaffolds: A versatile tool in regenerative medicine. *Organogenesis* **2013**, *9*, 239–244. [CrossRef]
6. Ramot, Y.; Haim-Zada, M.; Domb, A.J.; Nyska, A. Biocompatibility and safety of PLA and its copolymers. *Adv. Drug Deliv. Rev.* **2016**, *107*, 153–162. [CrossRef]
7. Hanon, M.M.; Dobos, J.; Zsidai, L. The influence of 3D printing process parameters on the mechanical performance of PLA polymer and its correlation with hardness. *Procedia Manuf.* **2020**, *54*, 244–249. [CrossRef]
8. Zaaba, N.F.; Jaafar, M. A review on degradation mechanisms of polylactic acid: Hydrolytic, photodegradative, microbial, and enzymatic degradation. *Polym. Eng. Sci.* **2020**, *60*, 2061–2075. [CrossRef]
9. Antoniac, I.; Popescu, D.; Zapciu, A.; Antoniac, A.; Miculescu, F.; Moldovan, H. Magnesium filled polylactic acid (PLA) material for filament based 3D printing. *Materials* **2019**, *12*, 719. [CrossRef]
10. Arockiam, A.J.; Subramanian, K.; Padmanabhan, R.G.; Selvaraj, R.; Bagal, D.K.; Rajesh, S. A review on PLA with different fillers used as a filament in 3D printing. *Mater. Today Proc.* **2021**, *50*, 2057–2064. [CrossRef]
11. Vidakis, N.; Petousis, M.; Velidakis, E.; Liebscher, M.; Tzounis, L. Three-Dimensional Printed Antimicrobial Objects of Polylactic Acid (PLA)-Silver Nanoparticle Nanocomposite Filaments Produced by an In-Situ Reduction Reactive Melt Mixing Process. *Biomimetics* **2020**, *5*, 42. [CrossRef]
12. Oksiuta, Z.; Jalbrzykowski, M.; Mystkowska, J.; Romanczuk, E.; Osiecki, T. Mechanical and thermal properties of polylactide (PLA) composites modified with Mg, Fe, and polyethylene (PE) additives. *Polymers* **2020**, *12*, 2939. [CrossRef]
13. Podstawczyk, D.; Skrzypczak, D.; Połomska, X.; Stargała, A.; Witek-Krowiak, A.; Guiseppi-Elie, A.; Galewski, Z. Preparation of antimicrobial 3D printing filament: In situ thermal formation of silver nanoparticles during the material extrusion. *Polym. Compos.* **2020**, *41*, 4692–4705. [CrossRef]
14. Leonés, A.; Lieblich, M.; Benavente, R.; Gonzalez, J.L.; Peponi, L. Potential applications of magnesium-based polymeric nanocomposites obtained by electrospinning technique. *Nanomaterials* **2020**, *10*, 1524. [CrossRef] [PubMed]
15. Manivasagam, G.; Suwas, S. Biodegradable Mg and Mg based alloys for biomedical implants. *Mater. Sci. Technol.* **2014**, *30*, 515–520. [CrossRef]
16. Sasaki, Y.; Sathi, G.A.; Yamamoto, O. Wound healing effect of bioactive ion released from Mg-smectite. *Mater. Sci. Eng. C* **2017**, *77*, 52–57. [CrossRef] [PubMed]
17. Bakhshi, R.; Mohammadi-Zerankeshi, M.; Mehrabi-Dehdezi, M.; Alizadeh, R.; Labbaf, S.; Abachi, P. Additive manufacturing of PLA-Mg composite scaffolds for hard tissue engineering applications. *J. Mech. Behav. Biomed. Mater.* **2023**, *138*, 105655. [CrossRef]

18. Ferrández-Montero, A.; Lieblich, M.; González-Carrasco, J.L.; Benavente, R.; Lorenzo, V.; Detsch, R.; Boccaccini, A.R.; Ferrari, B. Development of biocompatible and fully bioabsorbable PLA/Mg films for tissue regeneration applications. *Acta Biomater.* **2019**, *98*, 114–124. [CrossRef]
19. Cifuentes, S.C.; Frutos, E.; Benavente, R.; Lorenzo, V.; González-Carrasco, J.L. Assessment of mechanical behavior of PLA composites reinforced with Mg micro-particles through depth-sensing indentations analysis. *J. Mech. Behav. Biomed. Mater.* **2017**, *65*, 781–790. [CrossRef]
20. Brown, A.; Zaky, S.; Ray, H.; Sfeir, C. Porous magnesium/PLGA composite scaffolds for enhanced bone regeneration following tooth extraction. *Acta Biomater.* **2015**, *11*, 543–553. [CrossRef]
21. Shuai, C.; Li, Y.; Feng, P.; Guo, W.; Yang, W.; Peng, S. Positive feedback effects of Mg on the hydrolysis of poly-l-lactic acid (PLLA): Promoted degradation of PLLA scaffolds. *Polym. Test* **2018**, *68*, 27–33. [CrossRef]
22. Zhao, C.; Wu, H.; Ni, J.; Zhang, S.; Zhang, X. Development of PLA/Mg composite for orthopedic implant: Tunable degradation and enhanced mineralization, *Compos. Sci. Technol.* **2017**, *147*, 8–15. [CrossRef]
23. Li, X.; Chu, C.; Wei, Y.; Qi, C.; Bai, J.; Guo, C.; Xue, F.; Lin, P.; Chu, P.K. In vitro degradation kinetics of pure PLA and Mg/PLA composite: Effects of immersion temperature and compression stress. *Acta Biomater.* **2017**, *48*, 468–478. [CrossRef]
24. Cifuentes, S.C.; Gavilán, R.; Lieblich, M.; Benavente, R.; González-Carrasco, J.L. In vitro degradation of biodegradable polylactic acid/magnesium composites: Relevance of Mg particle shape. *Acta Biomater.* **2016**, *32*, 348–357. [CrossRef]
25. Thompson, C.; Biurrun, N.; Lizarralde, I. Processing and properties of PLA/Mg filaments for 3D printing of scaffolds for biomedical applications. *Rapid Prototyp. J.* **2022**, *28*, 884–894.
26. Barrasa, J.O.; Ferrández-Montero, A.; Ferrari, B.; Pastor, J.Y. Characterisation and modelling of pla filaments and evolution with time. *Polymers* **2021**, *13*, 2899. [CrossRef] [PubMed]
27. Suteja, T.J.; Soesanti, A. Mechanical Properties of 3D Printed Polylactic Acid Product for Various Infill Design Parameters: A Review. In *Journal of Physics: Conference Series*; IOP Publishing: Bristol, UK, 2020; Volume 1569, p. 042010. [CrossRef]
28. Peponi, L.; Navarro-Baena, I.; Báez, J.E.; Kenny, J.M.; Marcos-Fernández, A. Effect of the molecular weight on the crystallinity of PCL-b-PLLA di-block copolymers. *Polymer* **2012**, *53*, 4561–4568. [CrossRef]
29. Motoyama, T.; Tsukegi, T.; Shirai, Y.; Nishida, H.; Endo, T. Effects of MgO catalyst on depolymerization of poly-l-lactic acid to l,l-lactide. *Polym. Degrad. Stab.* **2007**, *92*, 1350–1358. [CrossRef]
30. Ansari, A.; Ali, A.; Asif, M. Shamsuzzaman Microwave-assisted MgO NP catalyzed one-pot multicomponent synthesis of polysubstituted steroidal pyridines. *New J. Chem.* **2018**, *42*, 184–197. [CrossRef]
31. Hsueh, M.H.; Lai, C.J.; Wang, S.H.; Zeng, Y.S.; Hsieh, C.H.; Pan, C.Y.; Huang, W.C. Effect of printing parameters on the thermal and mechanical properties of 3d-printed pla and petg, using fused deposition modeling. *Polymers* **2021**, *13*, 1758. [CrossRef]
32. McMurdie, H.F.; Morris, M.C.; Evans, E.H.; Paretzkin, B.; Wong-Ng, W.; Hubbard, C.R. Standard X-Ray Diffraction Powder Patterns from The JCPDS Research Associateship. *Powder Diffr.* **1986**, *1*, 265–275. [CrossRef]
33. Leonés, A.; Peponi, L.; Lieblich, M.; Benavente, R.; Fiori, S. In vitro degradation of plasticized PLA electrospun fiber mats: Morphological, thermal and crystalline evolution. *Polymers* **2020**, *12*, 2975. [CrossRef] [PubMed]
34. Migliaresi, C.; Fambri, L.; Cohn, D. A study on the in vitro degradation of poly(lactic acid). *J. Biomater. Sci. Polym. Ed.* **1994**, *5*, 591–606. [CrossRef] [PubMed]

Disclaimer/Publisher's Note: The statements, opinions and data contained in all publications are solely those of the individual author(s) and contributor(s) and not of MDPI and/or the editor(s). MDPI and/or the editor(s) disclaim responsibility for any injury to people or property resulting from any ideas, methods, instructions or products referred to in the content.

Article

Cohesive Zone Modeling of Pull-Out Test for Dental Fiber–Silicone Polymer

Ayman M. Maqableh [1,*] and Muhanad M. Hatamleh [2]

1 School of Electro-Mechanical Engineering, Luminus Technical University College (LTUC), Amman 11118, Jordan
2 Faculty of Applied Medical Science, Allied Dental Sciences Department, Jordan University of Science and Technology (JUST), Irbid 22110, Jordan; muhanad.hatamleh@gmail.com
* Correspondence: a.maqableh@ltuc.edu

Citation: Maqableh, A.M.; Hatamleh, M.M. Cohesive Zone Modeling of Pull-Out Test for Dental Fiber–Silicone Polymer. *Polymers* 2023, 15, 3668. https://doi.org/10.3390/polym15183668

Academic Editors: Jesús-María García-Martínez and Emilia P. Collar

Received: 2 August 2023
Revised: 2 September 2023
Accepted: 3 September 2023
Published: 6 September 2023

Copyright: © 2023 by the authors. Licensee MDPI, Basel, Switzerland. This article is an open access article distributed under the terms and conditions of the Creative Commons Attribution (CC BY) license (https://creativecommons.org/licenses/by/4.0/).

Abstract: Background: Several analytical methods for the fiber pull-out test have been developed to evaluate the bond strength of fiber–matrix systems. We aimed to investigate the debonding mechanism of a fiber–silicone pull-out specimen and validate the experimental data using 3D-FEM and a cohesive element approach. Methods: A 3D model of a fiber–silicone pull-out testing specimen was established by pre-processing CT images of the typical specimen. The materials on the scans were posted in three different cross-sectional views using ScanIP and imported to ScanFE in which 3D generation was implemented for all of the image slices. This file was exported in FEA format and was imported in the FEA software (PATRAN/ABAQUS, version r2) for generating solid mesh, boundary conditions, and material properties attribution, as well as load case creation and data processing. Results: The FEM cohesive zone pull-out force versus displacement curve showed an initial linear response. The Von Mises stress concentration was distributed along the fiber–silicone interface. The damage in the principal stresses' directions S11, S22, and S33, which represented the maximum possible magnitude of tensile and compressive stress at the fiber–silicone interface, showed that the stress is higher in the direction S33 (stress acting in the Z-direction) in which the lower damage criterion was higher as well when compared to S11 (stress acting in the XY plane) and S23 (stress acting in the YZ plane). Conclusions: The comparison between the experimental values and the results from the finite element simulations show that the proposed cohesive zone model accurately reproduces the experimental results. These results are considered almost identical to the experimental observations about the interface. The cohesive element approach is a potential function that takes into account the shear effects with many advantages related to its ability to predict the initiation and progress of the fiber–silicone debonding during pull-out tests. A disadvantage of this approach is the computational effort required for the simulation and analysis process. A good understanding of the parameters related to the cohesive laws is responsible for a successful simulation.

Keywords: FEA; FRC; fiber-reinforced composite

1. Introduction

Restoring significant facial defects often necessitates the use of a facial prosthesis, which can be constructed from silicone, acrylic resin, or a combination of both materials [1,2]. Various issues have been reported by different researchers that can affect the functionality and effectiveness of these facial prostheses. These problems encompass issues like silicone degradation, separation of silicone from the acrylic base, limitations in simulating natural facial expressions due to the stiffness of the acrylic base, and a reduction in the seamless integration of the prosthesis, resulting in visible gaps [3–6].

Efforts have been made to develop new polymeric materials that possess improved mechanical properties, including a higher tear strength, lower hardness, and reduced viscosity [6–8]. It has also been proposed that facial prostheses be fabricated in three layers, comprising a silicone rubber base material, an inner silicone gel layer, and a thin outer polymeric coating. However, attempts to modify materials to address specific shortcomings have often resulted in the compromise of other desirable characteristics. While existing materials for facial prostheses have seen enhancements, they are not yet considered ideal [9]. Ongoing research aims to create prostheses composed of two or more materials, laminated and bonded together, each possessing its own optimal properties [9].

Recently, there has been a growing interest in utilizing fiber-reinforced composite (FRC) as a biomaterial in dental and medical applications [10–13]. When applied to maxillofacial silicone prostheses, FRC tends to provide more stable margins and a strong bond between the silicone and the fibers. Moreover, the silicone component offers a realistic sensation when in contact with the skin, meeting the aesthetic requirements of patients and enhancing their quality of life [11]. The new technique reported indicated that the silicone polymer can be encapsulated in a retentive glass fiber-embedded framework as in fiber-embedded maxillofacial prostheses [11].

Fiber-reinforced composites, particularly those incorporating continuous glass fibers, offer a wide range of mechanical properties, dependent on factors such as the fiber type, orientation, quantity, and polymer matrix [14,15]. They have been proved to be suitable dental and medical biomaterials [16,17].

During clinical service, forces expected to influence the bond integrity between fibers and silicone polymers are likely to be generated when the patient holds the silicone to dislodge the prosthesis from the magnetic retentive sites or bars, or during cleaning of the prosthesis [18].

The pull out forces required to disrupt the bond integrity between fibers and silicone elastomer have been reported previously to be in the range of (12–20 N). However, earlier reports have indicated various difficulties with FRC bonding including the experimental handling of samples and/or the ability to observe the fracture events instrumentally or usually with enough detail [19–22].

Several analytical methods, such as the fragmentation test, microbond test, and fiber pull-out test have been developed to evaluate bond strength of fiber–matrix systems [20,23]. The single fiber pull-out test has been extensively used due to its importance in understanding the fiber–matrix interface, stress distribution, and strength within materials [15]. The bond strength is determined by measuring the force required to pull-out a fiber bundle embedded in the matrix [19,20,24–26]. Photoelasticity has been also used to experimentally measure the stress field at the fiber–matrix interface [27].

One of the most recent numerical methods to investigate fracture mechanics is the finite element method (FEM). This method is based on continuum damage theories, in which the difficulty in relating its parameters to well-defined experimental parameters (i.e., fracture energy) is pointed out. In addition, these methods do not consider the discrete nature of fracture mechanics [27,28]. A new FEM alternative for these theories has been related as a discrete approach based on the cohesive elements used at the interface between standard volume elements to nucleate cracks and propagate them following the deformation process [29–34]. The aim of the study was to investigate the debonding mechanism of a fiber–silicone pull-out specimen and validate the experimental data achieved using 3D-FEM and a cohesive element approach.

2. Material and Methods

2.1. Specimen Construction and Test Methodology

The construction process was previously described [15]. Three test samples were created by embedding unidirectional glass fiber bundles (manufactured by C&B Fibers, StickTech, Turku, Finland) with a diameter of 1.5 mm and an embedded length of 20 mm into a heat-polymerized silicone elastomer (Cosmesil M511, Principality Medical, Newport, UK). The construction of these specimens involved the use of a two-part sectional flask measuring 100 mm × 80 mm × 30 mm. The lower section of the flask had three holes with a diameter of 1.50 mm and a depth of 5 mm, into which the fiber bundles were securely fixed. Meanwhile, the upper section contained three cylindrical-shaped molds measuring 14.40 mm in diameter and 20 mm in length, where the silicone material was packed. The two parts of the flask were separated by a thin layer of sodium alginate (Hillier Dental, Kent, UK).

The glass fiber bundles were subjected to light polymerization for 4 min using a curing unit (ESPE visio ® Beta vario, 3M ESPE, Seefeld, Germany). Subsequently, the second part of the flask was assembled over the basal part, allowing the fiber bundles to protrude through the center of the cylindrical molds. The maxillofacial silicone elastomer was prepared according to the manufacturer's instructions, using a ratio of 10 g of rubber to 1 g of hardener, which was measured with a micro-balance. The silicone was manually mixed for 5 min and then mechanically mixed under vacuum conditions for 5 min using a Multi Vac 4 (Degussa, Munich, Germany).

After the mixing process was completed, the silicone was poured into the flask molds with the assistance of vibration. The flask contents were heat-polymerized in an oven (Gallenkamp, London, UK) in accordance with the manufacturer's guidelines, which involved heating at 100 °C for 1 h. Following this, the specimens were gently removed and allowed to cool at room temperature for 2 h. Subsequently, the specimens were carefully taken out and stored in a dry environment for 24 h.

A universal testing machine was utilized to undergo the mechanical test of pulling the fibers out the silicone matrices according to the international society organization standard (ISO 3501 Tensile Pull Out Test) [35]. A flat rectangular-shaped, metal plate (1.50 mm thick, 16 mm wide, and 20 mm long) was placed on the top of each specimen [11]. The plate had a middle opening (5.30 mm in diameter) through which the free parts of the fibers (which were 5 mm long) were pointing out. Then, the free part was fixed in the chuck of the upper member of the testing machine. The periphery of the metal plate was clasped between the grips of a holder which is fixed in the base of the testing machine. Since the metal plate was wider than the diameter of the specimen (16 mm and 14.20 mm), it clasped the specimen in place and prepared it for running the test without exerting any radial stresses on the silicone matrices (Figures 1 and 2). The specimens (n = 3) were tested and all specimens were fixed in alignment parallel to the long axis of the testing machine so that no bending movements were created in performing the test. The specimens were tested at room temperature (23 °C ± 1 °C).

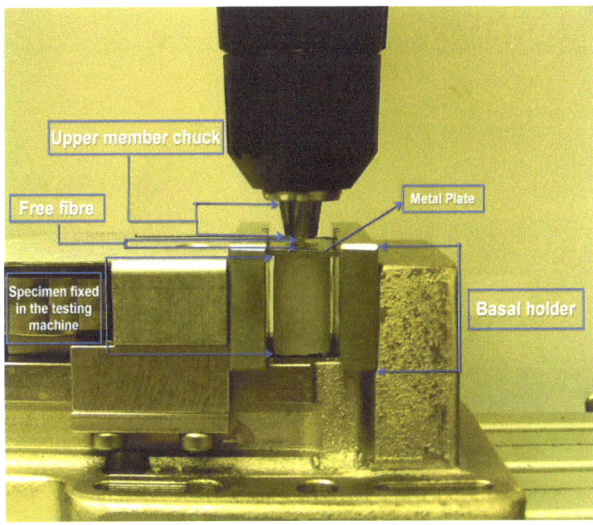

Figure 1. A graph showing how specimens were fixed between the grips of the testing machine.

Figure 2. A graph of an ongoing pulling of a fiber out a silicone body at a 1 mm/min pulling speed.

2.2. Finite Element Modeling: The Steps of FEM Modeling Are Shown in Figures 3 and 4

Pre-processing of CT image slices: The pre-processing of this study was performed to build a 3D model of a typical fiber–silicone pull-out testing specimen as shown in Figure 4. Firstly, 866 CT image slices (19.5 μm thick/each slice) were obtained by using a high resolution Micro-CT (μCT, SkyScan 1072, Aarstselaar, Belgium). Secondly, the materials on the scans were posted in three different cross-sectional views using ScanIP (Simpleware, Version U-2022.12) in which masks were applied for the different materials to allow extended visualization based on image density thresholding.

As a result of the segmentation of these masks, the ScanIP file was imported to ScanFE in which a 3D generation was implemented for all of the image slices. This file was exported in FEA format to be processed in ABAQUS. The file was imported in the finite element analysis and was re-meshed. The mesh was improved using Hypermesh 8.0 (Altair Hyperworks), because the congruence was lost during the previous remeshing process. The optimized STL files were imported in a finite element analysis software package (PATRAN/ABAQUS, version r2) for the generation of the solid mesh, boundary conditions, and material properties attribution, as well as load case creation and data processing.

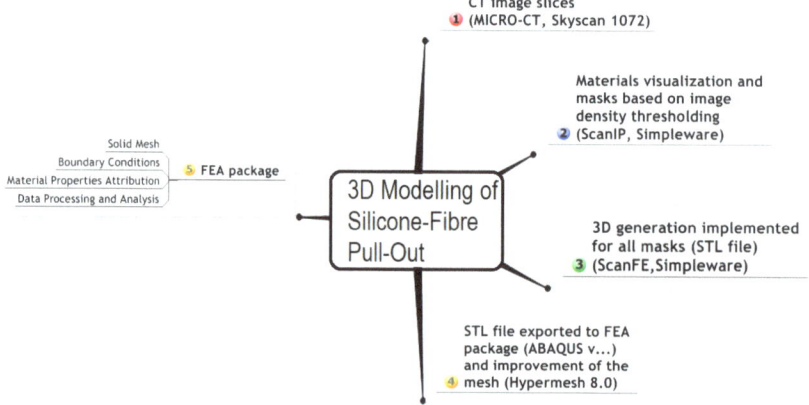

Figure 3. Steps of the 3D pull-out specimen modeling.

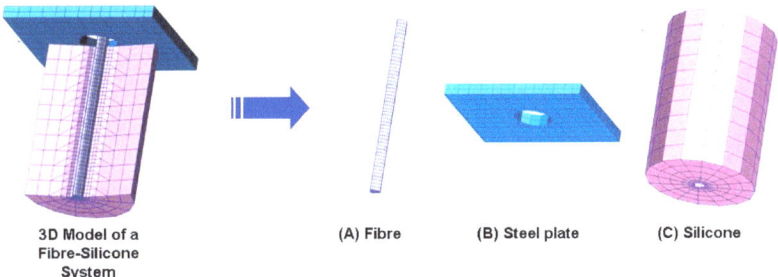

Figure 4. 3D model of a typical fiber–silicone pull-out testing specimen.

The solid mesh was performed by eight-node hexahedral elements and the debonding propagation was performed by using cohesive elements which are user-defined elements within ABAQUS. The total energy for a traction-separation damage evolution (G_c) was 0.11×10^{-3} J and the stress criteria for the cohesive approach S_{11}, S_{22}, and S_{33} were 5, 5, and 0.15 MPa, respectively (Table 1).

Table 1. Stress analysis in the Cartesian coordinate system.

Stress	Direction
S11	Normal stress component acting along the X-axis
S22	Normal stress component acting along the Y-axis
S33	Normal stress component acting along the Z-axis

2.3. Boundary Conditions, Material Properties and Data Processing

The boundary conditions applied to the model are represented in Figure 5. From the experimental pull-out test, the fiber displacement of load applied and the corresponding pull-out force are known for each load step.

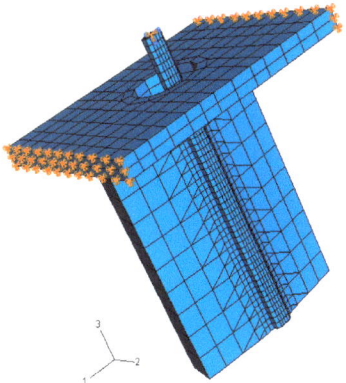

Figure 5. The boundary conditions applied to the model (appearing in orange colour).

No friction was assumed for the debonded region in the present study. Excluding the cohesive interface, which is anisotropic, it was assumed that all of the materials were homogeneous, isotropic, linear, and elastic and that all the interfaces were perfectly bonded and could transfer shear and normal stresses. The cohesive interface needs to be anisotropic because the cohesive elements need to be softened in the directions S11 and S22; decreasing the reaction force of them in these directions will enable the fiber to be pulled out from the silicone matrix. The properties of the materials are presented in Table 2.

Table 2. Materials properties.

Material	Elastic Modulus (MPa)	Poisson's Ratios
Silicone	1.90	0.40
Fiber	81.300	0.2
Steel	210.000	0.30
Cohesive interface	0.50 (E_1) 0.50 (E_2) 10.0 (E_3)	0.5

The Von Mises stress distribution of the fiber–silicone specimen was investigated and the maximum shear stress distribution was plotted to investigate the stress concentration during the debonding process at the cohesive interface (for the stress criteria of each direction: S11, S22, and S33).

3. Results

A typical experimental pull-out curve of 20 mm embedded length is shown in Figure 6. Force is plotted against displacement (linear region), exhibiting a single peak where the debonding took place, and then followed by lower peaks indicating frictional forces between the fiber and the matrix. The test terminated when the fiber was fully pulled out from the silicone matrix. In the initial quasi-linear region (A), the fiber extends as the force rises. At Point I, the shear damage initiates. When the maximum force D (debonding force) is reached, the fiber debonds from the matrix along the full embedded length. At Point II, a region of traction (T) initiates as a sudden drop of the force E (initial extraction force) required for pulling out the debonded fiber from the matrix. Finally, the force continues to decrease as successive lower peaks at the friction region (F), while the extracted length increases until the fiber is totally extracted from the matrix (Point III). The same curve is shown in Figure 7 with more details of the debonding region. For this curve, the displacement was plotted until a 1.8 mm implantation length of the fiber in the silicone matrix.

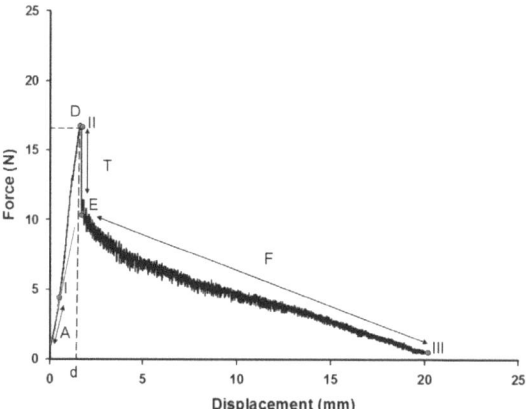

Figure 6. Experimental pull-out force-displacement curve of 20.0 mm embedded length. (A) is the quasi-linear region; (I) is the initial shear damage point; (D) is the maximum force; (d) is the displacement of debonding; (II) is the initial traction point; (T) is the region of traction; (E) is the initial extraction force; (F) is the region of friction; and (III) is the point of complete debonding.

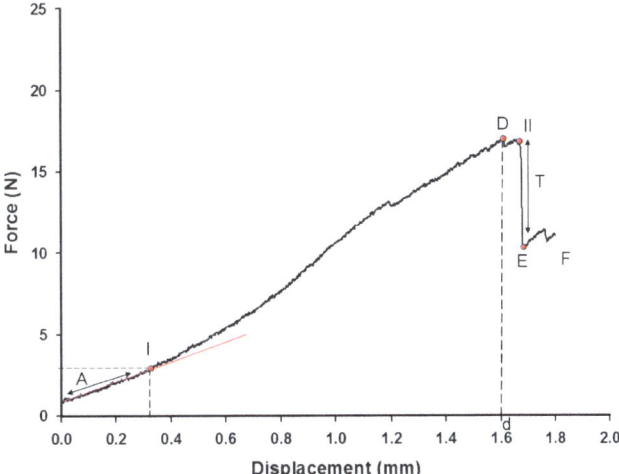

Figure 7. Experimental pull-out force-displacement curve for 1.8 mm implantation length. (A) is the quasi-linear region; (I) is the initial shear damage point; (D) is the maximum force; (d) is the displacement of debonding; (II) is the initial traction point; (T) is the region of traction; (E) is the initial extraction force; and (F) is the region of friction.

A FEM cohesive zone pull-out force versus displacement is shown in in Figure 8. The displacement was plotted until the debonding failure. The curve shows an initial linear response where the fiber extends as the force rises (quasi-linear region A) up to a peak load D (or in this case, to the apparent strength) when the fiber debonds from the silicone, followed by an exponential strain-softening region, after a drop of the curve to the initial extraction (I) (required for pulling out the debonded fiber from the silicone). The force continues to decrease at region T until the whole fiber is extracted from the silicone at 1.5 mm of displacement (Point III). After this region (T), the tensile stresses are higher than the shear and the displacement jumps across the interface. From the intrinsic cohesive failure model, as the interface separates, the magnitude tensile stress in the T region (the apparent strength) reaches a maximum, after which it progressively becomes null, meaning that the fiber–silicone is completely debonded.

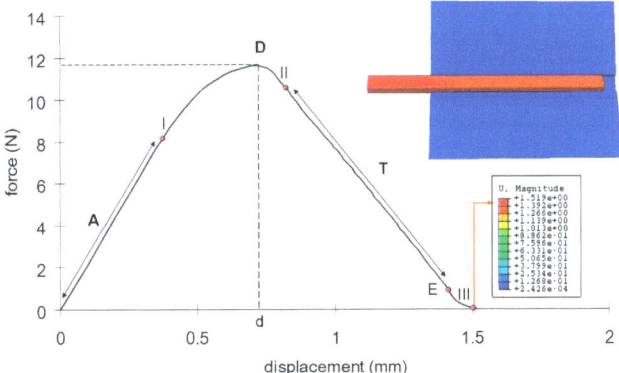

Figure 8. Apparent strength vs. displacement response of a fiber–silicone specimen simulated by the finite element method. (A) is the quasi-linear region; (I) is the initial shear damage point; (D) is the maximum force; (d) is the displacement of debonding; (II) is the initial traction point; (T) is the region of traction; (E) is the initial extraction force and (III) is the point of complete debonding.

The Von Mises stress concentration zone at the interface near the fiber is shown in Figure 9. At the region with the red fringe, the maximum shear stress concentration is represented. The stress is distributed along the interface and also at the end of the matrix. The damage in the three directions (S11, S22, and S33) for the cohesive zone (interface) is shown in Figures 10–12, respectively. The stress is higher in the direction S33 in which the lower damage criterion was higher as well (0.15 MPa) in comparison to the others (both 5 MPa). The initial debonding of the fiber from the silicone matrix is shown on Figure 13. The cohesive elements are located between the materials, representing the interface cohesive zone.

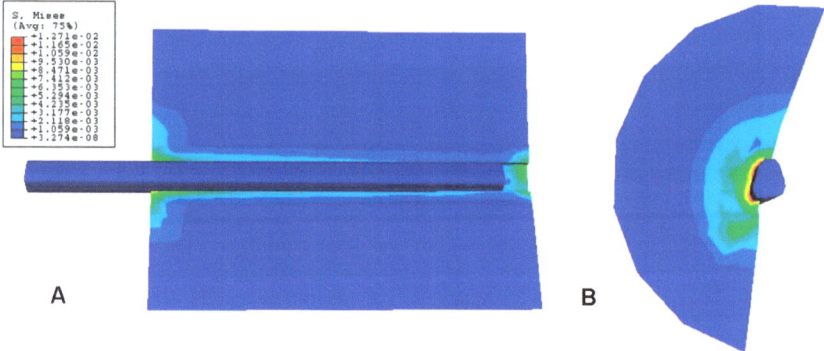

Figure 9. Von Mises stress of the fiber–silicone pull-out specimen simulation. (**A**) is a cross-sectional representation of the model (plane 1 0 3) and (**B**) is the top view of the model (plane 1 2 0).

Figure 10. Damage stress distribution at the cohesive zone (S11 direction).

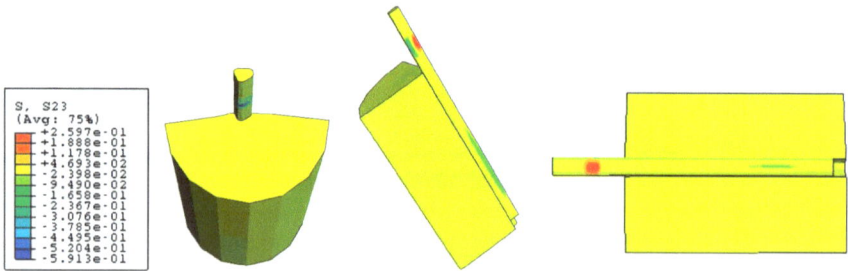

Figure 11. Damage stress distribution at the cohesive zone (S22 direction).

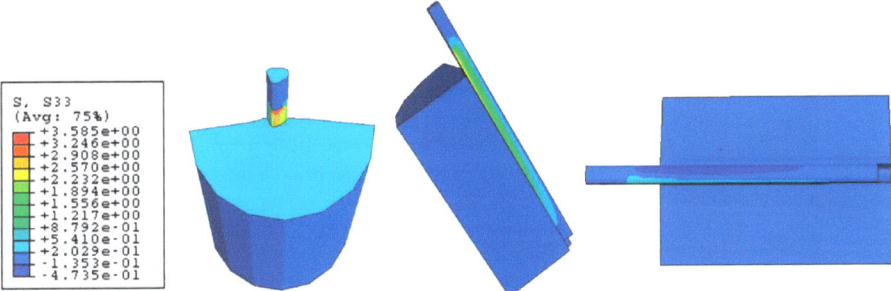

Figure 12. Damage stress distribution at the cohesive zone (S33 direction).

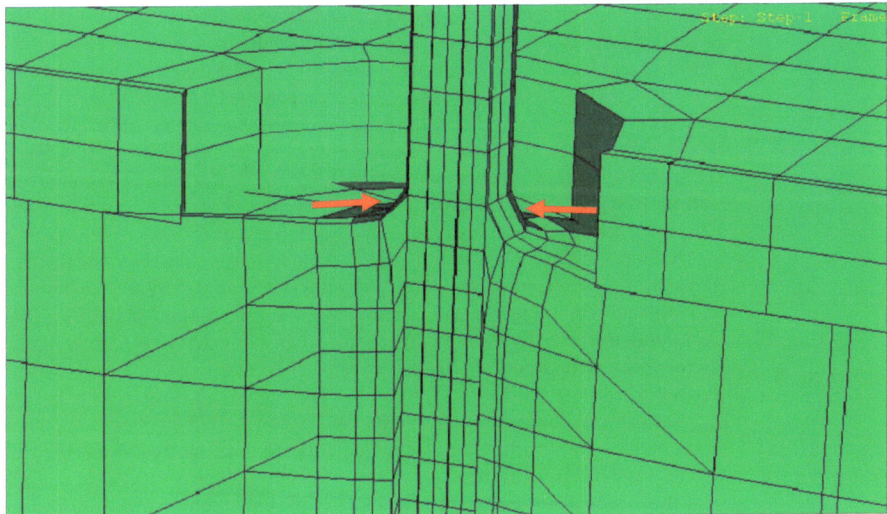

Figure 13. Cross-sectional virtual view showing the fiber–silicone interface. Initial debonding of the fiber from the silicone matrix (arrowed). The cohesive elements are located between the materials, representing the interface cohesive zone.

4. Discussion

The single-fiber pull-out test has been pointed out by many authors as a simple method that can be used to evaluate interfacial properties such as interfacial shear strength and frictional stress [20,24]. Most of them have used the interfacial shear strength as a criterion for fracture, i.e., a crack will propagate when the interfacial shear stress at a crack tip is greater than the shear strength of the interface. On the other hand, the debonding process can also be treated using the fracture energy as a failure criterion. According to Griffith's fracture criterion, the debonding will take place when the work done by the applied load minus the energy stored in the system is larger than the work of the interfacial detachment, which is the adhesive fracture energy. This energy is the amount of energy to separate the unit area of the interface and is used to characterize the bonding strength.

Some studies have pointed out that during the fiber pull-out tests, initial debonding, partial debonding, and complete debonding at the interface occur sequentially and the initial debonding and complete debonding are respectively the beginning and the end of the partial debonding [36,37]. They have reported that after initial debonding, the displacement increases with the increasing stress initially and before the displacement reaches its maximum value, the stress exhibits a decrease and the debonding occurs when the displacement reaches the maximum value. The experimental curves obtained in the present study are in agreement to these theories. In addition, they could indicate that the free part of the fiber outside the matrix is loaded to a point where partial debonding happened (I) and that part is free from the stress, causing an instantaneous drop of force to point E (initial extraction force) [38].

Recent studies have used numerical analysis to understand the stress distribution at the pull-out set-up. The photoelasticity technique has been used to investigate the stress field in a matrix at a fiber interface by calculating the shear stress profile and pointing out the location of micromechanical events [27], but this method cannot reach the complexity of the interface stress distribution in detail. The finite element method has also been used to investigate the single pull-out test; however, most of these studies were based on few parameters or on friction laws that still have known difficulties in understanding and simulating the interface behavior [23–25,28,39].

Most recently, the FEM was used to carry out the pull-out test simulation using the code ABAQUS® and its cohesive elements approach, which is characterized by interface elements with a cohesive constitutive model [30–32,40]. The interface elements were inserted between the fiber and the silicone matrix for representation of the bonded area and the debonding process. This cohesive approach includes the complex behavior of debonding which is a function of the fracture energy (critical energy release rate G_c or 'cohesive energy') and the strength of the interface (peak strength σ_c or 'cohesive strength'). For this type of simulation, an initial crack is not essential [32]. The separation of the cohesive interfaces is calculated from the difference of the displacements of the adjacent continuum elements [32]. There is a linear relationship that terminates when initial debonding occurs. After initial debonding, the stress increases at a slower rate with the increase in the displacement, and it reaches a maximum (in this study, this point is determined by the maximum shear stress, σ_c) when complete debonding occurs along the full embedded fiber length (one of the components of the displacement vector reaches a critical value (δ_c)) and the continuum elements initially connected by this cohesive element are disconnected, which means that the failure has occurred.

In Figure 6, Point I represents the failure in which the cohesive elements have lost their stiffness so that the continuum elements are disconnected. Since the solid elements overestimate the maximum shear stress, the failure of the cohesive elements is promoted and this was well represented by the simulation at around 0.75 mm of displacement, which confirms the potential of this method to assess the predictive sensitivities of structural damage tolerance. This prediction can be confirmed in Figure 5, which shows the point D with a higher value (1.60 mm) than the corresponding value in Figure 6. The maximum shear stress is achieved when the value of the normal stress (11, 22, and 33 direction), first shear stress, or second shear stress reaches the limit determined by the stress criteria (S_1: 5 MPa; S_2: 5 MPa, and S_3: 0.15 MPa). At this point, the stress drops to a new value and the defect responsible for fiber pull-out initiates. Then, the stress continuously decreases to zero until the fiber is fully pulled out of the silicone.

It is already known that the friction took place after the debonding between the fiber and matrix had happened, which supports the absence of this parameter in this study which is focused on the debonding that happens when the maximum shear stress is achieved. The differences between the two curves after the maximum shear stress are attributed to the fact that the friction was not simulated in this study. After the single peak at point D in Figure 4, where debonding took place, there were lower successive peaks indicating the presence of frictional forces between the fiber and the matrix. Since the cohesive interface model tends to "smear" the localized effects near the crack tip through gradual development of the displacement jumps and gradual decay of the tractions, these peaks are not described by the cohesive interface model [41]. In Figure 6, this region is represented by a quasi-linear region with some step-wise (T) where traction is acting almost without any obstacle and this is why the decrease is progressive until the full pull-out of the fiber from the silicone matrix instead of presenting successive peaks.

The step-wise characteristic at this region might happen as a function of the anisotropy of the cohesive interface which promotes a strain-softening region in the curve (Figure 6). The results from the shear stress distribution in the softened directions (1 and 2) of the present study in Figures 8 and 9 are in accordance with some authors that have shown that the damage in one direction is not independent of what happens in the other [40], which means that the fiber–silicone system represented could support shear and tensile stress in a combined way. In this type of simulation, the cohesive interface needs to be anisotropic because the cohesive elements need to be softened in Directions 1 and 2 to enable the fiber to be pulled out from the silicone matrix. The shear stress distribution in Direction 3 with the high elastic modulus (10 MPa) and the lowest damage criterion (0.15 MPa) is shown in Figure 10. And this represents the highest stress concentration, as Figure 11 shows that when the effective traction acting on the elements' facets reaches the cohesive strength of the material (maximum shear stress, σ_c), the cohesive elements are inserted adaptatively at their interior and this means that they are disconnected.

5. Conclusions

The comparison between the experimental values and the results from the finite element simulations show that the proposed cohesive zone model accurately reproduces the experimental results. These results are considered almost identical to the experimental observations about the interface. The FEA simulation is in good agreement with the experimental results in the part before the peak load, but the agreement is not as promising in the softening part. The cohesive element approach is a potential function that takes into account the shear effects with many advantages related to its ability to predict the initiation and progress of the fiber–silicone debonding during pull-out tests. Among the disadvantages of this approach are the computational efforts required for the simulation and analysis process. A good understanding of the parameters that are related to the cohesive laws is responsible for a successful simulation.

Author Contributions: Conceptualization, A.M.M. and M.M.H.; Software, A.M.M.; Writing—original draft, A.M.M. and M.M.H. All authors have read and agreed to the published version of the manuscript.

Funding: This research received no external funding.

Data Availability Statement: No new data were created.

Conflicts of Interest: The authors declare no conflict of interest.

References

1. Hatamleh, M.M.; Haylock, C.; Watson, J.; Watts, D.C. Maxillofacial prosthetic rehabilitation in the UK: A survey of maxillofacial prosthetists' and technologists' attitudes and opinions. *Int. J. Oral Maxillofac. Surg.* **2010**, *39*, 1186–1192. [CrossRef]
2. Hatamleh, M.M.; Watson, J. Construction of an implant-retained auricular prosthesis with the aid of contemporary digital technologies: A clinical report. *J. Prosthodont.* **2013**, *22*, 132–136. [CrossRef]
3. Hatamleh, M.M.; Watts, D.C. Effect of extraoral aging conditions on color stability of maxillofacial silicone elastomer. *J. Prosthodont.* **2010**, *19*, 536–543. [CrossRef] [PubMed]
4. Hatamleh, M.M.; Watts, D.C. Mechanical properties and bonding of maxillofacial silicone elastomers. *Dent. Mater.* **2010**, *26*, 185–191. [CrossRef] [PubMed]
5. Hatamleh, M.M.; Polyzois, G.L.; Silikas, N.; Watts, D.C. Effect of extraoral aging conditions on mechanical properties of maxillofacial silicone elastomer. *J. Prosthodont.* **2011**, *20*, 439–446. [CrossRef] [PubMed]
6. Hatamleh, M.M.; Hatamlah, H.M.; Nuseir, A. Maxillofacial prosthetics and digital technologies: Cross-sectional study of healthcare service provision, patient attitudes, and opinions. *J. Prosthodont.* **2023**. [CrossRef]
7. Hatamleh, M.M.; Watson, J.; Nuseir, A. Successful prosthetic salvage of a suboptimal autogenous auricular reconstruction with digital technologies: A report of 3 challenging treatments. *J. Prosthet. Dent.* **2022**, *128*, 1103–1108. [CrossRef]
8. Hatamleh, M.M.; Maqableh, A.M.; Al-Wahadni, A.; Al-Rabab'ah, M.A. Mechanical properties and bonding of maxillofacial silicone elastomer mixed with nano-sized anti-microbials. *Dent. Mater.* **2023**, *39*, 677–681. [CrossRef]
9. Hatamleh, M.M.; Polyzois, G.L.; Nuseir, A.; Hatamleh, K.; Alnazzawi, A. Mechanical Properties and Simulated Aging of Silicone Maxillofacial Elastomers: Advancements in the Past 45 Years. *J. Prosthodont.* **2016**, *25*, 418–426. [CrossRef]

10. Hatamleh, M.M.; Watts, D.C. Fibre reinforcement enhances bonding of soft lining to acrylic dental and maxillofacial prostheses. *Eur. J. Prosthodont. Restor. Dent.* **2008**, *16*, 116–121.
11. Kurunmaki, H.; Kantola, R.; Hatamleh, M.M.; Watts, D.C.; Vallittu, P.K. A fiber-reinforced composite prosthesis restoring a lateral midfacial defect: A clinical report. *J. Prosthet. Dent.* **2008**, *100*, 348–352. [CrossRef] [PubMed]
12. Watson, J.; Hatamleh, M.M. Complete integration of technology for improved reproduction of auricular prostheses. *J. Prosthet. Dent.* **2014**, *111*, 430–436. [CrossRef] [PubMed]
13. Nuseir, A.; Hatamleh, M.M.; Alnazzawi, A.; Al-Rabab'ah, M.; Kamel, B.; Jaradat, E. Direct 3D Printing of Flexible Nasal Prosthesis: Optimized Digital Workflow from Scan to Fit. *J. Prosthodont.* **2019**, *28*, 10–14. [CrossRef] [PubMed]
14. Hatamleh, M.M.; Maryan, C.J.; Silikas, N.; Watts, D.C. Effect of net fiber reinforcement surface treatment on soft denture liner retention and longevity. *J. Prosthodont.* **2010**, *19*, 258–262. [CrossRef]
15. Hatamleh, M.M.; Watts, D.C. Effects of accelerated artificial daylight aging on bending strength and bonding of glass fibers in fiber-embedded maxillofacial silicone prostheses. *J. Prosthodont.* **2010**, *19*, 357–363. [CrossRef]
16. Lassila, L.V.; Nohrstrom, T.; Vallittu, P.K. The influence of short-term water storage on the flexural properties of unidirectional glass fiber-reinforced composites. *Biomaterials* **2002**, *23*, 2221–2229. [CrossRef]
17. Miettinen, V.M.; Narva, K.K.; Vallittu, P.K. Water sorption, solubility and effect of post-curing of glass fibre reinforced polymers. *Biomaterials* **1999**, *20*, 1187–1194. [CrossRef]
18. Adisman, I.K. Prosthesis serviceability for acquired jaw defects. *Dent. Clin. N. Am.* **1990**, *34*, 265–284. [CrossRef]
19. Zucchini, A.; Hui, C. Detailed analysis of the fibre pull-out test. *J. Mater. Sci.* **1996**, *31*, 5631–5641. [CrossRef]
20. Wang, C. Fracture mechanics of single-fibre pull-out test. *J. Mater. Sci.* **1997**, *32*, 483–490. [CrossRef]
21. Tsai, K.; Kim, K. The Micromechanics of Fiber Pull-Out. *J. Mech. Phys. Solids* **1996**, *44*, 1147–1177. [CrossRef]
22. DiFrancia, C.; Ward, T.; Claus, R. The single-fibre pull-out test. 1:Review and interpretation. *Compos. Part A* **1996**, *27A*, 597–612. [CrossRef]
23. Kharrat, M.; Dammak, M.; Trabelsi, M. Finite element analysis of load transfer at a fibre-matrix interface during pull-out loading. *J. Adhes. Sci. Technol.* **2007**, *21*, 725–734. [CrossRef]
24. Beckert, W.; Lauke, B. Fracture mechanics finite element analysis of debonding crack extension for a single fibre pull-out specimen. *J. Mater. Sci. Lett.* **1995**, *14*, 333–336. [CrossRef]
25. Sun, W.; Lin, F. Computer Modeling and FEA Simulation for Composite Single Fiber Pull-out. *J. Thermoplast. Compos. Mater.* **2001**, *14*, 327–343. [CrossRef]
26. Banholzer, B.; Brameshuber, W.; Jung, W. Analytical evaluation of pull-out tests-The inverse problem. *Cem. Concr. Compos.* **2006**, *28*, 561–571. [CrossRef]
27. Zhao, F.; Hayes, S.; Patterson, E.; Jones, F. Phase-stepping photoelasticity for the measurement of interfacial shear stress in single fibre composites. *Compos. Part A* **2006**, *37*, 216–221. [CrossRef]
28. Lin, G.; Geubelle, P.; Sottos, N. Simulation of fiber debonding with friction in a model composite pushout test. *Int. J. Solids Struct.* **2001**, *38*, 8547–8562. [CrossRef]
29. Han, T.; Ural, A.; Chen, C.; Zehnder, A.; Ingraffea, A.; Billington, S. Delamination buckling and propagation analysis of honeycomb panels using a cohesive element approach. *Int. J. Fract.* **2002**, *115*, 101–123. [CrossRef]
30. Zhou, F.; Molinari, J. Dynamic crack propagation with cohesive elements: A methodology to address mesh dependency. *Int. J. Numer. Methods Eng.* **2004**, *59*, 1–24. [CrossRef]
31. Tsai, K.; Patra, A.; Wetherhold, R. Finite element simulation of shaped ductile fiber pullout using a mixed cohesive zone/friction interface model. *Compos. Part A* **2005**, *36*, 827–838. [CrossRef]
32. Scheider, I.; Brocks, W. Cohesive elements for thin-walled structures. *Comput. Mater. Sci.* **2006**, *37*, 101–109. [CrossRef]
33. Pandolfi, A.; Guduru, P.; Ortiz, M.; Rosakis, A. Three dimensional cohesive-element analysis and experiments of dynamic fracture in C300 steel. *Int. J. Solids Struct.* **2000**, *37*, 3733–3760. [CrossRef]
34. Ortiz, M.; Pandolfi, A. Finite-Deformation Irreversible Cohesive Elements for Three-Dimensional Crack-Propagation Analysis. *Int. J. Numer. Methods Eng.* **1999**, *44*, 1267–1282. [CrossRef]
35. ISO 3501:2021; Plastics Piping Systems—Mechanical Joints between Fittings and Pressure Pipes—Test Method for Resistance to Pull-Out under Constant Longitudinal Force. International Standards Organisation: Geneva, Switzerland,, 2021. Available online: https://www.iso.org/standard/82023.html (accessed on 2 September 2023).
36. Hsueh, C. Interfacial debonding and fiber pull-out stresses of fiber-reinforced composites. Part VI. Interpretation of fiber pull-out curves. *Mater. Sci. Eng.* **1991**, *A149*, 11–18. [CrossRef]
37. Hsueh, C. Interfacial Debonding and Fiber Pull-out Stresses of Fiber-reinforced Composites II: Non-constant Interfacial Bond Strength. *Mater. Sci. Eng.* **1990**, *A125*, 67–73. [CrossRef]
38. Gray, R. Analysis of the effect of embedded fibre length on fibre debonding and pull-out from an elastic matrix. Part 1—Review of theories. *J. Mater. Sci.* **1984**, *19*, 861–870. [CrossRef]
39. Quek, M.; Yue, C. An improved analysis for axisymetric stress distributions in the single fibre pull-out test. *J. Mater. Sci.* **1997**, *32*, 5457–5465. [CrossRef]

40. Rabinovitch, O. Debonding Analysis of Fiber-Reinforced-Polymer Strengthned Beams: Cohesive Zone Modeling Versus a Linear Elastic Fracture Mechanics Approach. *Eng. Fract. Mech.* **2008**, *75*, 2842–2859. [CrossRef]
41. Moreo, P.; Pérez, M.; García-Aznar, J.; Doblaré, M. Modelling the mixed-mode failure of cement-bone interfaces. *Eng. Fract. Mech.* **2006**, *73*, 1379–1395. [CrossRef]

Disclaimer/Publisher's Note: The statements, opinions and data contained in all publications are solely those of the individual author(s) and contributor(s) and not of MDPI and/or the editor(s). MDPI and/or the editor(s) disclaim responsibility for any injury to people or property resulting from any ideas, methods, instructions or products referred to in the content.

Article

Dry Friction and Wear Behavior of Laser-Sintered Graphite/Carbon Fiber/Polyamide 12 Composite

Abdelrasoul Gadelmoula [1,2,*] and Saleh Ahmed Aldahash [1]

1. Department of Mechanical and Industrial Engineering, College of Engineering, Majmaah University, Al-Majmaah 11952, Saudi Arabia; saldahash@mu.edu.sa
2. Department of Mechanical Design and Production Engineering, Faculty of Engineering, Assiut University, Assiut 71515, Egypt
* Correspondence: a.gadelmoula@mu.edu.sa

Abstract: Carbon fiber-reinforced polymers (CFRPs) are being used extensively in modern industries that require a high strength-to-weight ratio, such as aerospace, automotive, motorsport, and sports equipment. However, although reinforcement with carbon fibers improves the mechanical properties of polymers, this comes at the expense of abrasive wear resistance. Therefore, to efficiently utilize CFRPs in dry sliding contacts, solid lubricant is used as a filler. Further, to facilitate the fabrication of objects with complex geometries, selective laser sintering (SLS) can be employed. Accordingly, in the present work, graphite-filled carbon fiber-reinforced polyamide 12 (CFR-PA12) specimens were prepared using the SLS process to explore the dry sliding friction and wear characteristics of the composite. The test specimens were aligned along four different orientations in the build chamber of the SLS machine to determine the orientation-dependent tribological properties. The experiments were conducted using a pin-on-disc tribometer to measure the coefficient of friction (COF), interface temperature, friction-induced noise, and specific wear rate. In addition, scanning electron microscopy (SEM) of tribo-surfaces was conducted to specify the dominant wear pattern. The results indicated that the steady-state COF, contact temperature, and wear pattern of graphite-filled CFR-PA12 are orientation-independent and that the contact temperature is likely to approach an asymptote far below the glass transition temperature of amorphous PA12 zones, thus eliminating the possibility of matrix softening. Additionally, the results showed that the Z-oriented specimen exhibits the lowest level of friction-induced noise along with the highest wear resistance. Moreover, SEM of tribo-surfaces determined that abrasive wear is the dominant wear pattern.

Keywords: selective laser sintering; polyamide 12; carbon fibers; graphite; friction and wear; dry sliding

Citation: Gadelmoula, A.; Aldahash, S.A. Dry Friction and Wear Behavior of Laser-Sintered Graphite/Carbon Fiber/Polyamide 12 Composite. *Polymers* **2023**, *15*, 3916. https://doi.org/10.3390/polym15193916

Academic Editors: Jesús-María García-Martínez and Emilia P. Collar

Received: 30 August 2023
Revised: 24 September 2023
Accepted: 26 September 2023
Published: 28 September 2023

Copyright: © 2023 by the authors. Licensee MDPI, Basel, Switzerland. This article is an open access article distributed under the terms and conditions of the Creative Commons Attribution (CC BY) license (https://creativecommons.org/licenses/by/4.0/).

1. Introduction

Selective laser sintering (SLS), one of the most popular 3D printing techniques, is a layer-by-layer manufacturing process in which laser energy is used to fuse the particles of a thermoplastic powder in selected areas that match the layer cross-section of the object 3D model. The unfused powder provides support for the object being printed, thus eliminating the need for additional supporting structure, which makes SLS a favorable technique for additive manufacturing of objects with complex geometries [1]. In addition to being used for a long time as a rapid prototyping technique, SLS is currently being used to produce small-to-medium-sized patches of polymeric objects.

Among the few pure materials suitable for the SLS process, polyamide 12 (PA12) is considered the most popular thermoplastic powder because of its favorable sintering properties, which include good powder flowability, low melt viscosity, and a wide range between melting and crystallization temperatures [2]. Even though PA12 parts prepared by the SLS process have high chemical/environmental stability, reasonable mechanical properties, high impact resistance, and acceptable wear resistance, the application of SLS to modern industries calls for composite materials based on PA12 with tailored mechanical and tribological properties [3–5]. Therefore, fillers in the form of either particulates, fibers, or a combination of both have been introduced to enhance the mechanical and/or tribological properties of laser-sintered functional parts. The most common particulate fillers include graphite, molybdenum disulfide (MoS_2), polytetrafluoroethylene (PTFE), boron nitride (BN), and glass beads [6,7], while carbon fibers, glass fibers, and aramid fibers are the most widely used synthetic fibers for the reinforcement of polymer matrix [8].

Recently, polymeric composites have proven to be good replacements for traditional materials in modern applications that require a high strength-to-weight ratio, such as aerospace, automotive, motorsport, and sports equipment. Hence, due care must be paid when designing the polymer composite for a specific application. Proper selection of particulate/fiber material, shape and size, weight ratio, and surface treatment is extremely important. The properties of fiber-reinforced polymer composites are extremely sensitive to fiber weight fraction, fiber length, fiber orientation, fiber preprocessing, and fiber/matrix adhesion [9,10]. As the properties of SLS-fabricated parts are orientation-dependent [11–13], it was found that reinforcement with either particles or short fiber enhances the homogeneity of polymer composites prepared by the SLS process [14,15], while introducing long or continuous fibers as a reinforcement results in directional properties of the fibrous composite [9]. Further, the ease with which short fibers are processed saves manufacturing costs considerably, making them most suitable for the reinforcement of polymer matrices in additive manufacturing processes [16]. Among the short fibers used for reinforcement purposes, carbon fibers (CF) are the most widely used fibers to improve the mechanical properties of polymer composites used in the automotive and motorsport industries. Furthermore, as neat polymers are thermal insulators, reinforcement with carbon fibers enhances the thermal conductivity of polymeric composites significantly. Nevertheless, although reinforcement of the thermoplastic matrix with short carbon fibers improves the tensile strength and wear resistance, decreases the coefficient of friction and frictional heating, and enhances the thermal conductivity of thermoplastic composites, it was found that the elongation at break has decreased significantly [17–19]. Accordingly, the abrasive wear resistance of carbon fiber-reinforced polymers (CFRPs) is unsatisfactory. Moreover, in dry sliding conditions, it was reported that fiber crushing results in fiber thinning followed by fiber pull-out of the matrix, which in turn results in unstable frictional properties [20,21]. Therefore, reinforcement with short carbon fibers can efficiently promote the mechanical rather than the tribological properties of thermoplastic polymers. As short carbon fiber-reinforced polymers are being used in applications where they are prone to direct contact with metallic/polymeric counterparts, it becomes inevitable to improve the friction and wear characteristics of such composites. For this purpose, solid lubricants in the form of particulate fillers have been employed to boost the lubricity of carbon fiber-reinforced polymers.

Graphite is a popular solid lubricant that has been extensively used to improve the lubricity of tribo-surfaces. The lubricating effect of graphite stems from its distinctive layered crystal structure, in which carbon atoms are bonded covalently in a hexagonal pattern within each layer, and layers are bonded to each other by means of weak Van der Waals forces. As a result, the bulk shear strength of graphite is always lower than the interfacial shear resistance, allowing graphene layers to be easily removed and transferred to the counter-surface in sliding contact, thereby reducing the coefficient of friction and improving the wear resistance of the matrix [22,23]. In addition, graphite as a filler enhances the mechanical properties and thermal conductivity of the polymer matrix [22]. Hence, to improve the tribological properties of the thermoplastic matrix and maintain prominent mechanical properties, graphite powder can be introduced as a filler for carbon fiber-reinforced polymers. By combining the effects of graphite as a solid lubricant and short carbon fibers as a reinforcement of the thermoplastic matrix, the resulting composite can have enhanced tribological as well as mechanical properties.

With PA12 as a thermoplastic matrix, graphite-filled carbon fiber-reinforced PA12 composite can meet the increasing demand for a stiff, lightweight, thermally conductive, low-friction, wear-resistant, and chemically stable material for aircraft, automotive, and Motorsport industries. The composite can be ideal for applications with fixed and movable joints operating at low temperatures (less than the glass transition temperature of PA12) and under a wide range of bearing pressures in harsh environments. Moreover, with graphite powder, short carbon fibers, and PA12 as the constituents, the composite is most suitable for SLS processing, which facilitates the fabrication of complex functional geometries based on this composite. However, so far, very little has been reported regarding the tribological characteristics of a reinforced thermoplastic filled with a solid lubricant, and very few articles among them are concerned with composites prepared by the SLS technique [15,24–26]. The few reported results showed that inclusion of either PTFE or MoS_2 could effectively reduce the coefficient of friction of laser-sintered PA12 by forming a homogenous transfer film on the counter-surface [15,27], and that a 5–10% weight fraction of graphite was sufficient to significantly improve the tribological characteristics of PA6 [24].

Hence, this experimental study is dedicated to exploring the tribological features of graphite-filled carbon fiber-reinforced PA12 (CFR-PA12) composites prepared by the SLS process. Given that the properties of SLS parts are orientation-dependent, the test specimens are aligned along four orientations in the build chamber of the SLS machine. The tribological characteristics of graphite-filled CFR-PA12 composites are examined using a pin-on-disc tribometer in dry sliding conditions. Frictional heating as well as the friction-induced noise are measured, and the wear resistance of the composite is evaluated. Finally, scanning electron microscopy (SEM) is used to investigate the wear patterns that dominate the worn tribo-surface of test specimens. The results from this study, when compiled with relevant findings in the open literature, can contribute to a better understanding of the tribological characteristics of graphite-filled CFR-PA12 composite prepared by SLS under dry sliding conditions; this can facilitate widening the applicability of this composite in demanding industries.

2. Materials and Methods

2.1. Selective Laser Sintering of Test Specimens

Selective laser sintering was used to fabricate the test specimens because the SLS process has proven to be the most reliable 3D printing technology suitable for the manufacturing of objects with complex geometries from a wide range of thermoplastic powders. Further, during the SLS process, the unsintered powder supports the part being manufactured, thus eliminating the need for supporting structures. Moreover, the SLS system is advantageous because it is able to manufacture parts with tailored properties through the proper selection of fabrication parameters (laser power, scanning speed, scan spacing, and layer thickness). Another prominent feature of the SLS system is its relatively low manufacturing cost, as the unsintered powder can be reused in subsequent fabrication processes. Hence, the test specimens were prepared by the SLS technique using a powder mixture composed of polyamide 12 (PA12) powder as a matrix, a 5% weight ratio of graphite powder, and a 2.0% weight ratio of short carbon fibers. The powder mixture was developed in-house by Graphite Additive Manufacturing Ltd. (Aylesbury, UK) and supplied by EOS GmbH (Düsseldorf, Germany). Hence, micro-sized graphite powder was used as a particulate filler, while short carbon fibers with 8–10 µm diameter and 20–60 µm in length were used as reinforcement of PA12 matrix; it was reported that carbon fiber length of 10–100 µm is optimum for the SLS process [28].

Specimens in the form of pins with a diameter of 4 mm and a length of 25 mm were fabricated in the SLS build chamber in four different orientations. Fabrication parameters are given in Table 1. The used SLS system (3D Systems sPro 60 HD-HS) is equipped with a CO_2 laser source of 70 W and a wavelength of 10,600 nm. The specimens were oriented along the X-axis, Y-axis, Z-axis, and at 45° to the X-axis (X45) in the build chamber, as shown in Figure 1a, and were built along the positive Z-axis. The tribo-surfaces of test specimens are perpendicular to each orientation. It is worth noting that the SLS fabrication parameters outlined in Table 1 were selected by the manufacturer for optimum part density, dimensional accuracy, and mechanical properties. Cosmetic finishing (light media blasting) was applied to remove the unsintered powder particles from the specimen surface after drawing them out of the build chamber. Figure 1b shows clear marks of SLS build layers in the X-oriented specimen; however, this was not visible on the surface of other specimens.

Table 1. SLS parameters of graphite-filled CF-PA12 composite.

	Graphite/CF/PA12 Composite
Powder color	Black color
SLS system	3D Systems SLS sPro 60 HD-HS
Outline power (W)	11
Hatching power (W)	34
Scanning speed (m/s)	12
Scan spacing (mm)	0.15
Layer thickness (mm)	0.1

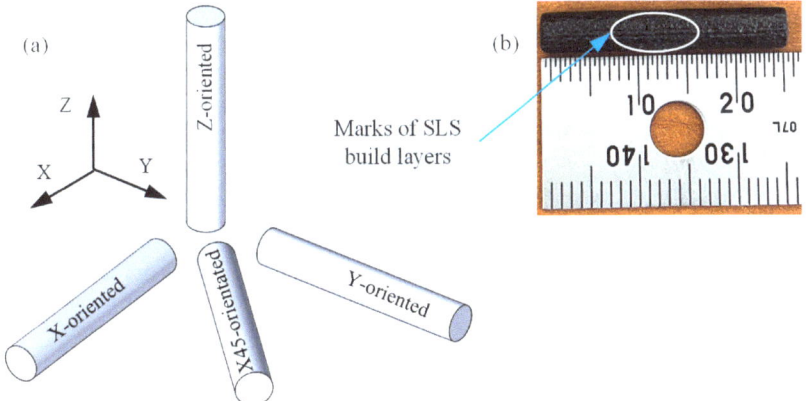

Figure 1. (a) Schematics of specimens' orientation in the SLS build chamber. (b) X-oriented test specimen showing marks of build layers.

2.2. Pin-on-Disc Tribometer

The dry sliding characteristics of graphite-filled CFR-PA12 composite were investigated using a pin-on-disc tribometer, at which the pin-shape test specimen is inserted into a pin holder located in the middle of a loading lever. Then, the pin is loaded against the steel disc by means of a hanging deadweight at the end of the loading lever (see Figure 2). The stainless steel disc is rotated by means of a drive unit while the specimen is fixed. To maintain dry sliding conditions at the interface, 99.9% ethanol was used to clean the disc surface to remove any dirt, oily substances, or adsorbed matter, and the experiments were conducted in a dry environment (air humidity was less than 10%). Several stainless steel discs were manufactured with close surface roughness so that every test specimen is examined against fresh disc tribo-surface; this was adopted to rule out the effect of the formed transfer film on friction and wear behavior at the interface. The surface roughness of steel discs was measured using portable surface roughness tester (SURFTEST SJ-210, Mitutoyo, Kanagawa, Japan), and the average surface roughness (Ra) value in the radial direction (i.e., perpendicular to the sliding direction) of all disc surfaces was 0.34 ± 0.03 microns.

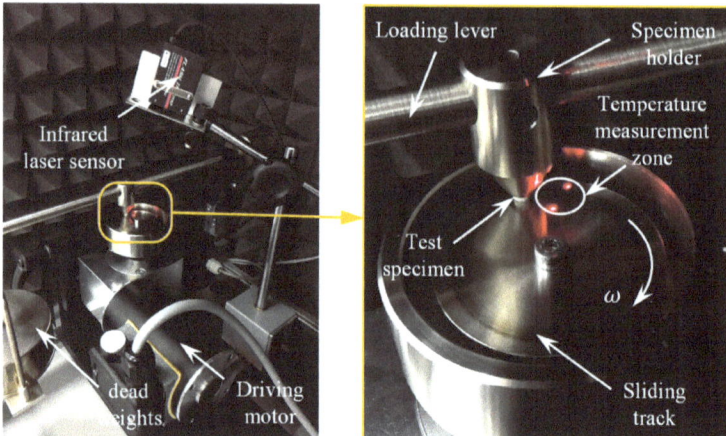

Figure 2. Experimental test rig using a pin-on-disc tribometer [7,20].

2.3. Experimental Measurements

The measurements were conducted to evaluate the coefficient of friction (COF), approximate contact temperature, friction-induced noise, specific wear rate, and to investigate the dominant wear pattern along each build orientation. To evaluate the COF at the composite/disc interface, the friction force was measured by means of a double-bending beam force sensor along with an NI-9237 module and an NI-compactDAQ controller [7,20]. A Labview code was developed to manage the sensor signal, in which a moving average function was used to return the average value of the COF every 0.5 s.

Since measurement of the actual contact temperature is not straightforward, a close approximation of contact temperature was obtained by measuring the disc temperature very close to the trailing edge of the pin (see Figure 2). For this purpose, an infrared laser temperature sensor (FT-H30, KEYENCE, Itasca, IL, USA) and a digital amplifier unit (FT-50AW, KEYENCE, Itasca, IL, USA) were used to detect the frictional heat emitted from a circular area of 6 mm in diameter in the vicinity of the contact zone. To measure the friction-induced noise, the pin-on-disc tribometer was arranged inside a semi-anechoic chamber lined with high-density polyurethane acoustic foam. The noise level was measured using a digital sound level meter (GM 1357, Benetech co., Shenzhen, China). An Arduino Uno microcontroller board (SMD R3, Arduino.cc, Ivrea, Italy) was used to manage the output signals of the infrared laser temperature sensor and the digital sound level meter, and a moving average function was applied to calculate the average values of contact temperature and frictional noise level every 0.5 s.

Finally, a thin film of gold (less than 100 Å) was deposited on the worn tribo-surfaces of graphite-filled CFR-PA12 specimens in a plasma sputtering device (sec-MCM-100P ion sputtering coater) to render them electrically conductive, and then scanned with a scanning electron microscope (SNE-4500M Plus, Sec Co., Suwon, Republic of Korea) to determine the wear patterns that predominated the tribo-surface.

2.4. Experimental Conditions

The results presented in this work outline the tribological characteristics of a graphite-filled CFR-PA12 composite prepared using SLS under dry sliding conditions against a steel counter surface. Tribological characterization was conducted using a pin-on-disc tribometer, where the pin is a rod with a diameter of 4 mm and length of 25 mm and made of laser-sintered graphite-filled CFR-PA12 composite, while the disc is made of stainless steel. The applied load on the pin was 50 N, thus the apparent contact pressure at the pin/disc interface was about 4 MPa; however, the actual contact pressure is much higher than the apparent value as the real contact area is much smaller than the apparent one [29].

The sliding track radius was 20 mm, and the disc rotation velocity was 120 rpm; thus, the pin sliding speed was about 250 mm/s. A dry sliding experiment was conducted for a duration of 45 min to explore the steady-state performance of the graphite-filled CFR-PA12 composite. The current experimental conditions are given in Table 2.

Table 2. Pin-on-disc experimental conditions of graphite-filled CFR-PA12 composite.

	Specifications
Normal load (N)	50
Disc rotation velocity (rpm)	120
Sliding track radius (mm)	20
Sliding speed (mm/s)	250
Test duration (min)	45
Disc initial temperature (°C)	29–30
Background noise level (dBA)	35–37
Humidity (%)	7–10

3. Results and Discussions

The results presented in the following sections address the potential effects of part orientation on the coefficient of friction, friction-induced noise, contact temperature, and specific wear rate of laser-sintered graphite-filled CFR-PA12 composites under dry sliding conditions against steel countersurfaces.

3.1. Coefficient of Friction (COF)

There are two components of friction: the mechanical component representing the material resistance to ploughing/microcutting by the asperities of the counter surface, and the adhesion component representing the resistance to shearing of adhesive junctions at spots of real contact. Indeed, the adhesion component contributes most to the frictional resistance of polymers and polymer composites [30,31]. Figure 3 shows the variations of the COF over sliding time for graphite-filled CFR-PA12 specimens with different orientations. The results from Figure 3a reveal that there has been a sharp increase in the COF of X-oriented, Y-oriented, and X45-oriented specimens during the running-in stage, while the Z-oriented specimen shows a gradual increase in COF during the same process. The running-in process lasts for a few minutes (less than 10 min), then a steady state is attained at which the COF of all specimens is remarkably close. Interestingly, the steady-state COF of all specimens is about 0.26 and is likely to fall behind this value as sliding continues. Such low COF indicates that abrasive wear dominates the tribo-surface of graphite-filled CFR-PA12 specimens, regardless of part build orientation. Indeed, reinforcing PA12 with short carbon fibers decreases the elongation at break considerably and, consequently, reduces the abrasive wear resistance of CFR-PA12. Additionally, introducing graphite powder as a filler further decreases the elongation at the beak and hence the abrasive wear resistance of the specimens, according to the Ratner-Lancaster correlation [31]. Due to the lamellar structure of graphite, upon sliding against a steel surface, the weak Van der Waals forces of attraction between layers allow graphite layers to be easily sheared away and transferred into the disc surface, thus increasing the interface lubricity and decreasing the steady-state COF. Figure 4 shows evidence of a transfer film attached to the sliding track on the disc surface after testing graphite-filled CFR-PA12 samples. Consequently, transferred graphite layers on the disc surface decrease the interfacial shear strength of adhesive junctions, rendering all junctions to be sheared away at the interface rather than at the polymer subsurface, and hence the frictional heating, friction-induced noise, and wear volume are decreased significantly.

The gradual increase of the COF for Z-oriented specimens at the running-in stage, shown in Figure 3b, is attributed to the corresponding gradual increase in real contact area. At the initial stages of the running-in process, carbon fibers contribute to supporting the normal load; however, as rubbing continues, the PA12 layer that encapsulates the fibers and graphite particles is continuously removed, rendering the fibers and graphite particles susceptible to direct contact with the disc surface. Consequently, thinning of carbon fibers occurs and more PA12 spots get into contact with the disc surface, resulting in a corresponding increase in adhesive resistance [20], hence the COF of the Z-oriented specimen increases accordingly.

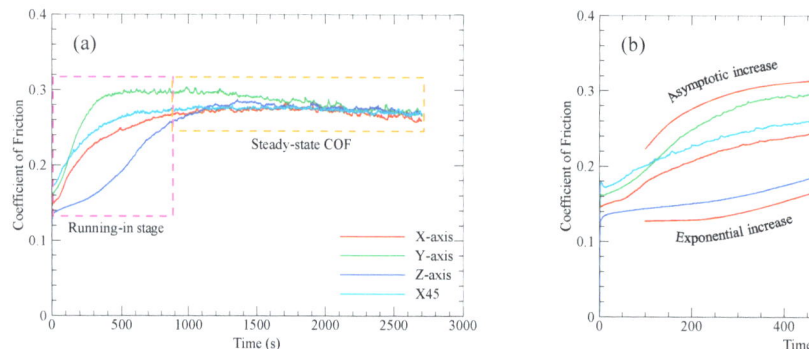

Figure 3. (**a**) Variations of the COF with sliding time. (**b**) COF increase patterns during the running-in stage.

Figure 4. Formation of transfer film on the disc surface (optical microscopy).

Contrarily, the steep increase in COF of specimens built along X, Y, and X45 directions during the running-in process is attributed to the frequent formation and shearing out of adhesive junctions at the sliding interface since it was reported that most frictional resistance is associated with adhesion at the interface until a stable transfer film of graphite layers is formed on the disc surface, then a steady state COF is reached and manifested by a uniform wear rate. Indeed, the highest wear volume of polymer composites is associated with the running-in process [32].

3.2. Friction-Induced Noise

In polymer tribology, friction-induced noise is attributed to the adhesive component of friction at the sliding interface. Adhesion bonds are formed either at surface or subsurface zones of real contact. Surface adhesive bonds are formed when the gap size between polymer and countersurface is less than 100 Å by means of weak Van der Waals molecular forces. Meanwhile, subsurface adhesive bonds are formed at both sides of ploughing/cutting asperities of the harder surface. In both cases, the breaking of adhesive bonds results in what is termed friction-induced noise, which is a ruling parameter in polymer tribology. Therefore, by reducing the COF, the friction-induced noise is reduced significantly. The most effective way to reduce the adhesive component of friction is to lubricate the interface, either using liquid or solid lubricant. Figure 5 shows frictional noise variations with sliding time for graphite-filled CFR-PA12 specimens of various orientations. It is worth noting that the background noise at the experimentation site was 35–37 dBA.

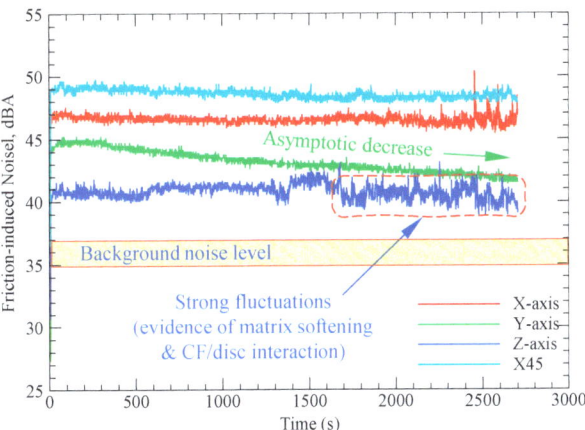

Figure 5. Variations of friction-induced noise with sliding time for specimens with different orientations.

Although the steady-state COF of all specimens is remarkably close, the results from Figure 5 depict that the test specimens have diverse levels of friction-induced noise throughout the test duration. A possible explanation for this inconsistency might be that the pattern of friction at the running-in stage has a prolonged effect on the friction-induced noise. The repeated formation/breaking of adhesive junctions between X-, Y-, and X45-oriented specimens and disc surfaces results in steep growth of both the COF and friction-induced noise at the running-in stage. It is apparent from Figures 3b and 5 that, during the running-in stage, the COF increases in an asymptotic manner and so does the friction-induced noise until steady state conditions are attained, and then the friction noise remains at the same level throughout the experiment duration. Another interesting finding from Figures 4 and 5 is the remarkable synchronous pattern of evolution of friction-induced noise and COF for the Y-oriented specimen, at which the friction-induced noise asymptotically decreases as the steady-state COF decreases. Another important remark from Figure 5 is that the Z-oriented specimen has the lowest friction noise level, but with severe fluctuations. Since the COF of the Z-oriented specimen is still near the lowest value amongst all orientations throughout the test duration, this behavior may be attributed to the interaction between carbon fibers and disc surface [20]; however, this explanation does not rule out the prominent effect of interfacial adhesion as the real area of contact increases with increasing sliding time. To emphasize the role of interfacial adhesion, Figure 6 shows clear marks of strong tearing of the PA12 matrix, manifested by the appearance of stringy tongues unfirmly attached to the surface. For tearing to occur, strong interfacial adhesion along with anisotropic cohesion properties must exist. Tearing results in the formation of PA12 wear flakes that may roll between surfaces or be attached to a sliding surface [33–35]. Similarly, the friction-induced noise of an X-oriented specimen tends to exhibit strong fluctuations near the end of testing time. This may be influenced by the softening of the PA12 matrix because of the cumulative frictional heating that fosters the formation of adhesive bonds, which results in an increase in frictional noise (see Figure 6).

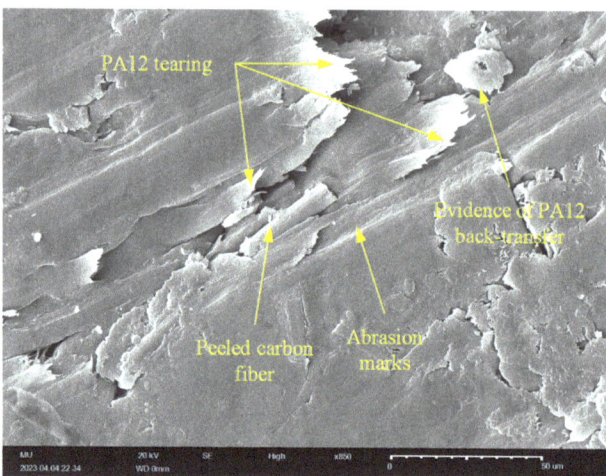

Figure 6. SEM (850×) of the worn surface Z-oriented specimen shows marks of adhesive wear.

3.3. Contact Temperature

When a thermoplastic slides against a hard counter surface, the interfacial interaction results in frictional energy losses in one or more of the following forms: (1) losses due to subsurface plastic deformations by means of hard asperities (ploughing and/or microcutting); (2) losses due to surface adhesion; (3) losses due to subsurface adhesion in ploughing and/or microcutting; and (4) losses due to polymer hysteresis (difference between induced elastic deformation energy and the retained back-pushing elastic energy). Indeed, ploughing and microcutting are principal sources of subsurface heating, while adhesion can be a major source of surface as well as subsurface heating. Meanwhile, hysteresis losses contribute to subsurface heating. Moreover, repeated formation/rupture of adhesive bonds causes severe vibrations of subsurface polymer molecules that contribute to subsurface heating [36]. Figure 7 shows the variations in approximate contact temperature with sliding time for specimens built along the four different orientations. It is apparent from Figure 7 that the approximate contact temperature increases in an asymptotic manner with increasing sliding time; however, there is no significant difference between the contact temperatures of different specimens. Further, the cumulative increase in contact temperature is only 13 °C after 45 min of dry sliding. Indeed, the enhanced thermal conductivity of graphite and carbon fibers contributes to the thermal stability of graphite-filled CFR-PA12 composites, regardless of build orientation. Additionally, the transfer film of graphite layers that formed on the disc surface acts to decrease surface and subsurface adhesion, which is the major source of frictional heating in dry sliding conditions. The most interesting result from Figure 7 is that increasing the sliding time is unlikely to cause a corresponding increase in contact temperature, as the contact temperatures tend to reach an asymptote far below the glass transition temperature (Tg) of the amorphous regions of the PA12 matrix. In fact, introducing CFs as a reinforcement is expected to increase the glass transition temperature of PA12. This can be explained as follows: upon sliding, the CFs support the majority of the applied load, and hence a high contact pressure is developed at the disc/CFs interface, which in turn is transmitted to the PA12 matrix. As the crystallinity percent of laser-sintered PA12 is 24.5% [37], high levels of contact pressure result in a corresponding increase in glass transition temperature in amorphous regions of the PA12 matrix. Accordingly, since the glass transition temperature of PA12 is around 50 °C [38,39], including graphite and carbon fibers is likely to increase the glass transition temperature beyond this value. As the contact temperature of graphite-filled CFR-PA12 increases in an asymptotic manner that is far below 50 °C, the result from Figure 7 demonstrates the potential of this composite to replace conventional bearings operating in harsh environmental conditions.

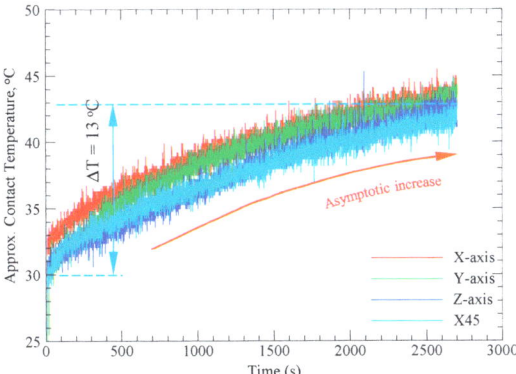

Figure 7. Asymptotic increase of contact temperature with sliding time.

3.4. Specific Wear Rate

Evaluation of a material's specific wear rate is an efficient method to visualize its wear resistance; wear resistance is proportional to the reciprocal of the specific wear rate. The specific wear rate of polymer composites (K_s) depends on the wear volume (ΔV), normal load (F_N), and sliding distance (L) and is expressed as [40]:

$$K_s = \frac{\Delta V}{F_N \times L} \quad (1)$$

Figure 8 shows the specific wear rates of graphite-filled CFR-PA12 specimens built along the four different orientations. The results reveal that the Z-oriented specimen has the lowest specific wear rate and hence the highest wear resistance amongst all other orientations. Accordingly, the results from Figure 8 are in good agreement with those obtained from Figures 3–7, as the Z-oriented specimen exhibited the lowest levels of COF and friction-induced noise. However, it is somewhat surprising that the X45-oriented specimen shows a comparatively high specific wear rate and hence a lower wear resistance, although it has a comparable COF to other specimens. A possible explanation for this behavior might be the continuous removal of the transfer film from the disc surface; it was found that the wear rate of polymers is dictated by the rate of removal of the transfer film from the countersurface rather than by the rate of polymer transfer into the film [31]. The transfer film can be removed from the disc surface by the peeling action of protuberant carbon fibers [20]; this is manifested by the detected high level of frictional noise of the X45-oriented specimen (see Figure 5). Likewise, the high frictional noise can be evidence of strong surface and/or subsurface adhesion that leads to localized tearing of the PA12 matrix and subsequent buildup of the transferred layer on the disc surface. Eventually, the repeated rubbing causes such lumps of wear debris to be detached and replaced by a fresh migrating film. Such poorly attached transferred film is the main reason behind the increased wear rate of the X45-oriented specimen.

Nevertheless, it is important to keep in mind that the tribological characteristics of thermoplastic composites are susceptible to small variations in experimental conditions. Therefore, the abovementioned results cannot be extrapolated to conditions other than those mentioned in Table 2. Hence, thorough investigations for a wide range of experimental conditions combined with a proper machine learning (ML) model can help in fabricating components with customized tribological properties based on graphite-filled CFR-PA12 composite [5]. Additionally, the results from Figures 3–8 suggest that functional sliding bearings based on graphite-filled CFR-PA12 composite where the bearing surface is built normal to the Z-orientation can perform reliably under low sliding speeds and high contact pressure; such prominent features render the laser-sintered graphite-filled CFR-PA12 composite ideal for aerospace, motorsport, and electric vehicle applications.

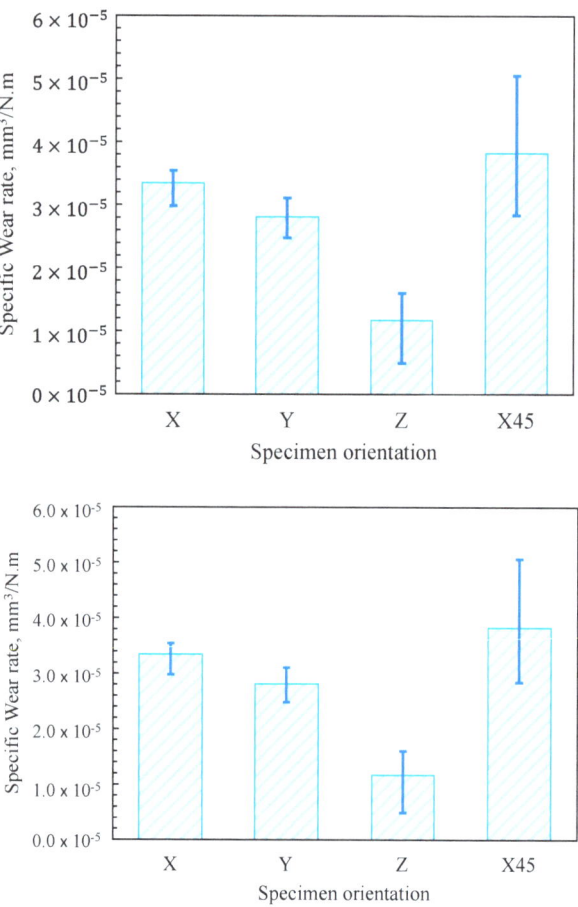

Figure 8. Specific wear rate of graphite-filled CFR-PA12 specimens with different orientations.

3.5. Scanning Electron Microscopy (SEM)

The worn tribo-surfaces of the specimens were investigated using SEM to determine the prevailing wear patterns for the purpose of suggesting microstructural solutions to enhance the wear resistance of the composite. Figure 9 shows an SEM picture of the worn surface of the X-oriented specimen. The tribo-surface shows evidence of weak adhesion and is characterized by clear marks of abrasive wear in the form of deep microcutting grooves. Further, the tribo-surface of the X-oriented specimen features two types of microcracks: (1) boundary microcracks at the CF/PA12 interface due to poor bonding that necessitates proper surface treatment of CF to enhance adhesion with the PA12 matrix; and (2) fatigue microcracks (aligned normal to the sliding direction) due to repeated shearing of adhesive junctions at the interface. Additionally, fiber thinning was detected, which supports the hypothesis that disc/fiber interaction contributes to the friction-induced noise of carbon fiber-reinforced polymers (CFRPs). Moreover, the existence of tiny rolls of PA12 wear debris (i.e., roll formation) that are collected in deep microcuts is evidence of weak interfacial adhesion. Furthermore, the tribo-surface of the X-oriented specimen shows a loosely attached PA12 lump as evidence of delamination. The delamination of the PA12 matrix occurs as a result of the propagation and subsequent coalescence of surface microcracks (either fatigue microcracks or boundary cracks).

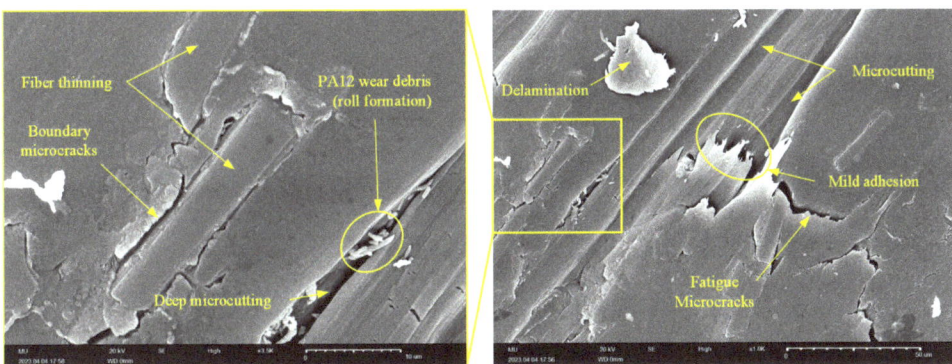

Figure 9. SEM of the worn tribo-surface of the X-oriented graphite-filled CFR-PA12 specimen.

Similarly, Figure 10 shows an SEM picture of the worn surface of the Y-oriented specimen. The results from Figure 10 indicate that abrasive wear is the dominant wear pattern of the Y-oriented tribo-surface, which is free from any signs of interfacial adhesion. Another interesting feature of the Y-oriented CF-PA12 tribo-surface is fiber thinning by the pulverization action of the steel countersurface, which releases fine graphite particles and further lubricates the composite/disc interface. Eventually, released graphite particles from both sources (i.e., shearing of graphite inclusions and pulverization of CFs) act to decrease the COF, friction-induced noise, and interface temperature. This may explain the remarkable asymptotic decrease of both the COF and the friction-induced noise of the Y-oriented specimen. Further analysis of Figure 10 reveals that upon coalescence of boundary cracks at the CF/PA12 interface, delamination occurs and releases lumps of PA12 that may initiate a state of three-body abrasion [41,42].

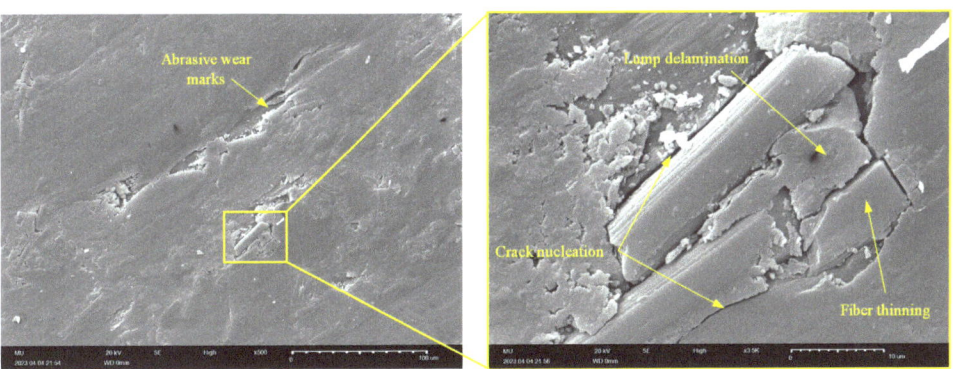

Figure 10. SEM of the worn tribo-surface of the Y-oriented graphite-filled CFR-PA12 specimen.

Likewise, Figure 11 further confirms that abrasive wear is the dominant wear pattern of the Z-oriented graphite-filled CFR-PA12 specimen; the tribo-surface is characterized by a continuous microcutting groove. In addition, the tribo-surface shows straightforward evidence of CF pulverization and shearing/chopping of graphite filler. Although thinning/pulverization of CFs produces further graphite powder that effectively reduces the COF and wear rate, this comes at the expense of frictional noise as the prolonged interaction between CFs and disc surface eventually increases the friction-induced noise of Z-oriented specimens (see Figure 5).

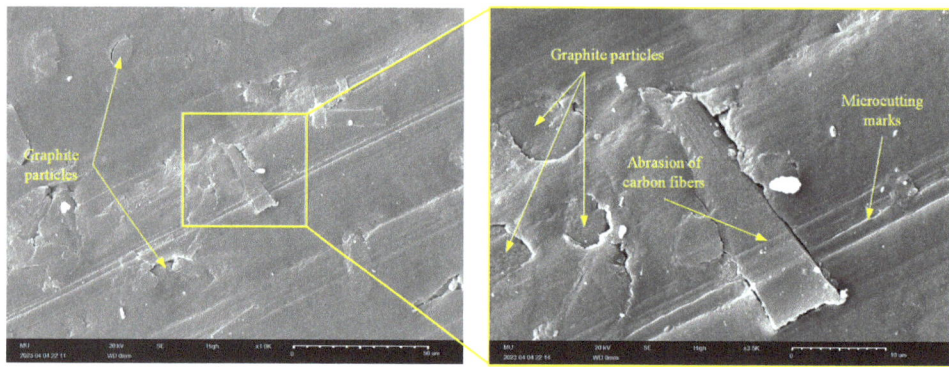

Figure 11. SEM of the worn tribo-surface of the Z-oriented graphite-filled CFR-PA12 specimen.

Similarly, Figure 12 shows an SEM picture of the tribo-surface of an X45-oriented specimen, where the worn surface features clear marks of deep microcutting grooves. Additionally, tiny graphite debris, which formed as a result of either CF crushing or shearing of graphite filler particles, is noticeable. However, the most interesting features of the tribo-surface of the X45-oriented specimen are the presence of protruded CFs and the deep microcutting grooves; these two features were the reason behind the friction and wear characteristics of the X45 specimen, as the protuberant carbon fibers that are aligned parallel to the sliding direction act to remove the transfer film from the disc surface, hence increasing the specific wear rate of the X45-oriented specimen. Further, the interaction of CF with the disc surface increases the friction-induced noise as CFs are stiff amorphous materials [20]. Meanwhile, the deep microcutting groove affirms the possibility of subsurface adhesion at either side of the cutting asperities and hence contributes to the elevated level of frictional noise.

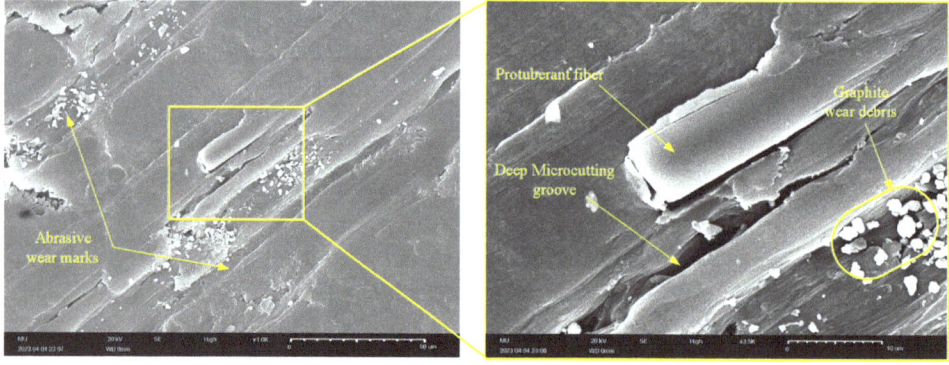

Figure 12. SEM of the worn tribo-surface of the X45-oriented graphite-filled CFR-PA12 specimen.

4. Conclusions

The present experimental study was conducted to explore the dry friction and wear characteristics of graphite-filled CFR-PA12 composites that were prepared by the SLS process. The effect of part-build orientation on tribological properties was investigated. For this purpose, the COF, contact temperature, friction-induced noise, and wear rate were measured for specimens built along four different orientations (X, Y, Z, and X45 orientations). The experiments were conducted using a pin-on-disc tribometer, and scanning electron microscopy (SEM) was used to investigate the wear patterns that dominated the tribo-surfaces. The results revealed that the steady-state COF, contact temperature, and wear pattern of graphite-filled CFR-PA12 are orientation-independent and that abrasive wear is the dominant wear pattern regardless of build orientation. In addition, the results showed that the Z-oriented specimen possesses the highest wear resistance combined with the lowest level of frictional noise among all other orientations. Unlike other orientations, the COF of Z-oriented specimens increases in an exponential manner during the running-in stage as a result of CFs/disc interactions. As a result of CFs/disc interaction, fiber crushing and/or thinning occurs, which eventually releases fine graphite particles between the rubbing surfaces and thus enhances the interface lubricity.

Furthermore, analysis of tribo-surfaces using SEM determined that shearing of graphite inclusions and fibers thinning release graphite debris at the interface, thus decreasing the COF, contact temperature, and friction-induced noise. Optical microscopy of the disc surface showed that a stable transfer film was formed on the disc surface that acts to decrease the COF, surface and subsurface adhesion, and specific wear rate of the Z-oriented test specimen.

However, despite the abovementioned promising features of the laser-sintered graphite-filled CFR-PA12 composite, further research is needed to explore its tribological properties under a wide range of operating conditions. With this in hand, artificial intelligence can be employed to help fabricate functional bearings with customized tribological properties.

Author Contributions: Conceptualization, A.G. and S.A.A.; methodology, A.G. and S.A.A.; software, A.G. and S.A.A.; validation, A.G. and S.A.A.; formal analysis, A.G. and S.A.A.; investigation, A.G. and S.A.A.; resources, A.G. and S.A.A.; data curation, A.G. and S.A.A.; writing—original draft preparation, A.G. and S.A.A.; writing—review and editing, A.G. and S.A.A.; visualization, A.G. and S.A.A.; supervision, A.G. and S.A.A.; project administration, A.G. and S.A.A.; funding acquisition, S.A.A. All authors have read and agreed to the published version of the manuscript.

Funding: This research was funded by the Deanship of Scientific Research at Majmaah University, grant number [R-2023-628].

Institutional Review Board Statement: Not applicable.

Data Availability Statement: The data are contained within the article.

Acknowledgments: The authors would like to thank the Deanship of Scientific Research at Majmaah University for supporting this work under project number [R-2023-628]. The authors are grateful to Graphite Additive Manufacturing Ltd. (Aylesbury, UK) for the careful preparation of test specimens.

Conflicts of Interest: The authors declare no conflict of interest.

References

1. Goodridge, R.D.; Tuck, C.J.; Hague, R.J.M. Laser sintering of polyamides and other polymers. *Prog. Mater. Sci.* **2012**, *57*, 229–267. [CrossRef]
2. Zhao, M.; Wudy, K.; Drummer, D. Crystallization Kinetics of Polyamide 12 during Selective Laser Sintering. *Polymers* **2018**, *10*, 168. [CrossRef] [PubMed]
3. Zárybnická, L.; Petrů, J.; Krpec, P.; Pagáč, M. Effect of Additives and Print Orientation on the Properties of Laser Sintering-Printed Polyamide 12 Components. *Polymers* **2022**, *14*, 1172. [CrossRef] [PubMed]
4. Yu, G.; Ma, J.; Li, J.; Wu, J.; Yu, J.; Wang, X. Mechanical and Tribological Properties of 3D Printed Polyamide 12 and SiC/PA12 Composite by Selective Laser Sintering. *Polymers* **2022**, *14*, 2167. [CrossRef]
5. Aldahash, S.A.; Salman, S.A.; Gadelmoula, A.M. Towards selective laser sintering of objects with customized mechanical properties based on ANFIS predictions. *J. Mech. Sci. Technol.* **2020**, *34*, 5075–5084. [CrossRef]
6. Randhawa, K.S.; Patel, A.D. A review on tribo-mechanical properties of micro- and nanoparticulate-filled nylon composites. *J. Polym. Eng.* **2021**, *41*, 339–355. [CrossRef]
7. Gadelmoula, A.; Aldahash, S.A. Tribological Properties of Glass Bead-Filled Polyamide 12 Composite Manufactured by Selective Laser Sintering. *Polymers* **2023**, *15*, 1268. [CrossRef]
8. Ogin, S.; Brøndsted, P.; Zangenberg, J. Composite materials: Constituents, architecture, and generic damage. In *Modeling Damage, Fatigue and Failure of Composite Materials*; Woodhead Publishing Series in Composites Science and Engineering; Elsevier: Amsterdam, The Netherlands, 2016; pp. 3–23. [CrossRef]
9. Kausar, A. Advances in Carbon Fiber Reinforced Polyamide-Based Composite Materials. *Adv. Mater. Sci.* **2019**, *19*, 67–82. [CrossRef]

10. Jing, W.; Hui, C.; Qiong, W.; Hongbo, L.; Zhanjun, L. Surface modification of carbon fibers and the selective laser sintering of modified carbon fiber/nylon 12 composite powder. *Mater. Des.* **2017**, *116*, 253–260. [CrossRef]
11. Aldahash, S.A.; Gadelmoula, A.M. Orthotropic properties of cement-filled polyamide 12 manufactured by selective laser sintering. *Rapid Prototyp. J.* **2020**, *26*, 1103–1112. [CrossRef]
12. Caulfield, B.; McHugh, P.E.; Lohfeld, S. Dependence of mechanical properties of polyamide components on build parameters in the SLS process. *J. Mater. Process. Technol.* **2007**, *182*, 477–488. [CrossRef]
13. Bai, J.; Yuan, S.; Chow, W.; Chua, C.K.; Zhou, K.; Wei, J. Effect of surface orientation on the tribological properties of laser sintered polyamide 12. *Polym. Test.* **2015**, *48*, 111–114. [CrossRef]
14. Yan, C.; Hao, L.; Xu, L.; Shi, Y. Preparation, characterisation and processing of carbon fibre/polyamide-12 composites for selective laser sintering. *Compos. Sci. Technol.* **2011**, *71*, 1834–1841. [CrossRef]
15. Nar, K.; Majewski, C.; Lewis, R. Evaluating the effect of solid lubricant inclusion on the friction and wear properties of Laser Sintered Polyamide-12 components. *Wear* **2023**, *522*, 204873. [CrossRef]
16. Zhang, H.; Zhang, Z.; Friedrich, K. Effect of fiber length on the wear resistance of short carbon fiber reinforced epoxy composites. *Compos. Sci. Technol.* **2007**, *67*, 222–230. [CrossRef]
17. Srinath, G.; Gnanamoorthy, R. Effect of Short Fibre Reinforcement on the Friction and Wear Behaviour of Nylon 66. *Appl. Compos. Mater.* **2005**, *12*, 369–383. [CrossRef]
18. Liu, Y.; Zhu, L.; Zhou, L.; Li, Y. Microstructure and mechanical properties of reinforced polyamide 12 composites prepared by laser additive manufacturing. *Rapid Prototyp. J.* **2019**, *25*, 1127–1134. [CrossRef]
19. Zhou, S.; Zhang, Q.; Wu, C.; Huang, J. Effect of carbon fiber reinforcement on the mechanical and tribological properties of polyamide6/polyphenylene sulfide composites. *Mater. Des.* **2013**, *44*, 493–499. [CrossRef]
20. Gadelmoula, A.; Aldahash, S.A. Effect of Reinforcement with Short Carbon Fibers on the Friction and Wear Resistance of Additively Manufactured PA12. *Polymers* **2023**, *15*, 3187. [CrossRef]
21. Friedrich, K. Wear of Reinforced Polymers by Different Abrasive Counterparts. *Compos. Mater. Ser.* **1986**, *1*, 233–287. [CrossRef]
22. Gilardi, R.; Bonacchi, D.; Spahr, M.E. Graphitic Carbon Powders for Polymer Applications. In *Encyclopedia of Polymers and Composites*; Palsule, S., Ed.; Springer: Berlin/Heidelberg, Germany, 2014; pp. 1–17. [CrossRef]
23. Meng, Y.; Xu, J.; Ma, L.; Jin, Z.; Prakash, B.; Ma, T.; Wang, W. A review of advances in tribology in 2020–2021. *Friction* **2022**, *10*, 1443–1595. [CrossRef]
24. Unal, H.; Esmer, K.; Mimaroglu, A. Mechanical, electrical and tribological properties of graphite filled polyamide-6 composite materials. *J. Polym. Eng.* **2013**, *33*, 351–355. [CrossRef]
25. Wu, H.; Chen, K.; Li, Y.; Ren, C.; Sun, Y.; Huang, C. Fabrication of Natural Flake Graphite/Ceramic Composite Parts with Low Thermal Conductivity and High Strength by Selective Laser Sintering. *Appl. Sci.* **2020**, *10*, 1314. [CrossRef]
26. Gadelmoula, A.M.; Aldahash, S.A. Effects of Fabrication Parameters on the Properties of Parts Manufactured with Selective Laser Sintering: Application on Cement-Filled PA12. *Adv. Mater. Sci. Eng.* **2019**, *2019*, 8404857. [CrossRef]
27. Johansson, P.; Elo, R.; Naeini, V.F.; Marklund, P.; Björling, M.; Shi, Y. Insights of the Ultralow Wear and Low Friction of Carbon Fiber Reinforced PTFE in Inert Trace Moisture Environment. *Tribol. Lett.* **2023**, *71*, 100. [CrossRef]
28. Goh, G.D.; Yap, Y.L.; Agarwala, S.; Yeong, W.Y. Recent Progress in Additive Manufacturing of Fiber Reinforced Polymer Composite. *Adv. Mater. Technol.* **2019**, *4*, 1800271. [CrossRef]
29. Myshkin, N.; Kovalev, A.; Spaltman, D.; Woydt, M. Contact mechanics and tribology of polymer composites. *J. Appl. Polym. Sci.* **2014**, *131*, 39870. [CrossRef]
30. Sinha Sujeet and Briscoe Brian, *Polymer Tribology*; Imperial College Press: London, UK, 2009.
31. Hutchings, I.M.; Shipway, P. *Tribology: Friction and Wear of Engineering Materials*, 2nd ed.; Elsevier: Amsterdam, The Netherlands, 2016.
32. Chang, L.; Zhang, Z.; Zhang, H.; Schlarb, A. On the sliding wear of nanoparticle filled polyamide 66 composites. *Compos. Sci. Technol.* **2006**, *66*, 3188–3198. [CrossRef]
33. Bahadur, S. The development of transfer layers and their role in polymer tribology. *Wear* **2000**, *245*, 92–99. [CrossRef]
34. Ye, J.; Khare, H.; Burris, D. Quantitative characterization of solid lubricant transfer film quality. *Wear* **2014**, *316*, 133–143. [CrossRef]
35. Rodiouchkina, M.; Lind, J.; Pelcastre, L.; Berglund, K.; Rudolphi, K.; Hardell, J. Tribological behaviour and transfer layer development of self-lubricating polymer composite bearing materials under long duration dry sliding against stainless steel. *Wear* **2021**, *484–485*, 204027. [CrossRef]
36. Lieng-Huang, L. *Polymer Science and Technology: Advances in Polymer Friction and Wear*; Plenum Press: New York, NY, USA; London, UK, 1974; Volume 5A. [CrossRef]
37. Rosso, S.; Meneghello, R.; Biasetto, L.; Grigolato, L.; Concheri, G.; Savio, G. In-depth comparison of polyamide 12 parts manufactured by Multi Jet Fusion and Selective Laser Sintering. *Addit. Manuf.* **2020**, *36*, 101713. [CrossRef]
38. Ma, N.; Liu, W.; Ma, L.; He, S.; Liu, H.; Zhang, Z.; Sun, A.; Huang, M.; Zhu, C. Crystal transition and thermal behavior of Nylon 12. *e-Polymers* **2020**, *20*, 346–352. [CrossRef]
39. Salmoria, G.; Leite, J.; Vieira, L.; Pires, A.; Roesler, C. Mechanical properties of PA6/PA12 blend specimens prepared by selective laser sintering. *Polym. Test.* **2012**, *31*, 411–416. [CrossRef]

40. Tewari, U.; Bijwe, J.; Mathur, J.; Sharma, I. Studies on abrasive wear of carbon fibre (short) reinforced polyamide composites. *Tribol. Int.* **1992**, *25*, 53–60. [CrossRef]
41. Li, J.; Xia, Y.C. The reinforcement effect of carbon fiber on the friction and wear properties of carbon fiber reinforced PA6 composites. *Fibers Polym.* **2009**, *10*, 519–525. [CrossRef]
42. Czelusniak, T.; Amorim, F.L. Influence of energy density on selective laser sintering of carbon fiber-reinforced PA12. *Int. J. Adv. Manuf. Technol.* **2020**, *111*, 2361–2376. [CrossRef]

Disclaimer/Publisher's Note: The statements, opinions and data contained in all publications are solely those of the individual author(s) and contributor(s) and not of MDPI and/or the editor(s). MDPI and/or the editor(s) disclaim responsibility for any injury to people or property resulting from any ideas, methods, instructions or products referred to in the content.

Article

Water Sorption in Hybrid Polyester/Glass/Jute Composites Processed via Compression Molding and Vacuum-Assisted Resin Transfer Molding

Rudá Aranha [1,*], Mario A. Albuquerque Filho [2], Cícero de Lima Santos [3], Viviane M. Fonseca [4], José L. V. Rivera [1], Antonio G. B. de Lima [3], Wanderley F. de Amorim, Jr. [3] and Laura H. Carvalho [2,*]

1. Escuela de Ingeniería Mecánica, Pontifícia Universidad Católica de Valparaíso 1, Valparaíso 2340025, Chile; jose.valin@pucv.cl
2. Post-Graduate Program in Materials Science and Engineering, Federal University of Campina Grande, Campina Grande 58429-900, Brazil; mario_alberto1910@hotmail.com
3. Mechanical Engineering Department, Federal University of Campina Grande, Campina Grande 58429-900, Brazil; cicero.santos@ufcg.edu.br (C.d.L.S.); antonio.gilson@ufcg.edu.br (A.G.B.d.L.); engenhariabrasileira1@gmail.com (W.F.d.A.J.)
4. Textile Engineering Department, Federal University of Rio Grande do Norte, Natal 59078-970, Brazil; fonseca.vmf@gmail.com
* Correspondence: ruda.aranha@pucv.cl (R.A.); heckerdecarvalho@yahoo.com.br (L.H.C.); Tel.: +55-83996156252 (R.A.); +55-83988211729 (L.H.C.)

Citation: Aranha, R.; Filho, M.A.A.; de Lima Santos, C.; Fonseca, V.M.; Rivera, J.L.V.; de Lima, A.G.B.; de Amorim, W.F., Jr.; Carvalho, L.H. Water Sorption in Hybrid Polyester/Glass/Jute Composites Processed via Compression Molding and Vacuum-Assisted Resin Transfer Molding. *Polymers* 2023, *15*, 4438. https://doi.org/10.3390/polym15224438

Academic Editors: Jesús-María García-Martínez and Emilia P. Collar

Received: 8 September 2023
Revised: 17 October 2023
Accepted: 19 October 2023
Published: 16 November 2023

Copyright: © 2023 by the authors. Licensee MDPI, Basel, Switzerland. This article is an open access article distributed under the terms and conditions of the Creative Commons Attribution (CC BY) license (https://creativecommons.org/licenses/by/4.0/).

Abstract: The aim of this work is to analyze water sorption in hybrid polyester/glass fabric/jute fabric composites molded via compression and VARTM (Vacuum-Assisted Resin Transfer Molding). The laminates were produced with five different stacking sequences and subjected to water sorption testing at room temperature, 50 °C and 70 °C. This study consisted of two stages: experimental and theoretical stages. The composites had a fiber volume content ranging from 30% to 40%. Water absorption and diffusion coefficient in the hybrid composites were intermediate to those reinforced with a single type of fiber. There were no significant differences in these properties based on fiber arrangement once the composites reached saturation. Diffusion coefficient values were higher for specimens with jute fiber on at least one of the outer surfaces. Water sorption rates increased with higher immersion temperatures. The water sorption at saturation point was not affected by the manufacturing process. Among the hybrid composites, those with jute on the surfaces showed the highest diffusion coefficient, while those with glass on the surface had the lowest values. Higher diffusion coefficient values were observed at temperatures of 50 °C and 70 °C. The main influencing factors on the absorbed moisture content for composites are the presence and content of jute fibers in the system and the immersion temperature. The manufacturing process does not affect the water sorption at saturation point.

Keywords: hybrid composites; jute fiber; VARTM; water sorption

1. Introduction

GFRP composites are increasingly being used as an alternative to traditional materials due to their low density, high strength, and stiffness. Aiming to reduce the environmental impact and preserve the environment, vegetable fibers are replacing synthetic fibers as reinforcement in composites, even if the mechanical properties of vegetable fibers are lower than synthetic ones. A solution to this problem is the partial substitution of synthetic fibers by plant fibers to create a hybrid composite [1]. Jute fibers have been used for this purpose because they are abundant, available in many countries, have low financial cost, and a good set of mechanical properties [2].

Fibrous hybrid composites are materials in which two or more different fibers—synthetic or natural—are used to reinforce a given matrix. Fibrous hybrid composites have intermediate mechanical properties between those of the same composite reinforced by each one of the fibers [3]. Therefore, hybrid composites reinforced by plant fibers or plant/synthetic hybrids cannot be intended for high-performance and costly applications because vegetable fibers have lower mechanical properties than synthetic fibers [1]. Even so, the use of these hybrid composites allows the material to reach suitable properties and meet project requirements, as well as improving a specific property of the material through the addition of a second reinforcement [4–6].

In fibrous hybrid composites, the fibers can be arranged in several different ways: interlaminar, intralaminar, and mixed [7–10]. Some other factors have an influence on the properties of hybrid composites: the mechanical properties and nature of the fibers, the length of the different types of fibers, and the quality of the interfacial connection between the different fibers and the matrix [10]. The synergistic effect is also reported for hybrid composites with two or more resin systems or additional constituents such as inserts, nanoparticles, and additives [11–13]. The fiber stacking sequence is another parameter that directly influences the mechanical properties of hybrid composites [14]. Some authors investigated the effects of adding glass fiber in composites reinforced with natural fibers [15–21]. All of them concluded that the addition of glass fibers increased the mechanical properties of these composites.

The properties of materials change over time, suffering degradation due to external effects. Most polymeric composites are sensitive to light, humidity, and heat, among other aggressive environments. The exposure of a material to aggressive environments generates changes over time that can be observed in engineering properties such as strength and rigidity; physical characteristics such as density; or in chemical characteristics such as reactivity to chemical products [22]. The degradation of composites can occur not only with the degradation of the material components, but also with the loss of interaction between them, that is, the deterioration of the fiber/matrix interface [15]. Glass fibers experience significant mechanical property losses due to corrosion mechanisms promoted by environmental exposure [23].

Aging tests are used to investigate the degradation of composites under service conditions. Accelerated or natural aging tests can be performed to produce physical and/or chemical degradation. For instance, in the water sorption test, the material is immersed in an aqueous solution, with or without controlled temperature, in order to degrade at an accelerated rate. This is a way of understanding the material's behavior when exposed to humidity. The glass transition temperature of thermosetting systems is modified when they are immersed in water at different temperatures until the saturation point is reached [24].

One-way composite materials absorb moisture into their structure through voids. The voids in the structure of the material will compromise the mechanical properties not only because of their existence, but also because they increase moisture absorption. Water absorption by composites does not occur only due to the existence of voids, but also due to the chemical affinity and type of matrix, temperature, polarity, diffusivity, and hydrogen bond formation, in addition to the nature, volumetric fraction, orientation, porosity, and geometry of fibers or fabrics [25–27].

Sorption is the phenomenon of mass transfer, where molecules from a liquid phase are associated with an immobile phase [28]. Water sorption in the composites followed a Fickian behavior [27,29]. Water sorption can be divided into adsorption that is related to the solid/fluid interface, where there is an accumulation of water molecules on the solid surface of the material, and absorption that occurs when water molecules penetrate the solid surface and settle inside the material. Water sorption in polymeric composites can be quite complex. When considering temperature variation, hygrothermal aging mainly damages the matrix, such as partial or total swelling and the formation of microcracks. This initial damage induces other damage mechanisms, such as interfacial debonding, delamination, fractures along the interface, resin particle loss, and fiber rupture [27,30–36]. Matrix swelling caused by water absorption is generally harmful to the fiber/matrix interface due to fiber detachment, which reduces the mechanical performance of the composite [27,32,37,38]. The diffusion of water molecules into polymeric networks will act as plasticizers. Plasticization tends to modify the glass transition temperature and decrease the strength of the composite [39]. On the other hand, moderate plasticization can also improve fracture toughness, preventing crack propagation in the material [40]. With a better adhesion between the reinforcement and the matrix, there was a reduction in water absorption, the diffusion coefficient, and swelling of the samples [29]. The fiber stacking sequence in hybrid composites should also be observed since it influences the water absorption of the composites [14,41,42].

When plant fibers are used as a reinforcement in polymer composites, the effects of water sorption on the composite are even more severe. Plant fibers, which are hydrophilic, are incompatible with thermoset resins, which are hydrophobic [31,32,43]. Therefore, the binding force becomes an impressive property to improve the adhesion between the fiber and the matrix. A decrease in the mechanical properties is expected due to the hygroscopic nature of these fibers and the poor fiber/matrix interface. Under constant environmental conditions, moisture content tends to increase with the volumetric fraction of plant fibers in the composite [32]. Such property losses were observed for polymeric composites reinforced by plant fibers of different nature, such as sisal [44], bamboo [45], flax [27,46], wood [47], hemp [33], and jute [48].

With the use of composite materials in various industries, understanding their behavior in real situations is essential. Therefore, correlating environmental effects such as temperature and humidity variations is important. However, due to the time and cost required for experimental studies in the actual conditions of material use, it is common to use different accelerated aging techniques as a more feasible alternative. The moisture content of fiber-reinforced polymer matrix composites has been extensively discussed in the literature [36,38,40,49–58]. However, investigations into the water sorption of hybrid composites reinforced with vegetable and synthetic fibers, such as the polyester/glass/jute composites, with different fiber stacking configurations on material properties, are still needed.

This study aims to investigate the effects of water sorption on polyester/fiberglass/jute fiber composites. The composites were processed using two methods: compression and Vacuum-Assisted Resin Transfer Molding (VARTM), and the laminates had five fiber stacking sequences. Water sorption tests were conducted, and it was possible to determine the water diffusion coefficient for these composites at different temperatures using the one-dimensional equation of Fick's second law for a flat plate.

After the tests, it was concluded that the jute fiber composites showed the highest moisture absorption and estimated diffusion coefficient at saturation point. The fiberglass composites exhibited the lowest absorption content, while the hybrid composites had an intermediate absorption rate. Higher water temperatures during the tests increased the moisture absorption rate for all composites. The hybridization of jute fibers with glass fibers reduced the amount of water absorbed by the composites compared to jute fiber composites. It was observed that the highest rate of water absorption by the composites occurred within the first 50 h of immersion. Regardless of the water sorption test condition, at saturation point, differences in the results were not observed when comparing the processing methods. The main influencing factors observed on the absorbed moisture content are the presence and content of jute fibers in the system and the processing method used in composite manufacturing. The novelty of this work is based on the following:

- The literature does not address the combinations and analyses used here when comparing hybrid composites reinforced with vegetable and synthetic fibers after immersion in water at different temperatures with different stacking sequences and manufactured using two different methodologies.
- This study allows a deeper understanding of the experimental water sorption behavior of this composite material under various working conditions to be obtained.

2. Materials and Methods

2.1. Materials

In this work, orthophtalic unsaturated polyester resin 10,316 produced for Reichhold, from Mogi das Cruzes, Brazil (Table 1), catalyzed with MEKP BUTANOX M-50 supplied by IBEX Chemical, Recife, Brazil, was used. The reinforcements used were type E glass fiber fabrics with a gramature of 330 g/m^2 as supplied by Redelease ltd, located in Sorocaba, Brazil, and jute plain weave fabric with a gramature of 330 g/m^2, manufactured by Cia. Têxtil, located in Castanhal, Brazil.

Table 1. Characteristics of resin—Reichhold guideline.

Characteristics	Values
Viscosidade Brookfield viscosity at 25 °C sp3: 60 rpm (CP)	250–350
Thixotropy index	1.30–2.10
Solid content—Reichhold method (%)	55–63
Density at 25 °C (g/cm^3)	1.07–1.11
Acidity index (mgKOH/g)	30 maximum
Exothermic Curve at 25 °C	
- Gel time (min)	5–7
- Single interval (min)	8–14
- Maximum temperature (°C)	140–180
Post-cure	60 °C

2.2. Manufacture of Composites

The composites investigated here were manufactured using two different methods: compression molding (Figure 1a) and VARTM (Figure 1b).

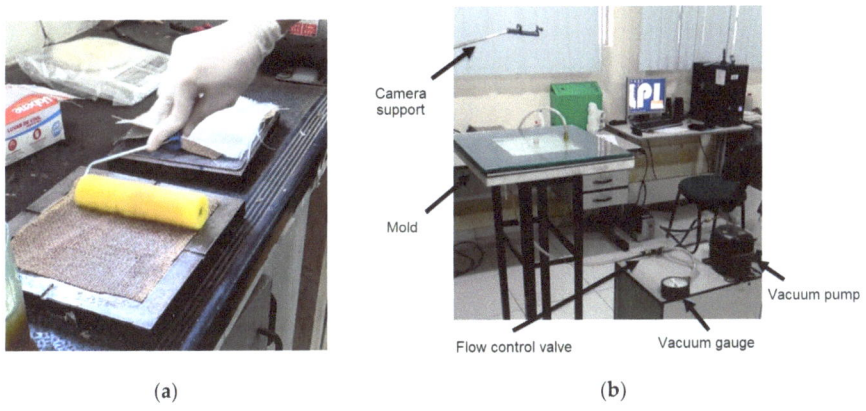

Figure 1. Manufacture of composite plates: (**a**) compression molding; (**b**) VARTM.

Regardless of the manufacturing method, each composite has four layers of glass fabric (G) and/or jute plain weave fabric (J) as the reinforcement. Hybrid composites were manufactured with two layers of each type of fabric arranged in different stacking sequences. The compression-molded composites were manufactured with the following layer stacking sequence: GGGG; JJJJ; GJGJ; JGGJ; GJJG. The composites manufactured by VARTM were manufactured with two stacking sequences: GJJG and JGGJ. Therefore, the effect of the fabrication method (compression or VARTM) can be compared between GJJG-C and GJJG-R composites, where 'C' stands for compression molding and 'R' stands for the VARTM process, and where 'C' stands for compression molding and 'R' stands for the VARTM process.

Compression molding of the plates was carried out in a metallic mold measuring 200 mm × 180 mm. The fabric layers were manually placed in the mold, laminated with a foam roller, and compressed at room temperature with 9 Ton in a Marconi uniaxial hydraulic press for 24 h before demolding and post-curing in an air circulation oven at 60 °C. VARTM processing was carried out in a rectangular mold with the same dimensions (200 mm × 180 mm), using a glass base and vacuum bag (Figure 1b), straight flow front, two entry points, and one exit point for resin with a diameter $\frac{1}{4}$" and a vacuum pressure of −0.3 bar.

At the end of the processing, the plates were weighed, and the volumetric fractions of the fibers could be determined. For this, the theoretical method (Equation (1)) was used [14].

$$V_f = \frac{w_j/\rho_j + w_g/\rho_g}{w_j/\rho_j + w_g/\rho_g + w_m/\rho_m}, \qquad (1)$$

where V_f is the volumetric fraction of the fibers of the laminate; w_j, w_g, and w_m are the masses of the jute fibers, glass fibers, and matrix, respectively; and ρ_j, ρ_g, and ρ_m are the densities of the jute fiber, glass fiber, and matrix, respectively.

The compression-molded and VARTM plates were cut on a CNC mill. The specimens obtained had dimensions of 20 mm × 20 mm for the water sorption tests according to the ASTM D570—81 Standard [59].

2.3. Water Sorption Test—Experimental

The effects of water sorption by the material were evaluated through the mass variation as a function of the immersion time of the composites. The samples submitted to the water sorption test had dimensions of 20 mm × 20 mm, as suggested by the ASTM D570—81 Standard [59]. The samples were sealed at the edges with resin as a way to prevent the fibers from having direct contact with water and absorbing by capillarity. The samples were dried in a vacuum oven at 60 °C for 4 h to reach a constant weight, ensuring that no residual moisture was present, before being placed in an aqueous medium. The specimens were weighed before and after the water absorption test at room temperature, T = 50 °C and T = 70 °C for up to 30 days (or until saturation). The water absorption curves for each type of composite as a function of immersion time were then plotted.

Equation (2) shows how to obtain water absorption as a function of time (M_t).

$$M_t = \frac{W_t - W_0}{W_0} 100(\%), \qquad (2)$$

where W_t is the mass of wet samples at time t and W_0 is the initial mass of dry samples.

2.4. Water Sorption Test—Theoretical

For the study of theoretical water sorption, a simplified Fickian diffusional model was used. It was assumed that the samples are flat plates, and the humidity flux is transient one-dimensional. Water sorption in transient regime is described by Fick's 2nd law (Equation (3)) [60].

$$\frac{\partial c}{\partial t} = D \frac{\partial^2 c}{\partial x^2}, \qquad (3)$$

Making the necessary adjustments and considering that M_t is the total amount of diffusing substance that entered the plate at time t, and M_∞ is the total amount of diffusing substance after infinite time, we have Equation (4).

$$\frac{M_t}{M_\infty} = 2 \left(\frac{Dt}{l^2} \right)^{1/2} \left(\pi^{-1/2} \right), \qquad (4)$$

In the initial stage of absorption, water absorption in time t (M_t) increases linearly with $\sqrt{t}$ and M_∞, which are associated with mass gain when the material approaches the saturation point. If we consider that the water absorption behavior follows the Fickian diffusion pattern, it can be described by Equation (5) [23,26,33,60–64].

$$\frac{M_t}{M_\infty} = 4\sqrt{\frac{Dt}{\pi h^2}}, \tag{5}$$

where h is the sample thickness. The average diffusion coefficient (D) is determined by the water absorption capacity (M_∞) and the kinetic constant of water absorption (k), obtained by the slope of the graph M_t versus $t^{1/2}$, described by Equation (6).

$$D = \pi \left(\frac{kh}{4M_\infty}\right)^2, \tag{6}$$

3. Results and Discussion

3.1. Volumetric Fraction of Fibers

Equation (1) was used to determine the fiber weight fraction and fiber volume fraction of the composites. The density of the fibers used for the theoretical calculation were 1.5 g/cm^3 (jute) and 2.54 g/cm^3 (glass). The values obtained for total and relative fiber weight and volume fractions are shown in Table 2.

Table 2. Total and relative fiber weight and volume fraction of manufactured composites.

Composites	Glass Fiber Weight Fraction (%)	Jute Fiber Weight Fraction (%)	Total Fiber Weight Fraction of Composites (%)	Fiber Volume Fraction of Composites (%)	Total Fiber Volume Fraction of Jute (%)
GGGG-C [1]	59.70 ± 3.27	0	59.70 ± 3.27	39.35 ± 3.21	0
JJJJ-C	0	44.10 ± 2.10	44.10 ± 2.10	36.87 ± 1.97	36.87 ± 1.97
GJGJ-C	15.23 ± 0.43	26.10 ± 0.74	41.33 ± 1.17	30.68 ± 1.02	24.77 ± 0.90
GJJG-C	15.23 ± 1.06	26.11 ± 1.82	41.34 ± 2.89	30.72 ± 2.55	24.81 ± 2.24
GJJG-R [1]	17.77 ± 1.77	30.46 ± 3.04	48.23 ± 4.82	36.99 ± 4.48	30.42 ± 4.06
JGGJ-C	15.28 ± 1.16	26.19 ± 2.00	41.47 ± 3.16	30.83 ± 2.76	24.91 ± 2.42
JGGJ-R	18.15 ± 1.18	31.12 ± 2.02	49.28 ± 3.20	37.94 ± 3.00	31.27 ± 2.73

[1] C and R indicate the manufacturing method: compression molding (C) and VARTM (R).

Glass fabric composites showed the highest values of fiber volume fraction, with an average of approximately 39%. The jute fabric composites showed an average value of 36% of the fiber volumetric fraction. The fiber volume fraction of hybrid composites was lower for composites manufactured using compression molding (30%) when compared with composites manufactured using VARTM (37%). These results were expected because the VARTM provides better compaction of the fibers during manufacturing compared to the compression molding method due to the use of a vacuum. The applied vacuum together with the resin flow front during manufacturing promotes more efficient air removal, thus reducing the void content in the composites.

3.2. Water Sorption: Experimental Analysis

Figure 2 shows the composites samples before and after the water sorption test at room temperature.

Surface changes were observed in the samples due to water absorption. The samples showed a change in color with a more "whitish" tone. Swelling of the samples at the end of the test caused by water absorption was also observed. Similar results were reported by [49].

Water absorption as a function of time was measured according to Equation (2). Figure 3 shows the water absorption of the composites at room temperature, Figure 4 shows the water absorption at T = 50 °C, and Figure 5 shows water absorption at T = 70 °C. The exposure time of the samples for all conditions was 696 h.

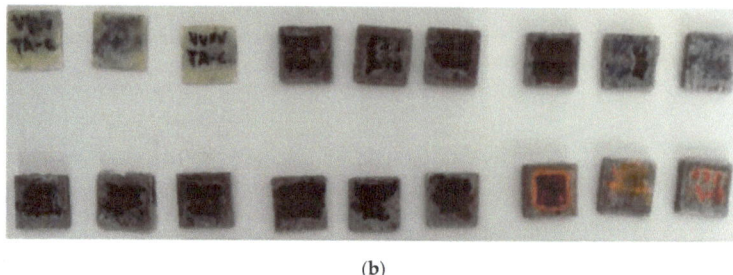

Figure 2. Water sorption test of the composites: (**a**) before the test; (**b**) at the end of the test (696 h).

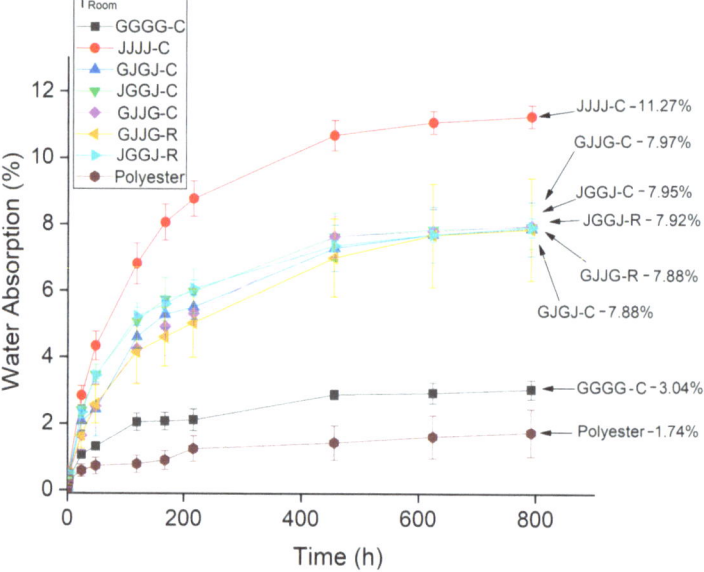

Figure 3. Water absorption at room temperature by the composites as a function of time (696 h).

It is clear that, for all exposure conditions (Figures 3–5), the jute fabric-reinforced composite showed higher water absorption, while the glass fabric composites showed the lowest water absorption. This result was expected considering the hydrophilic nature of jute.

The polyester samples exposed to water at room temperature exhibited an absorption of 1.74%, which was the lowest value observed. Despite being composed of two synthetic materials, the glass fabric composites (polyester/glass fiber) showed higher water absorption than the polyester samples. This difference in absorption results between the glass fabric composite and the polyester samples occurred due to the manufacturing process of the composites. The wetting process of the glass fibers by the resin is not perfect, resulting in difficulty in filling, mainly for the micropores [65]. This difficulty in resin wetting implies the presence of small voids in the laminate, and these voids are responsible for the penetration of moisture into the composites.

The hybrid composites exhibited intermediate absorption rates for all studied conditions. The absorption values observed in the hybrid composites (Figures 3–5) are closer to the values observed in the jute fabric composites. Due to its hydrophilic nature, jute is the main contributor to the total water absorption in hybrid systems. A slight variation in water absorption was found due to variations in the fiber stacking sequence and the method of manufacturing the hybrid composites (compression molding or VARTM). It should also be taken into account that the manufactured composites had a higher volumetric fraction of jute fiber than glass fiber, which corroborates these results.

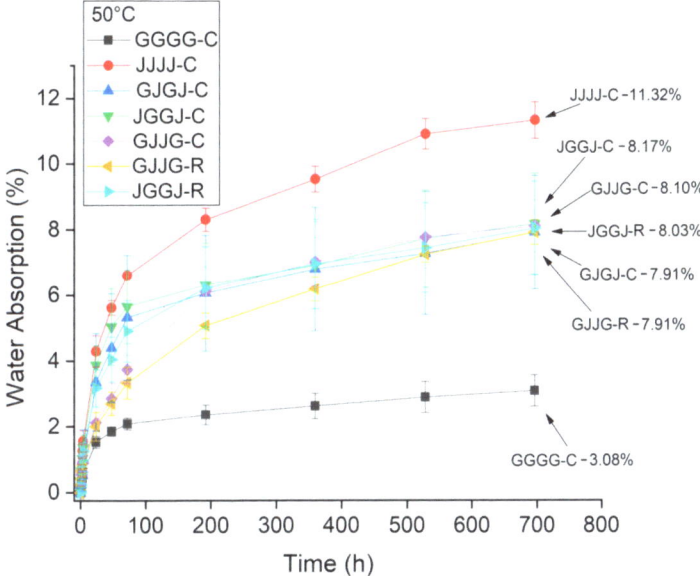

Figure 4. Water absorption at T = 50 °C by the composites as a function of time (696 h).

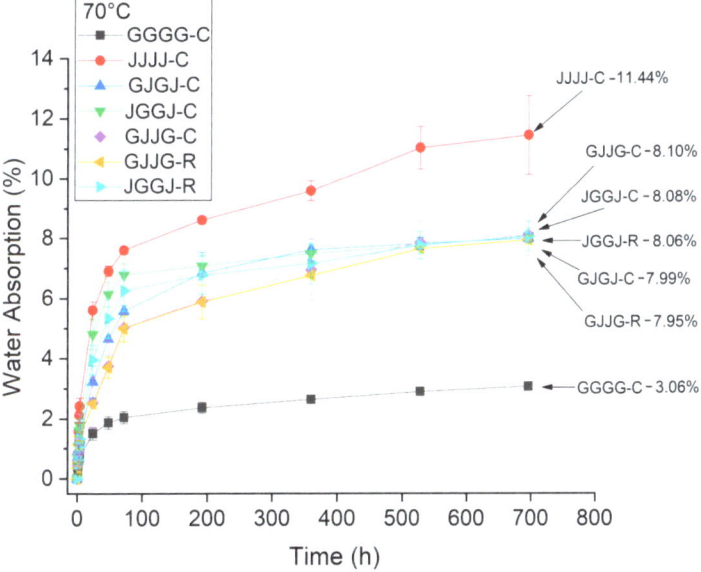

Figure 5. Water absorption at T = 70 °C by the composites as a function of time (696 h).

The hybridization of jute fibers with glass fibers reduced the amount of water absorbed by the composites compared to jute fiber composites. The results from Figures 3–5 indicate higher initial absorption in the hybrid composites with jute in the outer layers. Due to the hygroscopic nature and higher porosity of jute fibers, water diffuses through the resin and is absorbed in larger quantities when it comes into contact with these fibers. These results are consistent with the findings reported by others [9,66,67].

On the other hand, between 250 h and 350 h of immersion at room temperature, it was observed that the VJJV-C composites began to absorb more moisture among the hybrids and became the composite with the highest water absorption content. It is possible that within this time range (250–350 h), water particles reached the interior of the samples, as jute fibers absorbed more moisture than fiberglass. This behavior was not observed in the VJJV-R samples, which, although they showed an increase in water absorption, did not reach the same level as the samples manufactured via compression for the same immersion time. A higher rate of absorption for the VJJV-R samples occurred between 400 h and 600 h compared to the other hybrid composites. We can argue that this delay in the comparison between the samples is due to the fact that moisture has more difficulty penetrating the VARTM-manufactured samples. In longer immersion times (above 600 h), the small differences in water absorption between the hybrid composites were no longer evident.

For the conditions of T = 50 °C and T = 70 °C, the VJJV-C hybrid composites started to absorb more moisture than the other hybrids. However, this phenomenon was observed between 150 h and 200 h of immersion. We can consider that this acceleration in water absorption occurred due to the higher temperatures during the test. This behavior was not observed in the VJJV-R hybrid composite. Since there was a higher water absorption rate, after 500 h of testing, differences in absorption between the hybrid composites were not observed for these test conditions.

In Figures 6–8, it was observed that the highest rate of water absorption by the composites occurred within the first 50 h of immersion. The JVVJ-C hybrid composites exhibited values close to those of the jute fiber samples. Regardless of the manufacturing method, the JVVJ composites absorbed more moisture than the VJJV composites. This result was expected due to the stacking sequence of these composites, with jute fibers on the surfaces.

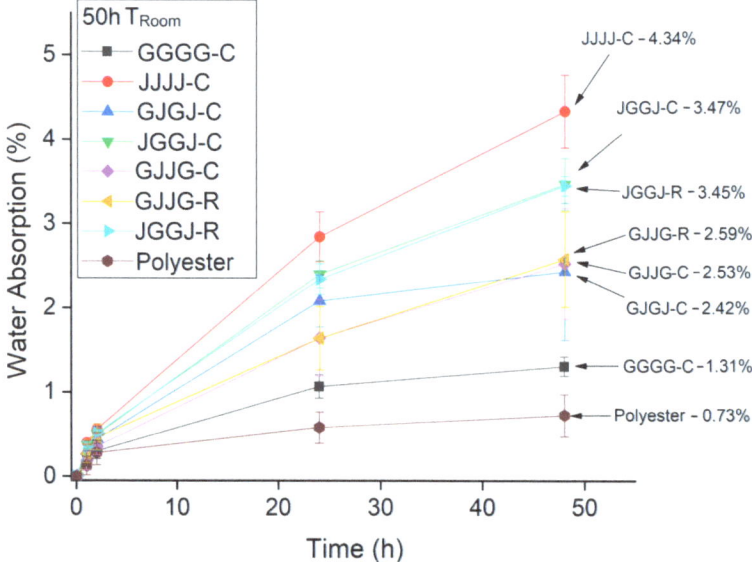

Figure 6. Absorption of water at room temperature by the composites in a time of 50 h.

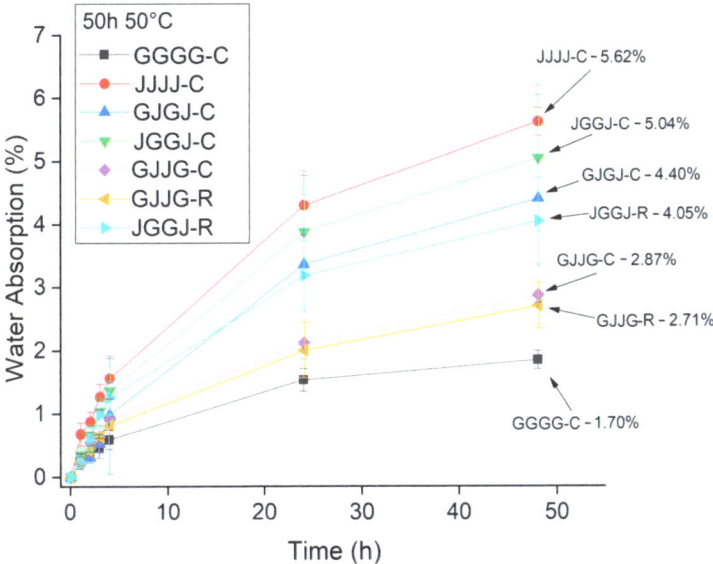

Figure 7. Absorption of water at T = 50 °C by the composites in a time of 50 h.

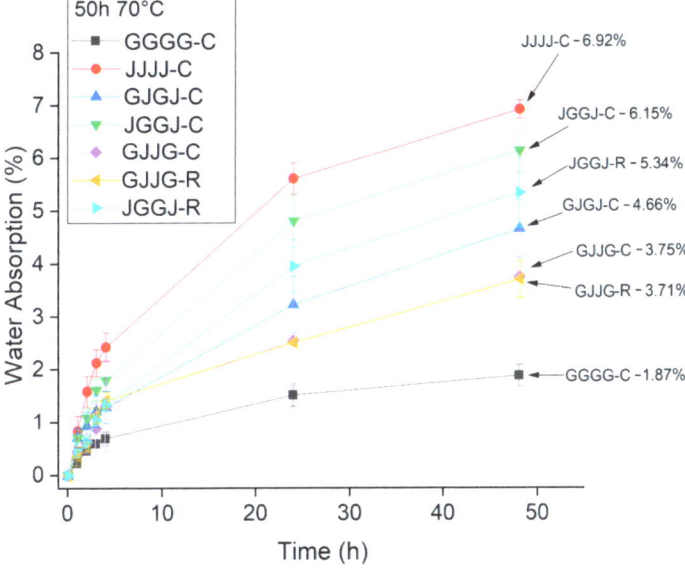

Figure 8. Absorption of water at T = 70 °C by the composites in a time of 50 h.

The VJVJ-C samples absorbed more moisture than the VJJV-C and VJJV-R samples in the initial time period. This occurred because the VJVJ-C composite had a jute layer on at least one side of the composite.

In Figure 7, an unexpected result was observed: the VJVJ-C composite showed a higher water absorption value than the JVVJ-R composite. Naturally, this result can be attributed to the different manufacturing methods used. However, other hypotheses were considered in addition to the manufacturing methods since this reason alone is not sufficient to justify such a result. The hypothesis is that there were micro-cracks in the matrix, causing an increased absorption rate due to moisture penetration through these areas. Water sorption and temperature act in two ways in composites: Firstly, through the plasticization of the macromolecular network of the resin, leading to irreversible property losses [23]. Secondly, through microcracks in the matrix and fiber/matrix debonding caused by matrix swelling and an increase in the amount of moisture absorption.

When comparing the effect of processing methods on the water sorption of hybrid composites at room temperature, the results showed equivalence: both the VJJV-C (7.97%) and VJJV-R (7.88%) composites, and the JVVJ-C (7.95%) and JVVJ-R (7.92%) composites, showed similar values to each other (Figure 6), even though the VARTM-fabricated composites (~37.50% total fiber content) had a higher total theoretical fiber volume fraction compared to the compression-fabricated composites (~30.75%) (Table 2). These results can be explained by the greater compaction in the VARTM laminate due to the application of vacuum during the resin infusion process. The use of vacuum allows for greater compaction of the system and generates a lower void content, reducing moisture absorption inside the panels. The results for temperatures T = 50 °C and T = 70 °C were similar to those observed at room temperature (Figures 7 and 8).

For the observed composite structures, the main influencing factors on the absorbed moisture content are the presence and content of jute fibers in the system and the processing method used in composite manufacturing, with the VARTM-fabricated composites showing similar absorption values for a higher jute fiber content compared to the compression-fabricated composites.

When comparing the absorption rate of composites at room temperature, T = 50 °C and T = 70 °C, an analysis was conducted for different immersion times (Figure 9). It was observed that regardless of the immersion time, samples immersed at room temperature showed the lowest water absorption rates, samples immersed at 50 °C showed intermediate rates, and samples immersed at 70 °C showed the highest absorption rates. This result was expected because an increase in temperature promotes an increase in molecular movement, favoring the penetration of moisture into the material.

It is also important to consider that the temperature of 70 °C is close to the glass transition temperature of the polyester matrix. The exposure of the material over time at 70 °C tends to cause surface degradation in the resin. In Figure 9d, it was observed that after 696 h, the composites reached saturation, showing very similar water absorption values regardless of the immersion temperature.

3.3. Water Sorption: Theoretical Analysis

The results shown in Table 3 indicate that the fiberglass composites had the lowest values for the diffusion coefficient (D), jute fiber composites had the highest values, while hybrid composites showed intermediate values for water sorption at room temperature. Among the hybrid composites, the JVVJ stacking sequence exhibited the highest diffusion coefficient values, VJVJ had intermediate values, and VJJV had the lowest values.

Table 3. Water absorption capacity (M_∞), kinetic constant (k), and diffusion coefficient (D) for composites immersed in water at room temperature.

Composites	M_∞ (%)	k (h^{-1})	D, $\times 10^{-13}$ (m^2/s)
VVVV-C	3.04	0.00189	0.76
JJJJ-C	11.27	0.00639	13.26
VJVJ-C	7.88	0.00437	7.40
VJJV-C	7.97	0.00373	5.27
VJJV-R	7.88	0.00370	5.56
JVVJ-C	7.95	0.00514	10.54
JVVJ-R	7.92	0.00508	10.37

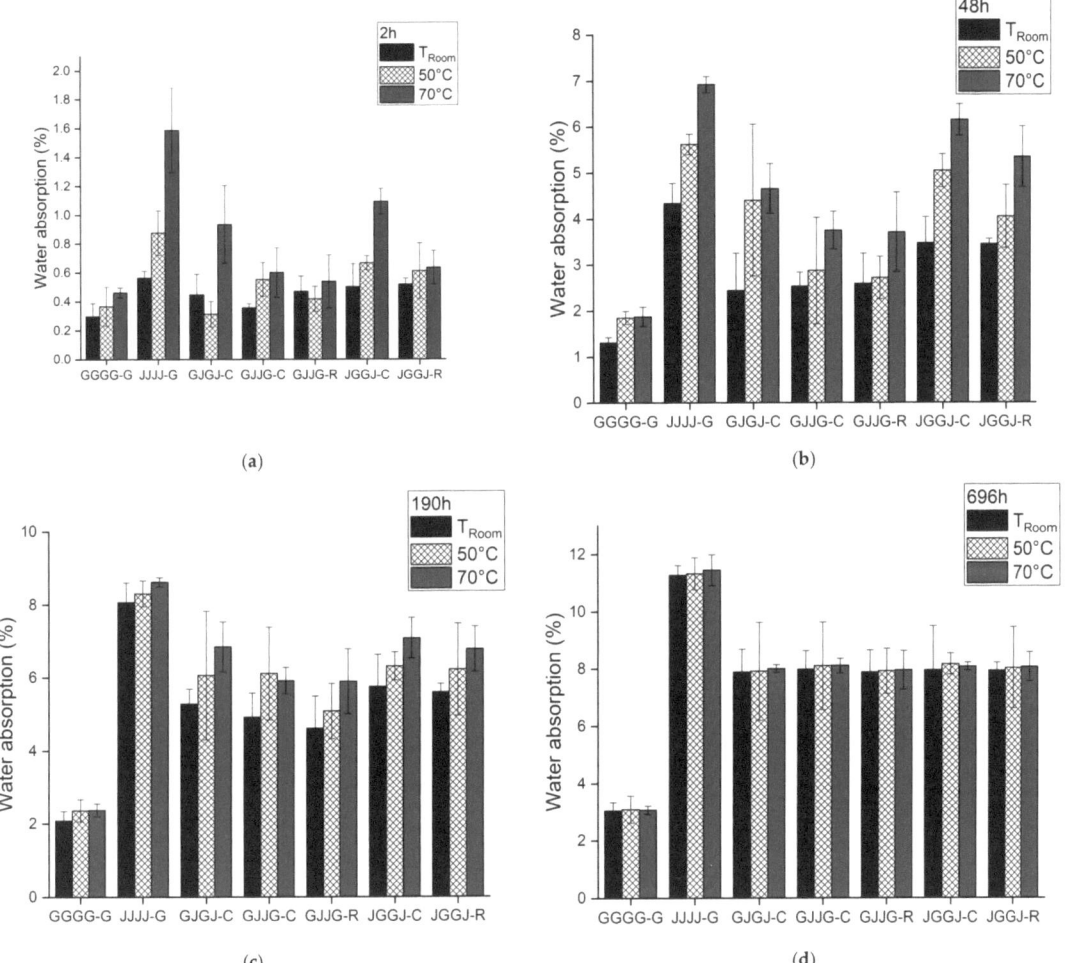

Figure 9. Comparison between sorption tests at room temperature, T = 50 °C and T = 70 °C for different immersion times: (**a**) 2 h; (**b**) 48 h; (**c**) 190 h; (**d**) 696 h.

These results for the hybrid composites indicate that the presence of fiberglass on the surface of the laminates initially delayed the diffusion of moisture into the material. On the other hand, the hybrid composite with alternating fibers (VJVJ-C) showed an intermediate diffusion value compared to the other hybrid types due to the presence of jute on only one surface of the laminate. Glass fiber, with its hydrophobic nature, when placed on the surface of the composite, acts as a protective barrier against the penetration of moisture from the interior, thus slowing down water absorption. No significant differences were observed between the manufacturing methods.

The diffusion coefficient of composites immersed at T = 50 °C and T = 70 °C (Tables 4 and 5) showed higher values than composites immersed at room temperature. The comparison of these values for all composites is shown in Figure 10. These results are expected because an increase in temperature tends to increase molecular movement, thereby increasing the rate at which composites absorb moisture. When immersed at T = 70 °C, the composites exhibited a higher diffusion coefficient compared to composites immersed under different conditions. As previously mentioned, this is due to the fact that the temperature of T = 70 °C is close to the glass transition temperature of the polyester, leading to matrix degradation and contributing to greater moisture diffusion within the material.

Table 4. Water absorption capacity (M_∞), kinetic constant (k), and diffusion coefficient (D) for composites immersed in water at T = 50 °C.

Composites	M_∞ (%)	k (h^{-1})	D, $\times 10^{-13}$ (m^2/s)
VVVV-C	3.08	0.00284	1.67
JJJJ-C	11.32	0.00756	18.40
VJVJ-C	7.91	0.00699	18.78
VJJV-C	8.10	0.00440	7.10
VJJV-R	7.91	0.00401	6.48
JVVJ-C	8.17	0.00662	16.55
JVVJ-R	8.03	0.00624	15.22

Table 5. Water absorption capacity (M_∞), kinetic constant (k), and diffusion coefficient (D) for composites immersed in water at T = 70 °C.

Composites	M_∞ (%)	k (h^{-1})	D, $\times 10^{-13}$ (m^2/s)
VVVV-C	3.06	0.00354	2.63
JJJJ-C	11.44	0.01244	48.77
VJVJ-C	7.99	0.0069	17.94
VJJV-C	8.10	0.00600	13.20
VJJV-R	7.95	0.00690	18.99
JVVJ-C	8.08	0.00918	32.54
JVVJ-R	8.06	0.00842	27.51

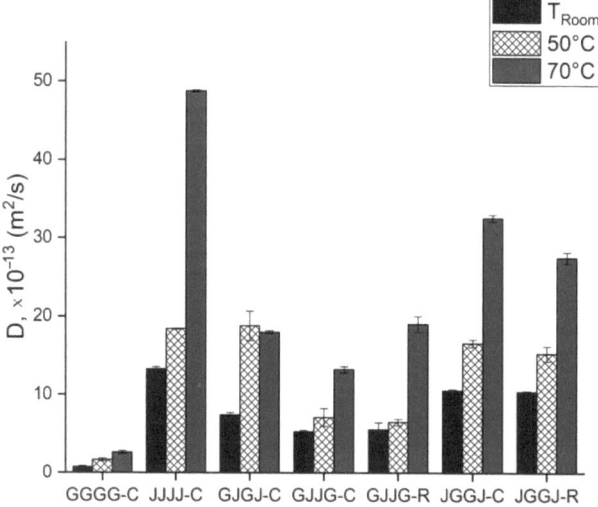

Figure 10. Comparative analysis of the diffusion coefficient of composites immersed in water at room temperature, 50 °C and 70 °C.

Attention should be paid to Equations (3)–(6), which only consider the initial stage of water absorption. During this stage, water absorption increases linearly, and the kinetic absorption constant (k) is obtained from the slope of the Mt versus t1/2 graph, which is used to calculate the diffusion coefficient (D). Although the water absorption of hybrid composites shows similarities in saturation, differences in the absorption curve behavior were observed during the initial hours of exposure due to the stacking sequence of the laminate layers. Regardless of the temperature condition during immersion, composites with jute fiber on the surface of the laminate exhibited a higher diffusion coefficient compared to composites with fiberglass on the surface.

4. Conclusions

Water sorption in hybrid polyester/fiberglass/jute fiber composites was analyzed. It was possible to manufacture the composites using both methods: compression molding and VARTM.

- The composites showed theoretical fiber volume fraction values ranging from approximately 30% to 40%, with the VARTM-produced hybrid composites showing higher values than the compression-molded hybrid composites.
- During the water sorption tests, the composites reached saturation after 696 h. Among them, the jute fiber composites showed the highest moisture absorption content after 696 h. The fiberglass composites exhibited the lowest absorption content, while the hybrid composites had an intermediate absorption rate. Using higher water temperatures during the tests increased the moisture absorption rate for all composites. The hybridization of jute fibers with glass fibers reduced the amount of water absorbed by the composites compared to jute fiber composites.
- It was observed that the highest rate of water absorption by the composites occurred within the first 50 h of immersion. Regardless of the test condition, higher moisture absorption rates were observed in the hybrid composites with jute layers on at least one of the surfaces during the initial 50 h.
- At long times, above 600 h, such differences were no longer observed for the composites immersed at room temperature (500 h for the composites immersed at 50 °C and 70 °C). Regardless of the water sorption test condition, at the time of 696 h, differences in the results were not observed when comparing the processing methods, as this is the time associated with sample saturation.
- When comparing the effect of processing methods on the water sorption of hybrid composites for all temperatures, the results showed equivalence. VJJV-C and VJJV-R composites, and the JVVJ-C and JVVJ-R composites, showed similar values to each other. VARTM-fabricated composites showed similar absorption values for a higher jute fiber content compared to the compression-fabricated composites.
- The estimation of the diffusion coefficient calculated through Fick's second law showed that the jute fiber composites exhibited the highest diffusion coefficient, the fiberglass composites showed the lowest values, and the hybrid composites presented intermediate values. The JVVJ composites had a higher diffusion coefficient among hybrids due to the presence of jute fiber on the surfaces. The VJJV composites had the lowest values for the diffusion coefficient. Higher values of the diffusion coefficient were observed at temperatures of 50 °C and 70 °C when compared to the composites immersed at room temperature.
- For the observed composite structures, the main influencing factors on the absorbed moisture content are the presence and content of jute fibers in the system, temperature increase, and the processing method used in composite manufacturing.

Author Contributions: Conceptualization, R.A.; methodology, R.A., C.d.L.S. and M.A.A.F.; validation, R.A., A.G.B.d.L., W.F.d.A.J. and L.H.C.; formal analysis, R.A., V.M.F., J.L.V.R., A.G.B.d.L., W.F.d.A.J. and L.H.C.; writing—original draft preparation, R.A.; writing—review and editing, R.A. and L.H.C.; supervision, J.L.V.R., A.G.B.d.L., W.F.d.A.J. and L.H.C. All authors have read and agreed to the published version of the manuscript.

Funding: This research was funded by the Brazilian Research Agencies: CAPES (Grant number 88882.455376/2019-01), CNPq (Grant number 305143/2020-4) and FAPESQ/PB (Grant number 3160/2021).

Institutional Review Board Statement: Not applicable.

Data Availability Statement: Not applicable.

Acknowledgments: The authors are indebted to LaMMEA (UFCG) and LCTF (UFCG) for laboratory support.

Conflicts of Interest: The authors declare no conflict of interest.

References

1. Sanjay, M.R.; Yogesha, B. Studies on Natural/Glass Fiber Reinforced Polymer Hybrid Composites: An Evolution. *Mater. Today Proc.* **2017**, *4*, 2739–2747. [CrossRef]
2. Singh, H.; Inder Preet Singh, J.; Singh, S.; Dhawan, V.; Kumar Tiwari, S. A Brief Review of Jute Fibre and Its Composites. *Mater. Today Proc.* **2018**, *5*, 28427–28437. [CrossRef]
3. Venkateshwaran, N.; ElayaPerumal, A.; Alavudeen, A.; Thiruchitrambalam, M. Mechanical and water absorption behaviour of banana/sisal reinforced hybrid composites. *Mater. Des.* **2011**, *32*, 4017–4021. [CrossRef]
4. Kretsis, G. A review of the tensile, compressive, flexural and shear properties of hybrid fibre-reinforced plastics. *Composites* **1987**, *18*, 13–23. [CrossRef]
5. Velmurugan, R.; Manikandan, V. Mechanical properties of palmyra/glass fiber hybrid composites. *Compos. Part A Appl. Sci. Manuf.* **2007**, *38*, 2216–2226. [CrossRef]
6. Swolfs, Y.; Gorbatikh, L.; Verpoest, I. Fibre hybridisation in polymer composites: A review. *Compos. Part A Appl. Sci. Manuf.* **2014**, *67*, 181–200. [CrossRef]
7. John, M.J.; Thomas, S. Biofibres and biocomposites. *Carbohydr. Polym.* **2008**, *71*, 343–364. [CrossRef]
8. Yu, H.; Longana, M.L.; Jalalvand, M.; Wisnom, M.R.; Potter, K.D. Pseudo-ductility in intermingled carbon/glass hybrid composites with highly aligned discontinuous fibres. *Compos. Part A Appl. Sci. Manuf.* **2015**, *73*, 35–44. [CrossRef]
9. Swolfs, Y.; Verpoest, I.; Gorbatikh, L. Recent advances in fibre-hybrid composites: Materials selection, opportunities and applications. *Int. Mater. Rev.* **2019**, *64*, 181–215. [CrossRef]
10. Rajpurohit, A.; Joannès, S.; Singery, V.; Sanial, P.; Laiarinandrasana, L. Hybrid Effect in In-Plane Loading of Carbon/Glass Fibre Based Inter- and Intraply Hybrid Composites. *J. Compos. Sci.* **2020**, *4*, 6. [CrossRef]
11. Kim, H.; Park, M.; Hsieh, K. Fatigue fracture of embedded copper conductors in multifunctional composite structures. *Compos. Sci. Technol.* **2006**, *66*, 1010–1021. [CrossRef]
12. Yao, H.; Sui, X.; Zhao, Z.; Xu, Z.; Chen, L.; Deng, H.; Liu, Y.; Qian, X. Optimization of interfacial microstructure and mechanical properties of carbon fiber/epoxy composites via carbon nanotube sizing. *Appl. Surf. Sci.* **2015**, *347*, 583–590. [CrossRef]
13. Paran, S.M.R.; Abdorahimi, M.; Shekarabi, A.; Khonakdar, H.A.; Jafari, S.H.; Saeb, M.R. Modeling and analysis of nonlinear elastoplastic behavior of compatibilized polyolefin/polyester/clay nanocomposites with emphasis on interfacial interaction exploration. *Compos. Sci. Technol.* **2018**, *154*, 92–103. [CrossRef]
14. Ahmed, K.S.; Vijayarangan, S. Tensile, flexural and interlaminar shear properties of woven jute and jute-glass fabric reinforced polyester composites. *J. Mater. Process. Technol.* **2008**, *207*, 330–335. [CrossRef]
15. Joseph, K.; Thomas, S.; Pavithran, C. Effect of chemical treatment on the tensile properties of short sisal fibre-reinforced polyethylene composites. *Polymer* **1996**, *37*, 5139–5149. [CrossRef]
16. Devireddy, S.B.R.; Biswas, S. Physical and mechanical properties of unidirectional banana–jute hybrid fiber-reinforced epoxy composites. *J. Reinf. Plast. Compos.* **2016**, *35*, 1157–1172. [CrossRef]
17. Latha, P.S.; Rao, M.V.; Kumar, V.K.; Raghavendra, G.; Ojha, S.; Inala, R. Evaluation of mechanical and tribological properties of bamboo–glass hybrid fiber reinforced polymer composite. *J. Ind. Text.* **2016**, *46*, 3–18. [CrossRef]
18. Sriranga, B.K.; Kirthan, L.J.; Ananda, G. The mechanical properties of hybrid laminates composites on epoxy resin with natural jute fiber and S-glass fibers. *Mater. Today Proc.* **2021**, *46*, 8927–8933. [CrossRef]
19. Mishra, C.; Ranjan Deo, C.; Baskey, S. Influence of moisture absorption on mechanical properties of kenaf/glass reinforced polyester hybrid composite. *Mater. Today Proc.* **2021**, *38*, 2596–2600. [CrossRef]
20. Braga, R.A.; Magalhaes, P.A.A. Analysis of the mechanical and thermal properties of jute and glass fiber as reinforcement epoxy hybrid composites. *Mater. Sci. Eng. C* **2015**, *56*, 269–273. [CrossRef]
21. Kaddami, H.; Arrakhiz, F.-e.; Hafs, O.; Assimi, T.E.; Boulafrouh, L.; Ablouh, E.-H.; Mansori, M.; Banouni, H.; Bouzit, S.; Erchiqui, F.; et al. Implementation and Characterization of a Laminate Hybrid Composite Based on Palm Tree and Glass Fibers. *Polymers* **2021**, *13*, 3444. [CrossRef]
22. White, J.R. Polymer ageing: Physics, chemistry or engineering? Time to reflect. *Comptes Rendus Chim.* **2006**, *9*, 1396–1408. [CrossRef]
23. Chateauminois, A.; Vincent, L.; Chabert, B.; Soulier, J.P. Study of the interfacial degradation of a glass-epoxy composite during hygrothermal ageing using water diffusion measurements and dynamic mechanical thermal analysis. *Polymer* **1994**, *35*, 4766–4774. [CrossRef]
24. Apicella, A.; Nicolais, L.; Astarita, G.; Drioli, E. Effect of thermal history on water sorption, elastic properties and the glass transition of epoxy resins. *Polymer* **1979**, *20*, 1143–1148. [CrossRef]
25. Andreopoulos, A.G.; Tarantili, P.A. Water sorption characteristics of epoxy resin–UHMPE fibers composites. *J. Appl. Polym. Sci.* **1998**, *70*, 747–755. [CrossRef]

26. Marcovich, N.E.; Reboredo, M.M.; Aranguren, M.I. Moisture diffusion in polyester–woodflour composites. *Polymer* **1999**, *40*, 7313–7320. [CrossRef]
27. Scida, D.; Assarar, M.; Poilâne, C.; Ayad, R. Influence of hygrothermal ageing on the damage mechanisms of flax-fibre reinforced epoxy composite. *Compos. Part B Eng.* **2013**, *48*, 51–58. [CrossRef]
28. Pignatello, J.J.; Xing, B. Mechanisms of Slow Sorption of Organic Chemicals to Natural Particles. *Environ. Sci. Technol.* **1996**, *30*, 1–11. [CrossRef]
29. Kaboorani, A. Characterizing water sorption and diffusion properties of wood/plastic composites as a function of formulation design. *Constr. Build. Mater.* **2017**, *136*, 164–172. [CrossRef]
30. Bond, D.A.; Smith, P.A. Modeling the Transport of Low-Molecular-Weight Penetrants Within Polymer Matrix Composites. *Appl. Mech. Rev.* **2006**, *59*, 249–268. [CrossRef]
31. Arbelaiz, A.; Cantero, G.; Fernández, B.; Mondragon, I.; Gañán, P.; Kenny, J.M. Flax fiber surface modifications: Effects on fiber physico mechanical and flax/polypropylene interface properties. *Polym. Compos.* **2005**, *26*, 324–332. [CrossRef]
32. El Hachem, Z.; Célino, A.; Challita, G.; Moya, M.-J.; Fréour, S. Hygroscopic multi-scale behavior of polypropylene matrix reinforced with flax fibers. *Ind. Crops Prod.* **2019**, *140*, 111634. [CrossRef]
33. Dhakal, H.N.; Zhang, Z.Y.; Richardson, M.O.W. Effect of water absorption on the mechanical properties of hemp fibre reinforced unsaturated polyester composites. *Compos. Sci. Technol.* **2007**, *67*, 1674–1683. [CrossRef]
34. Le Duigou, A.; Davies, P.; Baley, C. Seawater ageing of flax/poly(lactic acid) biocomposites. *Polym. Degrad. Stab.* **2009**, *94*, 1151–1162. [CrossRef]
35. Kafodya, I.; Xian, G.; Li, H. Durability study of pultruded CFRP plates immersed in water and seawater under sustained bending: Water uptake and effects on the mechanical properties. *Compos. Part B Eng.* **2015**, *70*, 138–148. [CrossRef]
36. Ghabezi, P.; Harrison, N.M. Hygrothermal deterioration in carbon/epoxy and glass/epoxy composite laminates aged in marine-based environment (degradation mechanism, mechanical and physicochemical properties). *J. Mater. Sci.* **2022**, *57*, 4239–4254. [CrossRef]
37. Esleman, E.A.; Önal, G. Effect of saltwater on the mechanical properties of basalt/carbon/glass-epoxy hybrid composites. *J. Compos. Mater.* **2022**, *56*, 3783–3799. [CrossRef]
38. Hu, Y.; Lang, A.W.; Li, X.; Nutt, S.R. Hygrothermal aging effects on fatigue of glass fiber/polydicyclopentadiene composites. *Polym. Degrad. Stab.* **2014**, *110*, 464–472. [CrossRef]
39. Xian, G.; Li, H.; Su, X. Effects of immersion and sustained bending on water absorption and thermomechanical properties of ultraviolet cured glass fiber-reinforced acylate polymer composites. *J. Compos. Mater.* **2013**, *47*, 2275–2285. [CrossRef]
40. Hu, Y.; Li, X.; Lang, A.W.; Zhang, Y.; Nutt, S.R. Water immersion aging of polydicyclopentadiene resin and glass fiber composites. *Polym. Degrad. Stab.* **2016**, *124*, 35–42. [CrossRef]
41. Tripathy, C.; Ranjan Deo, C.; Kumar Das, S. Performance studies of polyester-based hybrid composites reinforced with palmyra-palm leaf stalk and glass fibers. *Mater. Today Proc.* **2021**, *38*, 2671–2674. [CrossRef]
42. Velmurugan, G.; Natrayan, L. Experimental investigations of moisture diffusion and mechanical properties of interply rearrangement of glass/Kevlar-based hybrid composites under cryogenic environment. *J. Mater. Res. Technol.* **2023**, *23*, 4513–4526. [CrossRef]
43. Kabir, M.M.; Wang, H.; Lau, K.T.; Cardona, F. Chemical treatments on plant-based natural fibre reinforced polymer composites: An overview. *Compos. Part B Eng.* **2012**, *43*, 2883–2892. [CrossRef]
44. Chow, C.P.L.; Xing, X.S.; Li, R.K.Y. Moisture absorption studies of sisal fibre reinforced polypropylene composites. *Compos. Sci. Technol.* **2007**, *67*, 306–313. [CrossRef]
45. Chen, H.; Miao, M.; Ding, X. Influence of moisture absorption on the interfacial strength of bamboo/vinyl ester composites. *Compos. Part A Appl. Sci. Manuf.* **2009**, *40*, 2013–2019. [CrossRef]
46. Airinei, A.; Asandulesa, M.; Stelescu, M.D.; Tudorachi, N.; Fifere, N.; Bele, A.; Musteata, V. Dielectric, Thermal and Water Absorption Properties of Some EPDM/Flax Fiber Composites. *Polymers* **2021**, *13*, 2555. [CrossRef] [PubMed]
47. Krapež Tomec, D.; Straže, A.; Haider, A.; Kariž, M. Hygromorphic Response Dynamics of 3D-Printed Wood-PLA Composite Bilayer Actuators. *Polymers* **2021**, *13*, 3209. [CrossRef]
48. Akil, H.M.; Cheng, L.W.; Mohd Ishak, Z.A.; Abu Bakar, A.; Abd Rahman, M.A. Water absorption study on pultruded jute fibre reinforced unsaturated polyester composites. *Compos. Sci. Technol.* **2009**, *69*, 1942–1948. [CrossRef]
49. Komai, K.; Minoshima, K.; Shiroshita, S. Hygrothermal degradation and fracture process of advanced fibre-reinforced plastics. *Mater. Sci. Eng. A* **1991**, *143*, 155–166. [CrossRef]
50. Kim, H.J.; Seo, D.W. Effect of water absorption fatigue on mechanical properties of sisal textile-reinforced composites. *Int. J. Fatigue* **2006**, *28*, 1307–1314. [CrossRef]
51. Rezgani, L.; Madani, K.; Feaugas, X.; Touzain, S.; Cohendoz, S.; Valette, J. Influence of water ingress onto the crack propagation rate in a AA2024-T3 plate repaired by a carbon/epoxy patch. *Aerosp. Sci. Technol.* **2016**, *55*, 359–365. [CrossRef]
52. Kabbej, M.; Guillard, V.; Angellier-Coussy, H.; Thoury-Monbrun, V.; Gontard, N.; Orgéas, L.; Du Roscoat, S.R.; Gaucel, S. From 3D real structure to 3D modelled structure: Modelling water vapor permeability in polypropylene/cellulose composites. *Polymer* **2023**, *269*, 125672. [CrossRef]
53. Razavi-Nouri, M.; Karami, M. Water sorption kinetics of acrylonitrile-butadiene rubber/poly(ethylene-co-vinyl acetate)/organoclay nanocomposites. *Polymer* **2018**, *154*, 101–110. [CrossRef]

54. You, M.; Wang, B.; Singh, P.; Meng, J. Water and salt transport properties of the cellulose triacetate/reduced graphene oxide nanocomposite membranes. *Polymer* **2020**, *210*, 122976. [CrossRef]
55. Siedlaczek, P.; Sinn, G.; Peter, P.; Wan-Wendner, R.; Lichtenegger, H.C. Characterization of moisture uptake and diffusion mechanisms in particle-filled composites. *Polymer* **2022**, *249*, 124799. [CrossRef]
56. Glaskova-Kuzmina, T.; Aniskevich, A.; Papanicolaou, G.; Portan, D.; Zotti, A.; Borriello, A.; Zarrelli, M. Hydrothermal Aging of an Epoxy Resin Filled with Carbon Nanofillers. *Polymers* **2020**, *12*, 1153. [CrossRef]
57. Ghabezi, P.; Harrison, N.M. Indentation characterization of glass/epoxy and carbon/epoxy composite samples aged in artificial salt water at elevated temperature. *Polym. Test.* **2022**, *110*, 107588. [CrossRef]
58. Wang, Z.; Zhao, X.-L.; Xian, G.; Wu, G.; Singh Raman, R.K.; Al-Saadi, S.; Haque, A. Long-term durability of basalt- and glass-fibre reinforced polymer (BFRP/GFRP) bars in seawater and sea sand concrete environment. *Constr. Build. Mater.* **2017**, *139*, 467–489. [CrossRef]
59. *ASTM D 570-98*; Standard Test Method for Water Absorption of Plastics. American Society for Testing Materials (ASTM): West Conshohocken, PA, USA, 2022.
60. Crank, J. *The Mathematics of Diffusion*, 2nd ed.; Oxford University Press: Oxford, UK, 1975.
61. Alamri, H.; Low, I.M. Mechanical properties and water absorption behaviour of recycled cellulose fibre reinforced epoxy composites. *Polym. Test.* **2012**, *31*, 620–628. [CrossRef]
62. Muñoz, E.; García-Manrique, J.A. Water Absorption Behaviour and Its Effect on the Mechanical Properties of Flax Fibre Reinforced Bioepoxy Composites. *Int. J. Polym. Sci.* **2015**, *2015*, 390275. [CrossRef]
63. Lin, Y.C.; Chen, X. Moisture sorption–desorption–resorption characteristics and its effect on the mechanical behavior of the epoxy system. *Polymer* **2005**, *46*, 11994–12003. [CrossRef]
64. Manaila, E.; Craciun, G.; Ighigeanu, D. Water Absorption Kinetics in Natural Rubber Composites Reinforced with Natural Fibers Processed by Electron Beam Irradiation. *Polymers* **2020**, *12*, 2437. [CrossRef] [PubMed]
65. Lee, G.W.; Lee, N.J.; Jang, J.; Lee, K.J.; Nam, J.D. Effects of surface modification on the resin-transfer moulding (RTM) of glass-fibre/unsaturated-polyester composites. *Compos. Sci. Technol.* **2002**, *62*, 9–16. [CrossRef]
66. Silva, R.V.; Aquino, E.M.F.; Rodrigues, L.P.S.; Barros, A.R.F. Curaua/Glass Hybrid Composite: The Effect of Water Aging on the Mechanical Properties. *J. Reinf. Plast. Compos.* **2009**, *28*, 1857–1868. [CrossRef]
67. Saidane, E.H.; Scida, D.; Assarar, M.; Sabhi, H.; Ayad, R. Hybridisation effect on diffusion kinetic and tensile mechanical behaviour of epoxy based flax–glass composites. *Compos. Part A Appl. Sci. Manuf.* **2016**, *87*, 153–160. [CrossRef]

Disclaimer/Publisher's Note: The statements, opinions and data contained in all publications are solely those of the individual author(s) and contributor(s) and not of MDPI and/or the editor(s). MDPI and/or the editor(s) disclaim responsibility for any injury to people or property resulting from any ideas, methods, instructions or products referred to in the content.

Article

Effect of Water Absorption and Stacking Sequences on the Tensile Properties and Damage Mechanisms of Hybrid Polyester/Glass/Jute Composites

Rudá Aranha [1,*], Mario A. Albuquerque Filho [2], Cícero de L. Santos [3], Tony Herbert F. de Andrade [4], Viviane M. Fonseca [5], Jose Luis Valin Rivera [1,*], Marco A. dos Santos [3], Antonio G. B. de Lima [3], Wanderley F. de Amorim, Jr. [3] and Laura H. de Carvalho [2]

1. Escuela de Ingeniería Mecánica, Pontifícia Universidad Católica de Valparaíso, Valparaíso 2340025, Chile
2. Post-Graduate Program in Materials Science and Engineering, Federal University of Campina Grande, Campina Grande 58429-900, Brazil; mario_alberto1910@hotmail.com (M.A.A.F.); heckerdecarvalho@yahoo.com.br (L.H.d.C.)
3. Mechanical Engineering Department, Federal University of Campina Grande, Campina Grande 58429-900, Brazil; cicero.santos@ufcg.edu.br (C.d.L.S.); santos.marco@ufcg.edu.br (M.A.d.S.); antonio.gilson@ufcg.edu.br (A.G.B.d.L.); wanderley.ferreira@professor.ufcg.edu.br (W.F.d.A.J.)
4. Petroleum Engineering Department, Federal University of Campina Grande, Campina Grande 58429-900, Brazil; tonyhebert@uaepetro.ufcg.edu.br
5. Textil Engineering Department, Federal University of Rio Grande do Norte, Natal 59078-970, Brazil; viviane.muniz@ufrn.br
* Correspondence: ruda.aranha@pucv.cl (R.A.); jose.valin@pucv.cl (J.L.V.R.); Tel.: +55-83996156252 (R.A.); +56-993447848 (J.L.V.R.)

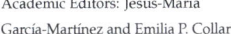

Citation: Aranha, R.; Filho, M.A.A.; Santos, C.d.L.; de Andrade, T.H.F.; Fonseca, V.M.; Rivera, J.L.V.; dos Santos, M.A.; de Lima, A.G.B.; de Amorim, W.F., Jr.; de Carvalho, L.H. Effect of Water Absorption and Stacking Sequences on the Tensile Properties and Damage Mechanisms of Hybrid Polyester/Glass/Jute Composites. *Polymers* **2024**, *16*, 925. https://doi.org/10.3390/polym16070925

Academic Editors: Jesús-María García-Martínez and Emilia P. Collar

Received: 30 January 2024
Revised: 15 March 2024
Accepted: 18 March 2024
Published: 28 March 2024

Copyright: © 2024 by the authors. Licensee MDPI, Basel, Switzerland. This article is an open access article distributed under the terms and conditions of the Creative Commons Attribution (CC BY) license (https:// creativecommons.org/licenses/by/ 4.0/).

Abstract: The aim of this work is to analyze the effect of water absorption on the mechanical properties and damage mechanisms of polyester/glass fiber/jute fiber hybrid composites obtained using the compression molding and vacuum-assisted resin transfer molding (VARTM) techniques with different stacking sequences. For this purpose, the mechanical behavior under tensile stress of the samples was evaluated before and after hygrothermal aging at different temperatures: TA, 50 °C, and 70 °C for a period of 696 h. The damage mechanism after the mechanical tests was evaluated using SEM analysis. The results showed a tendency for the mechanical properties of the composites to decrease with exposure to an aqueous ambient, regardless of the molding technique used to conform the composites. It was also observed that the stacking sequence had no significant influence on the dry composites. However, exposure to the aqueous ambient led to a reduction in mechanical properties, both for the molding technique and the stacking sequence. Damage such as delamination, fiber pull-out, fiber/matrix detachment, voids, and matrix removal were observed in the composites in the SEM analyses.

Keywords: hybrid composites; jute fiber; VARTM; water sorption; hybrid effect; tensile properties; water absorption; stacking sequence; temperature effect

1. Introduction

Materials undergo degradation due to external influences, leading to changes in their properties over time. Exposure to aggressive environments or factors such as light, humidity, and heat tends to alter the mechanical, chemical, and physical properties of materials [1]. Specifically for composite materials, the change in properties can occur not only through the degradation of the material components but also through the degradation of the interface formed by these components, also referred to as the matrix/reinforcement interface [2].

To assess the degradation of composites under real-service conditions, accelerated or natural aging tests are conducted, inducing physical and/or chemical degradation. One method to comprehend the material's behavior when exposed to humidity is through water sorption tests. In these tests, the material is immersed in an aqueous solution, sometimes with a controlled temperature, to expedite the degradation process.

Water absorption in composites is influenced by various factors, such as the chemical affinity and type of matrix, temperature, polarity, diffusivity, hydrogen bond formation, as well as the nature, volumetric fraction, orientation, porosity, and geometry of fibers or fabrics [3–5]. Another factor that has a significant influence on water absorption in composites is the voids present in the structure, which tend to increase water absorption. Water absorption, in turn, tends to compromise the mechanical properties of the material through complex mechanisms, including partial or total swelling of the matrix, plasticization of the matrix, material loss, and formation of cracks [6,7].

Matrix swelling can be detrimental to the fiber/matrix interface, promoting the delamination of fibers and matrix and thus reducing the mechanical properties of the material [5,8–10]. Similarly, plasticization can decrease the material's mechanical properties and alter the glass transition temperature [11]. However, there are reports where plasticization increases the fracture toughness of the material, delaying crack propagation through the material [12]. Hygrothermal aging primarily damages the matrix, with this initial damage triggering other mechanisms such as fiber/matrix debonding and fiber breakage as part of the deterioration induced by aging [5].

The type of fiber and its composition also affect water absorption, with some fibers absorbing more moisture than others. There is currently a trend to replace synthetic fibers with natural fibers as reinforcements in polymer composites to reduce environmental impact. Similarly, replacing petroleum-based plastic with starch-based film has been an alternative for producing green composites [13,14].

The effects of water absorption in compounds reinforced by natural fibers are intensified. This happens mainly due to the poor interface of these fibers with the polymeric matrix as to hydrophilic characteristics by nature, which means that they have a natural affinity to absorb water, either by capillarity or because of the percentage of cellulose present in the fibers. A common problem is that the hydrophilic nature of plant fibers makes them incompatible with most polymeric resins, which are generally hydrophobic [8,15,16]. This way, a decrease in mechanical properties and an increase in moisture content is expected with the volume fraction of the fiber. This behavior is generally associated with the poor fiber/matrix interface.

It is well-known that vegetable fibers have inferior mechanical properties compared to synthetic fibers [17–19]. Property losses have been observed in polymeric composites reinforced by plant fibers such as sisal [20], bamboo [21], flax [5], hemp [22,23], and jute [24]. The development of hybrid composites, where part of the fibers is synthetic and part is vegetable, is an alternative to overcome this issue [23–27]. Among the vegetable fibers, jute fibers excel in this application due to their widespread availability in numerous countries, cost-effectiveness, and possession of favorable mechanical properties [27,28]. However, the utilization of hybrid composites consisting exclusively of vegetable fibers or a combination of synthetic and vegetable fibers in high-performance applications should be conducted with prudence. Despite these factors, the use of vegetable fibers in hybrid composites can result in materials with suitable properties and be attractive in specific applications depending on project requirements [29–31].

Fibrous hybrid composites are materials that combine two or more distinct types of fibers to reinforce a particular matrix. They exhibit intermediate mechanical properties compared to the same composite reinforced by each fiber individually and offer various possibilities for fiber arrangements, including interlaminar, intralaminar, and mixed configurations [32–36]. Several factors play a role in influencing the properties of these hybrid composites, such as the mechanical properties and characteristics of the fibers, the length of the different fiber types, and the quality of the interfacial bonding between the fibers and the matrix [36].

Synergistic effects can be observed when the properties of hybrid composites are higher or lower than expected. The synergistic effect, also known as the hybrid effect, refers to the impact on the stress–strain response of mechanical loading in hybrid composites compared to non-hybrid composites [37]. The initial studies on the hybrid effect were reported between the early 1970s and 1980s [37–42]. It is defined in two different ways: one is based on the increase in the failure stress of the hybrid composite compared to non-hybrid fibers of low elongation [31,36], and the other is based on the rule of mixtures, used as a parameter to evaluate the deviation in mechanical behavior between hybrid and non-hybrid composites. However, in some cases, certain properties remain constant regardless of the amount of fibers added to the composite [35].

Although it is widely used, some caution must be exercised regarding the rule of mixtures. The rule of mixtures is not linear for all properties and, therefore, not suitable for estimating all mechanical properties. For instance, it is not recommended to use it for estimating the flexural strength of a composite [36]. The synergistic effect is also reported for hybrid composites with two or more resin systems or additional constituents, such as inserts, nanoparticles, and additives [43–45]. Another parameter that directly influences the mechanical properties of hybrid composites is the fiber stacking sequence. Moreover, there are situations where a property shows a negative hybrid effect for one characteristic while displaying a positive hybrid effect for another property [36]. For example, in carbon/glass hybrid composites, the ultimate stress at break displays a negative hybrid effect due to the positive hybrid effect on maximum strain [37].

The properties of jute composites can be enhanced by incorporating glass fiber as outer layers in the laminate, whereas jute should be used on the inside [46]. Several studies have explored the impact of incorporating glass fibers into composites reinforced with natural fibers [47–56]. In all cases, the consensus was that the inclusion of glass fibers led to a notable enhancement in the mechanical properties of these composites. However, a reduction in mechanical properties was correlated with moisture absorption [51]. Effects such as matrix cracks, delamination, fractures along the interface, resin particle loss, and fiber misalignment were observed due to the impact of moisture diffusion in hybrid composites [22]. When the glass fiber is used in the outer layer, lower water absorption is observed in comparison to other hybrid laminates with plant fabric on the outer layer [57].

The effect of water moisture on the mechanical performance of fiber-reinforced polymer matrix composites has been extensively studied in the literature [10,12,22,51,57–60]. However, there is still a need for further investigation regarding the sorption effect on the mechanical behavior of hybrid polyester/glass/jute composites, especially with different fiber stacking configurations, to fully understand its influence on material properties.

This study aims to assess how water absorption affects the tensile mechanical properties of composites at various temperatures. We investigate composites manufactured using different molding techniques and reinforcement stacking sequences. Two molding methods, compression molding and vacuum-assisted resin transfer molding (VARTM), were employed, producing laminates with five distinct fiber stacking sequences. Damage mechanisms were analyzed using SEM.

A previous study [61] conducted experimental and theoretical research on water absorption in these composites. The findings indicated that hybridization improved mechanical properties, with hybrid compounds showing intermediate results between glass fiber and jute fiber composites.

The novelty of our work lies in its unique approach. The existing literature lacks discussions on combinations and analyses used in evaluating hybrid compounds reinforced with plant and synthetic fibers after water immersion at varying temperatures, with diverse stacking sequences, and manufactured using two different processes. Hence, our study provides deeper insights into the mechanical behavior and damage mechanisms of dry and wet composite materials.

2. Materials and Methods

2.1. Materials

This study utilized orthophthalic unsaturated polyester resin 10316-10 produced for Reichhold, from Mogi das Cruzes, SP, Brazil (Table 1), catalyzed with MEKP BUTANOX M-50 supplied by IBEX Químicos e Compósitos Ltd.a., Recife, PE, Brazil. The reinforcements used were type E glass fiber fabrics with a gramature of 330 g/m^2, as supplied by Redelease Ltd., Sorocaba, Brazil, and jute plain weave fabric with a gramature of 330 g/m^2, manufactured by Cia. Têxtil Castanhal from Castanhal, PA, Brazil.

Table 1. Characteristics of resin—Reichhold guideline.

Characteristics	Values
Brookfield viscosity at 25 °C sp3: 60 rpm (CP)	250–350
Thixotropy index	1.30–2.10
Solid content—Reichhold method (%)	55–63
Density at 25 °C (g/cm^3)	1.07–1.11
Acidity index (mgKOH/g)	30 maximum
Exothermic curve at 25 °C	
- Gel time (min)	5–7
- Single interval (min)	8–14
- Maximum temperature (°C)	140–180
Post-cure	60 °C

2.2. Manufacturing of Composites

The composites were manufactured using two different methods: compression molding (Figure 1a) and VARTM (Figure 1b).

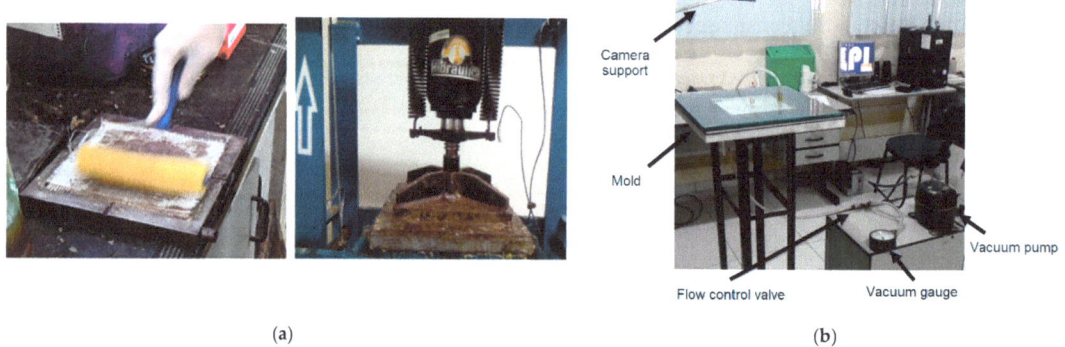

Figure 1. Manufacturing of composite plates: (**a**) compression molding; (**b**) VARTM.

Regardless of the method used for fabrication, each composite is composed of four layers of reinforcement, utilizing either glass fabric (G) or jute plain weave fabric (J) arranged in different stacking sequences (Figure 2): GGGG; JJJJ; GJGJ; JGGJ; GJJG. A comparison of fabrication methods (compression or VARTM) can be illustrated, for instance, by comparing GJJG-C and GJJG-R, where "C" denotes compression molding, and "R" denotes the VARTM process.

The laminates were manufactured with dimensions of 200 mm × 180 mm, both through compression molding and VARTM. In compression molding, the fabric layers were manually positioned in the metal mold, and lamination was carried out using a foam roller (Figure 1a). Compression molding was performed using hydraulic press produced for Marconi Equipamentos para Laboratórios Ltd.a, Piracicaba, Brasil, at room temperature with 9 Ton/24 h. In VARTM fabrication, the plates were produced in a mold with a glass base and vacuum bag (Figure 1b), employing a transverse flow front, two resin entry points, one exit point with a $\frac{1}{4}''$ diameter, and a vacuum pressure of −0.3 bar. Post-curing of the compression molding and VARTM laminates was conducted in an air circulation oven at a temperature of 60 °C.

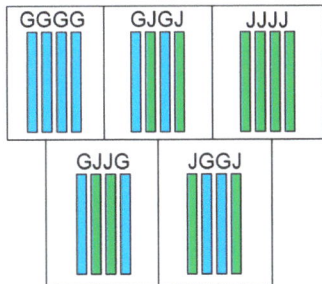

Figure 2. Stacking sequences; blue color represents glass fibers and green jute fibers.

After fabrication, the plates were weighed, and the theoretical method [14] was employed to determine the volumetric fractions of the fibers (Equation (1)).

$$V_f = \frac{w_j/\rho_j + w_g/\rho_g}{w_j/\rho_j + w_g/\rho_g + w_m/\rho_m}, \quad (1)$$

where the V_f is the volumetric fraction of the fibers of the laminate; w_j, w_g, and w_m are the masses of the jute fibers, glass fibers, and matrix, respectively; and ρ_j, ρ_g, and ρ_m are the densities of the jute fiber, glass fiber and matrix, respectively. Thus, the percentage of fiber volume in the hybrid composites manufactured by compression molding was around 31%, while for glass fiber it was 39%, and for jute fiber it was 37%. With VARTM molding technique, the fiber content was 37%.

The laminates were cut on a CNC mill. The specimens obtained had dimensions of 100 mm × 13 mm × 3.5 mm for the tensile tests, adapting the recommendations of the ASTM D3039 standard [62].

In a previous work, both experimental and theoretical water sorption tests were conducted for these composites, determining the required exposure time for them to reach saturation [61]. These absorption curves served as a foundation for establishing the immersion duration in water for the tensile samples analyzed in this study.

2.3. Tensile Tests

The tensile tests were carried out in the MTS universal testing machine, model 810 with a 100 KN load cell, following the recommendations of the ASTM D3039 standard [62]. The tests were carried out at room temperature, with a displacement rate of 1 mm/min. In order to determine the influence of water absorption on the mechanical properties of the composites, the test was carried out with samples that were immersed in water for 696 h (saturation time) and dry samples, i.e., those not immersed in water. Water immersion temperatures were as follows: room temperature, 50 °C, and 70 °C. Tensile tests were conducted on samples immersed in water for 696 h at the three specified temperatures, as well as on samples exposed to 70 °C for times that led to an estimated 3% absorption (1.5% for glass fiber composites). The average results of five samples are reported for each set of samples manufactured.

2.4. Hybrid Effect Methodology

For the calculation of the hybrid effect, two methodologies were employed: one based on the rule of mixtures and another using a theoretical value for the strain at break of hybrid composites [63,64]. In the first methodology, the rule of mixtures was employed to establish a theoretical value for the stress at break of hybrid composites. Equation (2) is used when the volumetric fraction of the composite with higher strength (V_{fH}) is lower than the critical volumetric fraction (V_{fcrit}) [63]. Equation (3) is applied when the volumetric fraction of the composite with higher strength (V_{fH}) is greater than the critical volumetric fraction (V_{fcrit}) [63]. The critical volumetric fraction is a theoretical value obtained by equating Equations (2) and (3).

$$\sigma_{htheoretical} = \left(1 - V_{fm}\right)\left(\sigma_H V_{fLE} + \varepsilon_{LE} E_H V_{fH}\right); V_{fH} \leq V_{fcrit}, \quad (2)$$

$$\sigma_{htheoretical} = \left(1 - V_{fm}\right)\left(\sigma_H V_{fH}\right); V_{fH} \geq V_{fcrit}, \quad (3)$$

where V_{fm} is the volume fraction of the matrix, V_{fLE} is the relative volume fraction of fibers with lower strength, and E_H is the elastic modulus of the composite with higher strength. The calculation of the hybridization effect for stress is determined by taking the difference between the experimental stress value and the theoretical value (Equation (4)).

$$\lambda_{stress} = \sigma_{hybrid} - \sigma_{theoretical}, \quad (4)$$

The hybrid effect can also be calculated from the theoretical value for the strain at break of hybrid composites (Equation (5)), and similar to stress, the calculation of the hybridization effect for strain is also performed by taking the difference between the experimental values for strain in hybrid composites and the theoretical value (Equation (6)) [63].

$$\varepsilon_{htheoretical} = \varepsilon_{LE} V_{fLE} + \varepsilon_H V_{fH}, \quad (5)$$

$$\lambda_{strain} = \varepsilon_{hybrid} - \varepsilon_{htheoretical}, \quad (6)$$

2.5. Scanning Eletron Microscope (SEM)

The fracture surfaces of tensile test samples were used to investigate sample morphology and to verify the damage mechanisms in the fracture region. The equipment used was a scanning electron microscope (SEM) Vega 3 microscope produced for Tescan, from Brno, Czech Republic. The fractured portions of the samples were cut and gold-coated uniformly over the surface for examination. The accelerating voltage used in this work was 20 kV. Only one sample each composition was tested.

3. Results and Discussion
3.1. Volumetric Fraction of Fibers

Table 2 presents the total and relative values of the weight and volumetric fractions of the fiber composites. For the theoretical calculation and results, fiber densities of 1.5 g/cm³ (jute) and 2.54 g/cm³ (glass) were utilized, as shown in Equation (1).

Table 2. Total and relative fiber weight and volume fraction of manufactured composites.

Composites	Glass Fiber Weight Fraction (%)	Jute Fiber Weight Fraction (%)	Total Fiber Weight Fraction of Composites (%)	Fiber Volume Fraction of Composites (%)	Total Fiber Volume Fraction of Jute (%)
GGGG-C [1]	59.70 ± 3.27	0	59.70 ± 3.27	39.35 ± 3.21	0
JJJJ-C	0	44.10 ± 2.10	44.10 ± 2.10	36.87 ± 1.97	36.87 ± 1.97
GJGJ-C	15.23 ± 0.43	26.10 ± 0.74	41.33 ± 1.17	30.68 ± 1.02	24.77 ± 0.90
GJJG-C	15.23 ± 1.06	26.11 ± 1.82	41.34 ± 2.89	30.72 ± 2.55	24.81 ± 2.24
GJJG-R [1]	17.77 ± 1.77	30.46 ± 3.04	48.23 ± 4.82	36.99 ± 4.48	30.42 ± 4.06
JGGJ-C	15.28 ± 1.16	26.19 ± 2.00	41.47 ± 3.16	30.83 ± 2.76	24.91 ± 2.42
JGGJ-R	18.15 ± 1.18	31.12 ± 2.02	49.28 ± 3.20	37.94 ± 3.00	31.27 ± 2.73

[1] C and R indicate the manufacturing method: compression molding (C) and VARTM (R).

The volume fraction of fibers in the composites ranged from approximately 31% to 39%, with the hybrid composites manufactured by compression molding showing the lowest values. Hybrid composites fabricated using VARTM had an average volumetric fiber fraction of 37%. The difference in results between the manufacturing methods was anticipated, as VARTM promotes better fiber compaction during composite fabrication due to the use of a vacuum, in contrast to compression molding. The use of a vacuum, along with the resin flow front during the manufacturing process, enhances the effective removal of air, consequently decreasing the void content in the composites.

3.2. Tensile Tests Results

Figure 3 and Tables 3 and 4 display the tensile test results of the samples for all analyzed conditions.

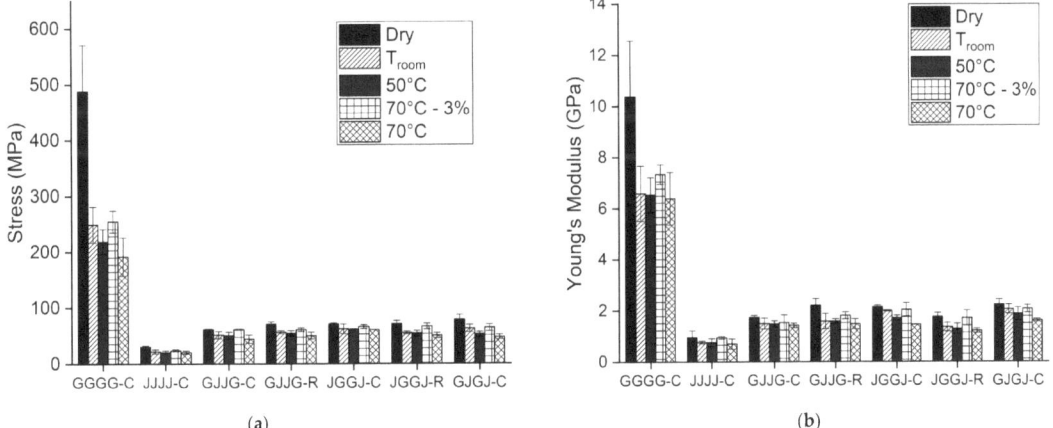

Figure 3. Mechanical properties of the composites under dry conditions and subjected to water sorption at different temperature conditions: (**a**) Stress at break; (**b**) Elastic modulus.

Table 3. Fracture stress of composites for all conditions.

Composites	Fracture Stress (MPa)				
	Dry	Room	50 °C	70 °C	70 °C–3%
GGGG-C	488.22 ± 82.21	249.29 ± 32.28	219.01 ± 22.36	191.78 ± 34.61	254.60 ± 19.43
JJJJ-C	30.36 ± 1.87	21.89 ± 3.67	20.65 ± 3.28	20.53 ± 2.74	23.83 ± 1.66
GJJG-C	61.39 ± 0.88	51.61 ± 6.46	50.78 ± 6.29	44.38 ± 7.44	61.23 ± 1.21
GJJG-R	70.72 ± 0.53	56.54 ± 2.67	54.25 ± 4.80	50.41 ± 6.20	60.92 ± 3.44
JGGJ-C	70.72 ± 2.04	62.23 ± 8.47	61.14 ± 0.72	60.70 ± 0.18	65.97 ± 3.39
JGGJ-R	71.09 ± 5.73	55.34 ± 2.64	54.15 ± 4.84	50.49 ± 4.92	66.87 ± 5.23
GJGJ-C	78.69 ± 8.76	62.90 ± 6.67	52.43 ± 4.37	47.38 ± 4.26	64.20 ± 6.03

Figure 3 shows that the stress and elastic modulus results of the composites meet the expected outcomes for all analyzed conditions: dry samples, water immersion at room temperature, 50 °C, 70 °C for 696 h, and 70 °C with 3% absorption. The fiberglass composites exhibited the highest mechanical strength values, jute fiber composites displayed the lowest values, and the hybrid composites demonstrated intermediate values between fiberglass and jute fiber composites.

Table 4. Elastic modulus of composites for all conditions.

Composites	Elastic Modulus (GPa)				
	Dry	Room	50 °C	70 °C	70 °C–3%
GGGG-C	10.38 ± 2.18	6.59 ± 1.08	6.53 ± 0.68	6.39 ± 1.03	7.33 ± 0.38
JJJJ-C	0.95 ± 0.26	0.76 ± 0.05	0.76 ± 0.15	0.70 ± 0.20	0.94 ± 0.05
GJJG-C	1.73 ± 0.08	1.49 ± 0.21	1.48 ± 0.12	1.42 ± 0.09	1.53 ± 0.30
GJJG-R	2.21 ± 0.25	1.58 ± 0.30	1.58 ± 0.09	1.47 ± 0.20	1.80 ± 0.14
JGGJ-C	2.14 ± 0.07	1.99 ± 0.03	1.71 ± 0.10	1.45 ± 0.01	2.03 ± 0.26
JGGJ-R	1.75 ± 0.16	1.35 ± 0.17	1.28 ± 0.21	1.21 ± 0.09	1.70 ± 0.30
GJGJ-C	2.22 ± 0.21	2.05 ± 0.18	1.86 ± 0.25	1.61 ± 0.05	2.06 ± 0.15

It was further noted that, despite different stacking sequences, the hybrid composites showed similar results for stress and modulus under all conditions and that the mechanical behavior of the hybrid composites was closer to that of jute fiber composites than fiberglass composites for all conditions. This was attributed to the higher volumetric fraction of jute fiber compared to fiberglass in the hybrid composites. The addition of glass fibers to jute fiber composites resulted in higher stress and modulus values, but the increase was marginal compared to fiberglass composites. The properties of hybrid composites depend on the relative volumetric fraction of each reinforcement, meaning that a higher relative volumetric fraction of a particular fiber will make the hybrid composite properties closer to those reinforced exclusively by that type of fiber [29,31].

Analyzing Table 2, it was found that the volumetric fraction of jute fibers was significantly higher than that of glass fibers in all hybrid composites. It is important to consider that the low adhesion between jute and glass fibers and the matrix created a low-quality interface, contributing to the low stress and modulus values exhibited by the hybrid composites. When moisture interacts with the fiber, it primarily penetrates through the cross-sectional area. This interaction between the hydrophilic fiber and the hydrophobic matrix causes the fiber to swell within the matrix. As a result, the bonding at the interface weakens, leading to dimensional instability, matrix cracking, and reduced mechanical properties of the composites [16].

Few differences were observed considering the manufacturing method, regardless of the test conditions (values were very close due to the identical composition of the composites; variables such as the stacking sequence and manufacturing method did not have a significant impact on the final results).

For the hybrid composites, statistical tests found no evidence to reject the null hypothesis, meaning that all hybrid composites showed equivalent results. This suggests that, although there are differences in water absorption at low rates for different stacking sequences [61], these differences do not reflect in mechanical behavior at saturation. Additionally, a reduction in stress at break values was observed for all composites exposed to moisture compared to dry composites. The more severe the conditions the composites were subjected to, the lower the observed stresses at break, indicating that the combination of moisture and temperature was highly detrimental to the materials. The higher the exposure temperature, the lower the observed stress at break, making it a determining factor in the mechanical behavior of composites, with 70 °C being the most severe condition adopted.

The choice of a fixed value for the amount of absorbed moisture by the samples was made to assess the effect of sorption on the mechanical properties of the composites, with equal water absorption for all samples over a different immersion time than the saturation time. The most severe condition (70 °C) was chosen, and the selection of the approximately 3% absorption value for the samples was based on the analysis of sorption graphs [61]. For fiberglass composites, tests were conducted with an absorption value of approximately 1.5%, as these composites reached saturation with about 3% water absorption.

For the maximum absorption condition of 3%, a considerable reduction in tensile strength was observed compared to the results of tests on dry composites, even for a shorter exposure period, highlighting the detrimental effect of moisture on the properties of this type of composite.

Dry composites exhibited better mechanical properties, followed by composites immersed at 70 °C with 3% absorption. This was followed by composites immersed in water for 696 h at room temperature, 50 °C, and 70 °C, respectively. These results were expected because higher moisture sorption leads to lower mechanical properties. The sorption kinetics increase with temperature, but the sorption level reached by the composites at the saturation time is similar for different conditions. Thus, after a certain time, composites immersed in water at room temperature will reach sorption values similar to those immersed at higher temperatures. Therefore, the higher the controlled temperature during immersion, the more rapidly the material will reach equilibrium sorption.

Fiberglass composites showed a trend of reduced properties with increasing temperature during water sorption tests until saturation. Only the fiberglass composite immersed at 70 °C with 1.5% absorption showed no variation in properties, considering standard deviations. Jute fiber composites maintained the trend of mechanical property loss with water sorption for all conditions. This behavior can be attributed to the degradation of the polyester matrix when subjected to water immersion. Thus, it was possible to observe that, except for these specific conditions, the temperature factor did not show significant variation in material stress.

Through the fiber stacking sequence, it was observed that there was a tendency for an increase in mechanical properties when fiberglass fibers were positioned in the center of the composite. Placing fiberglass fibers in the center of the composite improved the quality of the interface between the resin and these fibers. The same behavior was observed for the modulus values of all materials.

Few differences were observed between composites immersed in water for 696 h because, at this immersion time, the composites had already reached equilibrium sorption for all immersion temperatures. Larger differences were observed between dry composites compared to the results of tests under other conditions, highlighting the effect of moisture on mechanical properties.

Considering all the studied conditions, it was possible to conclude that, although some differences between the types of hybrid composites were observed under certain conditions, overall, the various fiber arrangements in the composites had little influence on their mechanical properties.

3.3. Hybrid Effect

When the strength values are higher than the theoretical values, this indicates that the composite in question exhibited a positive hybrid effect, just as strength values lower than the theoretical values signify a negative hybrid effect. The hybrid effect of the composites is depicted in Figures 4–8 for the dry (Figure 4), as well as room-temperature (Figure 5), 50 °C (Figure 6), 70 °C (Figure 7), and 70 °C–3% (Figure 8) conditions.

In Figure 4a, a positive hybridization effect for stress at break was observed for all composites, where the composite GJJG-C showed the lowest value (+17.36%), and the composite GJGJ-C exhibited the highest value (+50.43%). The other composites presented intermediate values: GJJG-R (+35.19%), JGGJ-C (+35.19%), and JGGJ-R (+35.90%). In Figure 3b, the hybridization effect for strain at break is negative for the composites GJJG-R (−6.51%) and JGGJ-C (−14.96%), both showing values below the expected value for these composites. The composites GJJG-C (+6.53%) and JGGJ-R (+22.24%) showed a positive hybridization effect for strain at break, while the strain at break of the composite GJGJ-C was equivalent to the theoretical deformation, showing neither a positive nor negative hybridization effect.

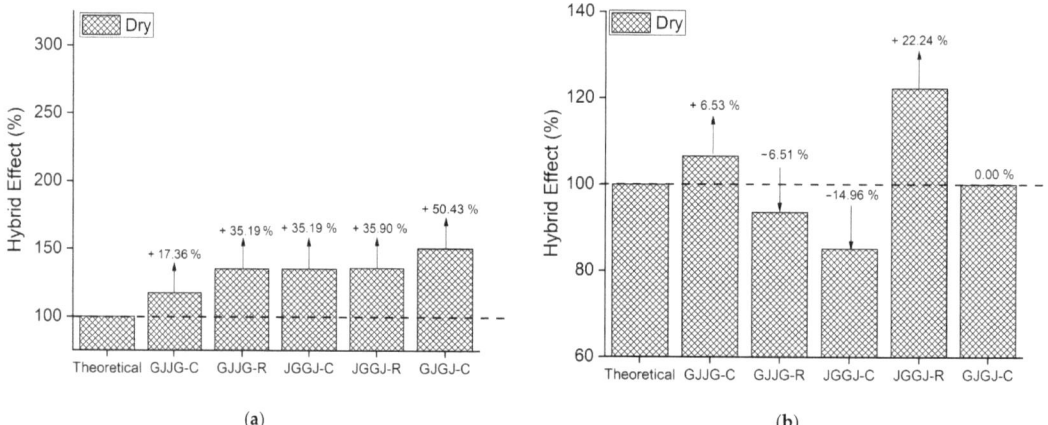

Figure 4. Hybrid effect of dry composites: (**a**) stress; (**b**) strain.

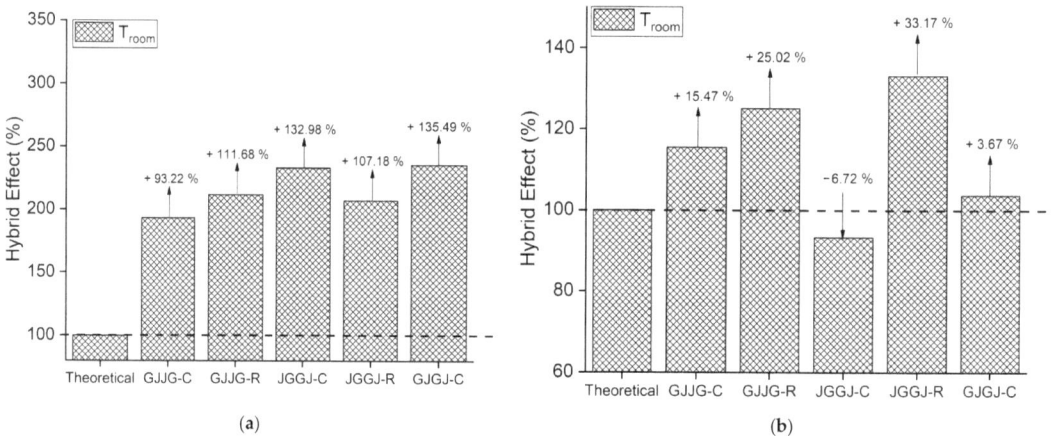

Figure 5. Hybrid effect of dry composites at room temperature: (**a**) stress at break; (**b**) strain at break.

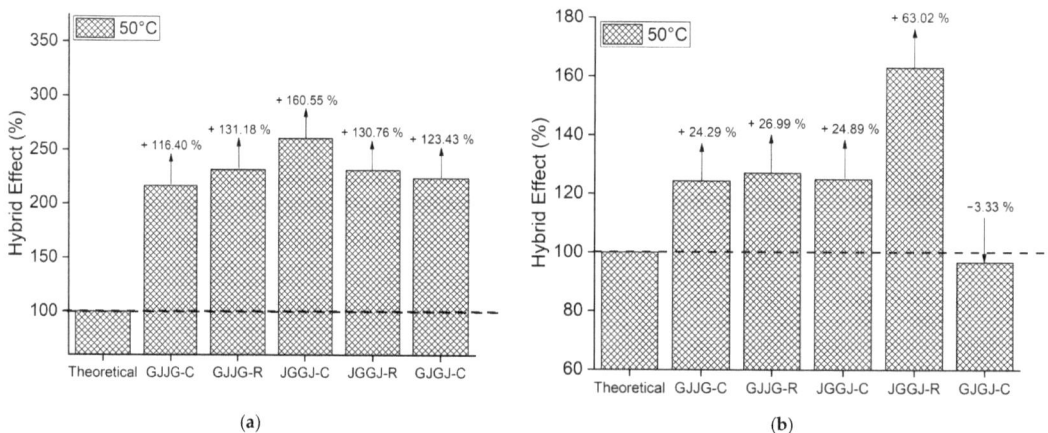

Figure 6. Hybrid effect of wet composites at 50 °C: (**a**) stress at break; (**b**) strain at break.

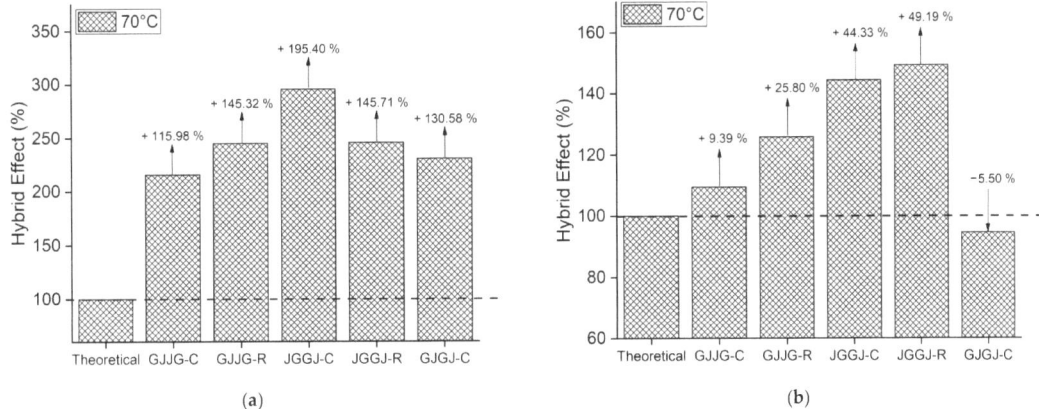

Figure 7. Hybrid effect of wet composites at 70 °C: (**a**) stress at break; (**b**) strain at break.

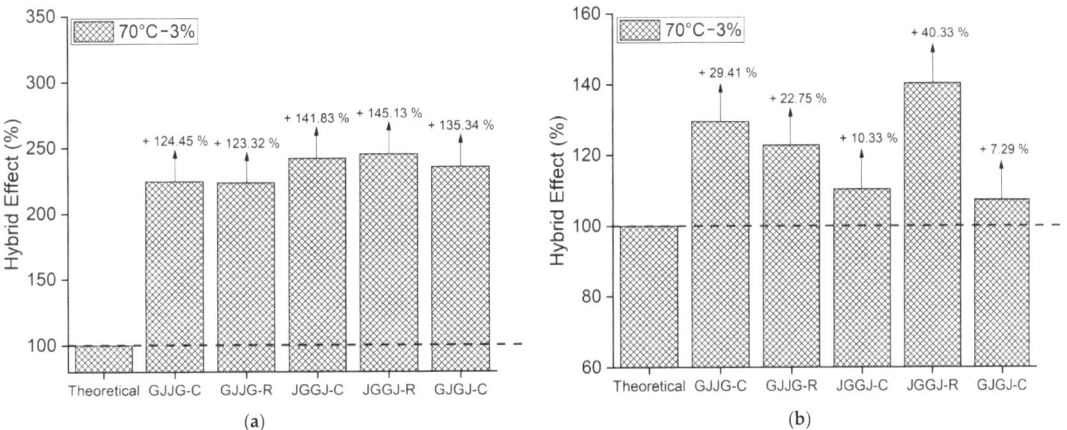

Figure 8. Hybrid effect of wet composites at 70 °C with 3% water absorption: (**a**) stress at break; (**b**) strain at break.

The variation in the hybridization effect in Figure 3b may be associated with the processing of the JGGJ composites, as the JGGJ-C composite exhibited a negative hybridization effect, while the JGGJ-R composite showed a positive hybridization effect. One consideration is that when fibers with lower strength have a higher volumetric fraction than fibers with higher strength, hybrid composites fail when the tensile strain reaches a value close to the failure strain of composites reinforced with the lower-strength fiber [63] (in our case, jute). However, if the higher-strength fiber has a greater volumetric fraction, the lower-strength fibers fail at the beginning, but the hybrid composites still maintain their integrity until the failure of the higher-strength fibers due to their greater failure strain.

Upon analyzing Figures 5a, 6a, 7a and 8a, it is evident that all composites exhibit a positive hybrid effect for stress during exposure to room temperature, 50 °C, and 70 °C for 696 h, as well as 70 °C–3%. The values for the hybridization effect observed for stress were considerably higher than those observed in dry composites. The reason for this difference in values occurred due to the presence of moisture and/or temperature, causing a significant reduction in the stress of the fiberglass composites compared to the dry composites, which were used as a reference to calculate the theoretical value. Equation (6) used for calculating the theoretical stress takes into account the stress value of the composite with higher strength (fiberglass composite). In Figure 3, it can be observed that fiberglass composites exhibited much lower average stress when subjected to water sorption/temperature, directly affecting the values obtained for the hybridization effect studied here.

As for strain (Figures 5b, 6b, 7b and 8b), the hybridization effect presented results that were similar to dry composites, although a slight increase in values was observed for all degradation conditions. An increase in the value of the hybridization effect was expected since Equation (5) uses the strain values of fiberglass composites and jute fiber composites to determine the theoretical strain of hybrid composites through the rule of mixtures. As there was a decrease in the strain values of fiberglass composites and jute fiber composites exposed to moisture compared to dry composites, these increases in the hybridization effect are justified. An increase in the strain of composites exposed to moisture is also expected due to the plasticization of the matrix caused by water absorption.

3.4. Damage Mechanism of Dry and Wet Composites

SEM images of the fracture surface of the composites were obtained for both dry and water-immersed samples at 70 °C for the hybrid composites. Figure 9 displays the SEM images of the fracture surface of the dry composites.

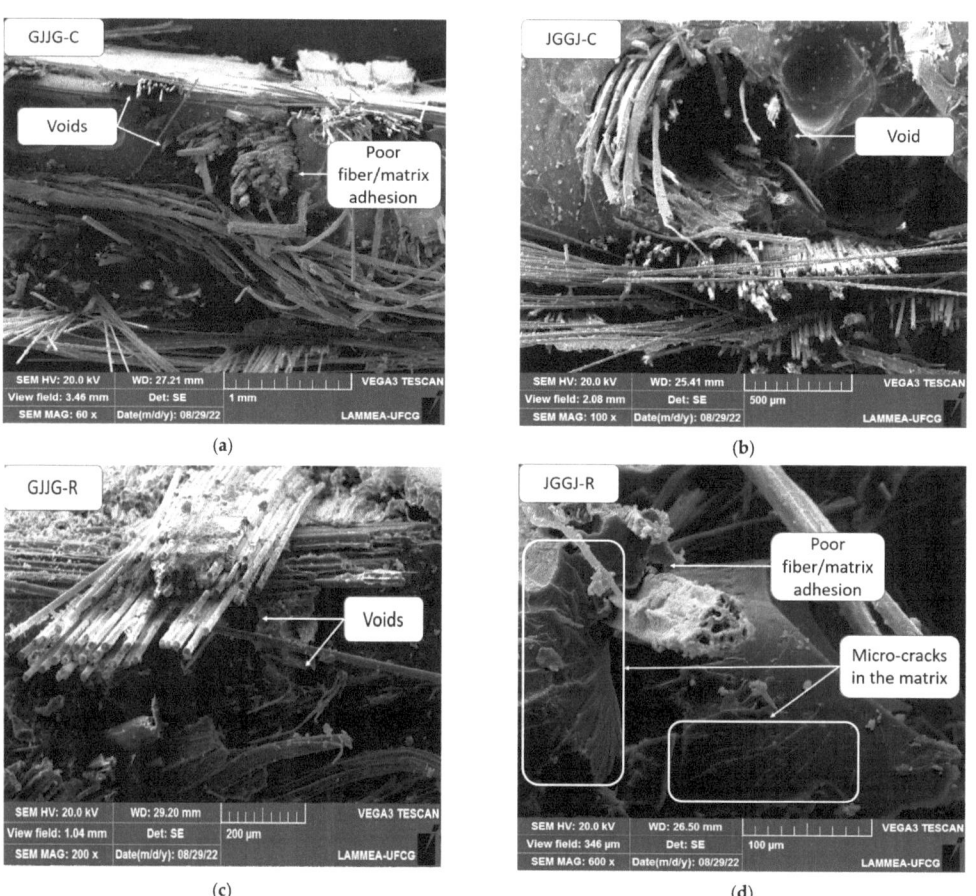

Figure 9. SEM images of dry hybrid composites with different stacking sequences: (**a**) composite GJJG-C; (**b**) JGGJ-C; (**c**) GJJG-R; (**d**) JGGJ-R.

Figure 9 reveals various failure mechanisms in the composites subjected to tensile testing. Fiber breakage, matrix fracture, fiber debonding, and fiber pull-out can be observed after the tensile test in all composites. It was also possible to clearly identify the different fibers and the matrix in the composites, and even after fracture, the glass fibers showed a certain regularity of arrangement, unlike the jute fibers, which appeared more irregular (Figure 9b,c). The difference in fiber arrangement within the composite has a direct influence on their mechanical behavior [65].

In Figure 9a–c, it is possible to identify voids in the composites, as well as fractures of both longitudinal glass and jute fibers along the length of the specimens. These voids were caused by fiber pull-out during tensile tests and were observed in all hybrid composites. In Figure 9b, complete removal of jute fiber bundles can be seen. In Figure 9a,d, debonding between jute fibers and the polymeric matrix is also observed, along with poor wetting of the fibers by the resin, resulting in a low-quality interface due to little or no adhesion between the phases. Fiber pull-out and debonding between the fibers and the matrix are the main factors responsible for the reduction in tensile and elastic modulus values [2,25,51,52,66]. One of the functions of a composite matrix is to transfer the load to the fibers through interfacial shear stress, so the fracture behavior also depends on the interfacial strength [25].

In Figure 9d, micro-cracks in the form of streaks that formed in the polymeric matrix can also be observed. One of the main failure mechanisms observed in hybrid composites was caused by the propagation of micro-cracks in the matrix [5]. The cracks propagate easily through the matrix, indicating that little resistance is offered, as evidenced by the poor interfacial bonding observed from the fracture [65]. In hybrid composites, when the load is applied, cracks in the matrix occur before the final failure [67].

Figure 10 shows that, in addition to the jute fibers, the glass fibers also exhibited poor adhesion with the matrix, where it was possible to identify the glass fibers, jute fibers, and the matrix phase. Voids caused by the pull-out of glass fibers at that location were observed. In Figure 10a, a highlighted region is magnified, as illustrated in Figure 10b, where it is evident that even the interface between the glass fibers and the resin shows low quality, with regions where the fibers were pulled out arranged in the longitudinal direction of the sample, and poor wetting of the glass fibers by the resin also observed.

Figure 10. Low adhesion of the glass fibers in the dry composites: (**a**) 150× zoom; (**b**) 500× zoom.

Figure 11 illustrates SEM images of the fracture surface of the composites subjected to water sorption at 70 °C. It can be observed that the issues in the dry composites related to fiber pull-out, voids, fiber fracture, and low adhesion between fibers and the matrix were also observed in these composites. In addition to the aforementioned problems, removal and fragmentation of the matrix were also observed due to its degradation caused by moisture (Figure 11a–c). Matrix degradation on the fracture surface was also observed (Figure 11d).

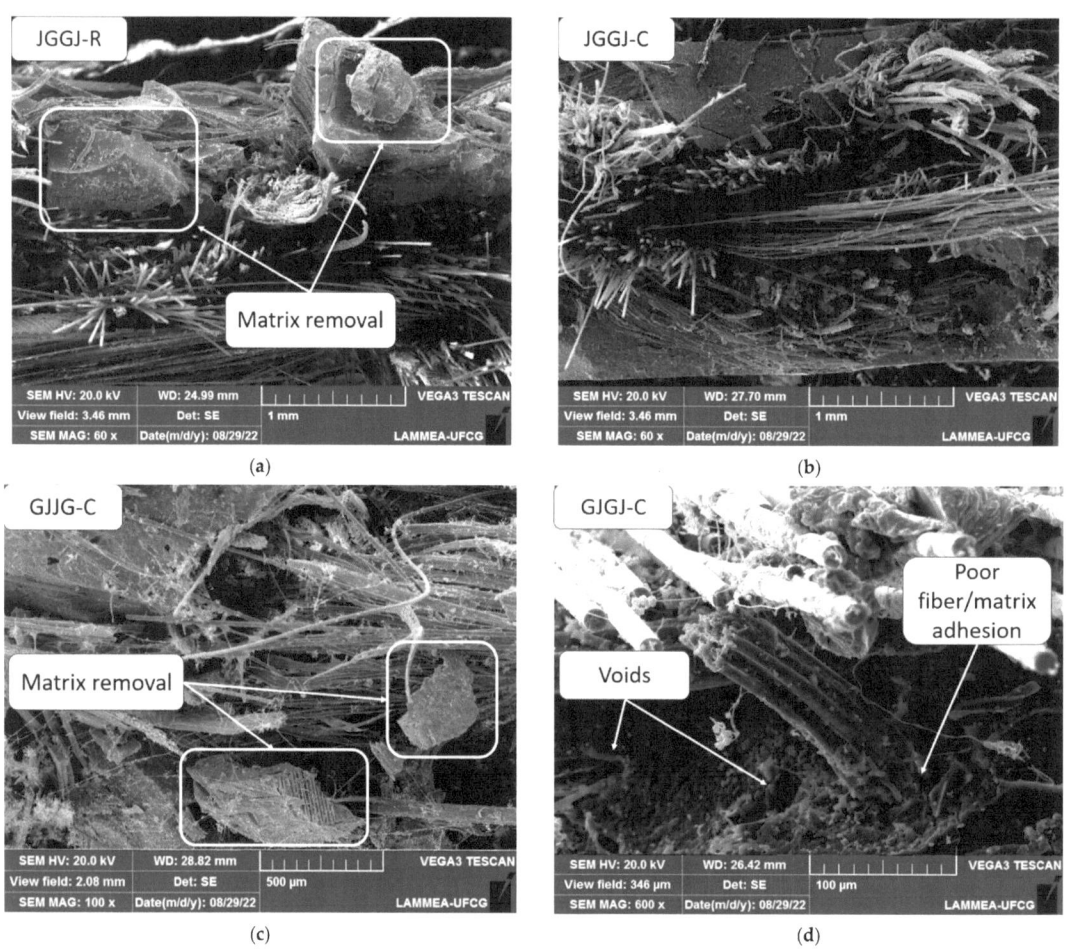

Figure 11. SEM images of wet hybrid composites with different stacking sequences at 70 °C: (**a**) composite JGGJ-R; (**b**) JGGJ-C; (**c**) GJJG-C; (**d**) GJJG-R.

In Figure 11c, the degradation of jute fibers can be further observed, as if they had "unraveled", generating a tangle of jute fibers with glass fibers and fragmented matrix particles. Based on this observation, Figure 12 was generated, showing a microscopy image of each hybrid composite sample that was exposed to moisture to assess whether this result was repeated for all composites.

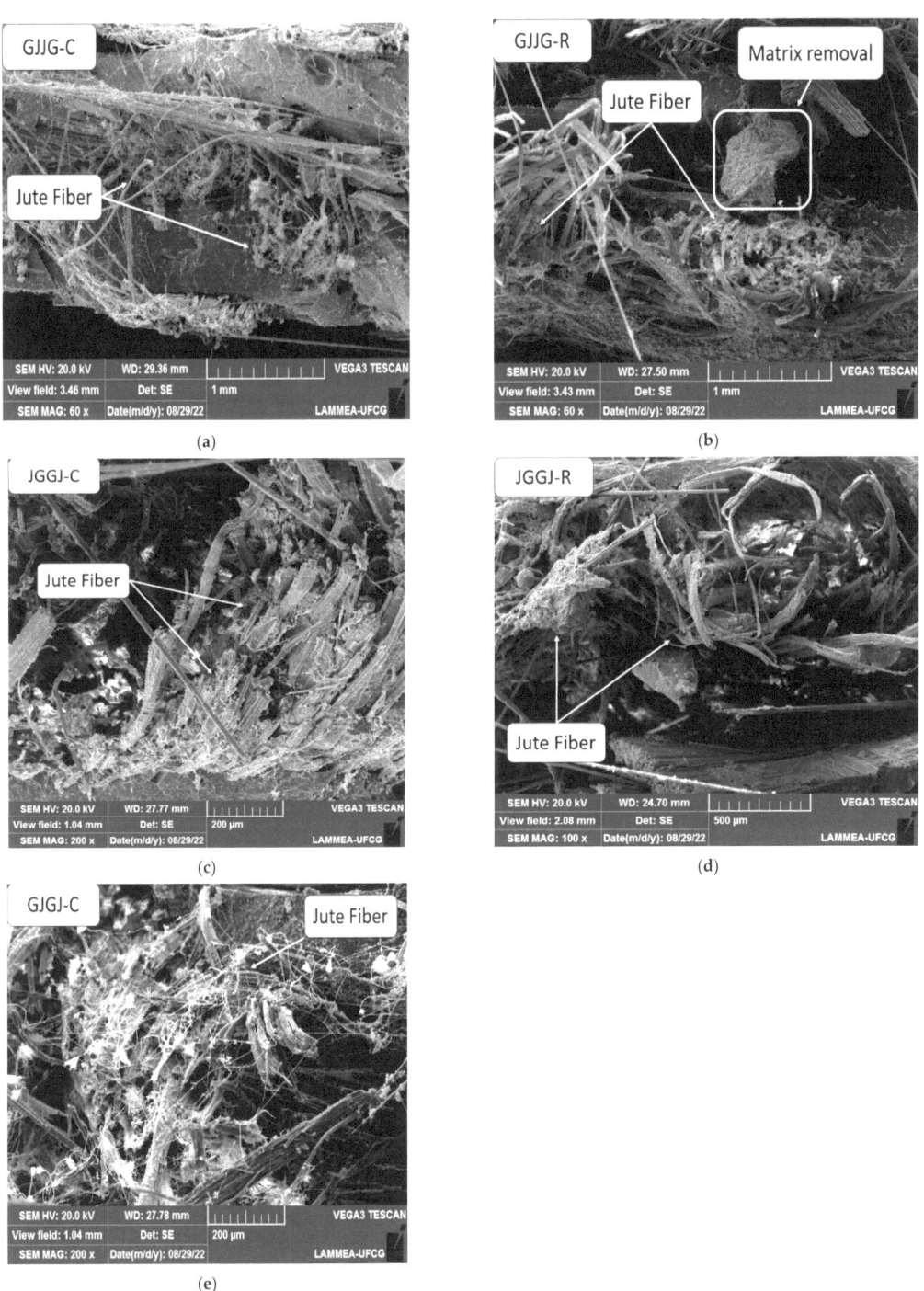

Figure 12. SEM images of the degradation mode of wet hybrid composites with different stacking sequences at 70 °C: (**a**) GJJG-C; (**b**) GJJG-R; (**c**) JGGJ-C; (**d**) JGGJ-R; (**e**) GJGJ-C.

The analysis of Figure 12 shows that this result was repeated for all types of composites analyzed, regardless of the manufacturing method or arrangement of jute fibers. This phenomenon can also be referred to as a microfibril explosion due to the action of moisture [5]. Hygrothermal aging, specifically at higher temperatures, does not induce other damage mechanisms to the composite; it merely accelerates degradation due to plasticization and reorientation of microfibrils in jute fibers [5]. Since the loss of mass in jute fibers only occurs at temperatures around 150 °C, we can currently consider that the microfibril explosion was caused by excess moisture absorbed by the jute fibers and not by an effect of temperature. In future work, it would be interesting to verify whether the same behavior of jute fibers occurs in samples subjected to other conditions studied in this work or only at 70 °C.

Observations in Figures 11 and 12 allow us to conclude that degradation, removal, and fragmentation of the matrix, as well as microfibrillation in jute fibers, were the main mechanisms (observed in SEM) responsible for the lower mechanical performance of composites subjected to water sorption at 70 °C compared to dry composites. This is because failure mechanisms such as fiber pull-out, voids, fiber fracture, and low adhesion between fibers and the matrix were observed in composites in both conditions. Although composite swelling due to moisture absorption is considered a factor of the reduction in mechanical properties due to interfacial degradation of phases, this was not observed in scanning electron microscopy.

4. Conclusions

This study investigated the effect of water absorption on the mechanical behavior and damage mechanisms in hybrid polyester/glass fiber/jute fiber composites with various stacking sequences. It was found that the addition of glass fiber enhanced the mechanical properties of jute fiber composites. Glass fiber composites consistently exhibited the highest mechanical properties, while jute fiber composites demonstrated the lowest mechanical properties, with hybrid composites falling between these extremes. Interestingly, hybrid composites displayed behavior more closely resembling that of jute fiber composites rather than glass fiber composites. Overall, the different fiber arrangements showed minimal influence on mechanical properties across the various conditions studied.

Furthermore, this study observed few disparities in mechanical properties between composites fabricated via compression molding and VARTM methods. Dry composites performed the best, followed by those aged at 70 °C with approximately 3% water absorption. Subsequently, composites immersed in water at room temperature for 696 h, in water at 50 °C for 696 h, and in water at 70 °C for 696 h showed progressively poorer mechanical properties. These results were expected and were attributed to the varying degrees of water absorption by the systems.

Moreover, positive hybridization effects on stress were noted across all conditions examined. However, certain composites exhibited negative effects on deformation under specific conditions. SEM images of fractured samples revealed a weak interface between the matrix and fibers, showcasing different fiber and matrix phases, fiber pull-out, voids, matrix cracks, moisture-induced matrix removal, and poor adhesion between fibers and the matrix. These findings offer valuable insights into the mechanical behavior and structural integrity of composite materials under different environmental conditions.

Author Contributions: Conceptualization, R.A.; methodology, R.A., C.d.L.S. and M.A.A.F.; validation, R.A., A.G.B.d.L., W.F.d.A.J. and L.H.d.C.; formal analysis, R.A., T.H.F.d.A., V.M.F., J.L.V.R., M.A.d.S., A.G.B.d.L., W.F.d.A.J. and L.H.d.C.; writing—original draft preparation, R.A.; writing—review and editing, V.M.F., A.G.B.d.L. and L.H.d.C.; supervision, J.L.V.R., A.G.B.d.L., W.F.d.A.J. and L.H.d.C. All authors have read and agreed to the published version of the manuscript.

Funding: This research was funded by the Brazilian Research Agencies: CAPES (grant number 88882.455376/2019-01), CNPq (grant number 305143/2020-4) and FAPESQ/PB (grant number 3160/2021).

Institutional Review Board Statement: Not applicable.

Data Availability Statement: Data are contained within the article.

Acknowledgments: The authors are indebted to LaMMEA (UFCG), LCTF (UFCG) and Escuela de Ingeniería Mecánica (PUCV) for laboratory support.

Conflicts of Interest: The authors declare no conflicts of interest.

References

1. White, J.R. Polymer ageing: Physics, chemistry or engineering? Time to reflect. *Comptes Rendus Chim.* **2006**, *9*, 1396–1408. [CrossRef]
2. Joseph, P.V.; Rabello, M.S.; Mattoso, L.H.C.; Joseph, K.; Thomas, S. Environmental effects on the degradation behaviour of sisal fibre reinforced polypropylene composites. *Compos. Sci. Technol.* **2002**, *62*, 1357–1372. [CrossRef]
3. Andreopoulos, A.G.; Tarantili, P.A. Water sorption characteristics of epoxy resin–UHMPE fibers composites. *J. Appl. Polym. Sci.* **1998**, *70*, 747–755. [CrossRef]
4. Marcovich, N.E.; Reboredo, M.M.; Aranguren, M.I. Moisture diffusion in polyester–Woodflour composites. *Polymer* **1999**, *40*, 7313–7320. [CrossRef]
5. Scida, D.; Assarar, M.; Poilâne, C.; Ayad, R. Influence of hygrothermal ageing on the damage mechanisms of flax-fibre reinforced epoxy composite. *Compos. Part B Eng.* **2013**, *48*, 51–58. [CrossRef]
6. Bond, D.A.; Smith, P.A. Modeling the Transport of Low-Molecular-Weight Penetrants Within Polymer Matrix Composites. *Appl. Mech. Rev.* **2006**, *59*, 249–268. [CrossRef]
7. Kafodya, I.; Xian, G.; Li, H. Durability study of pultruded CFRP plates immersed in water and seawater under sustained bending: Water uptake and effects on the mechanical properties. *Compos. Part B Eng.* **2015**, *70*, 138–148. [CrossRef]
8. El Hachem, Z.; Célino, A.; Challita, G.; Moya, M.-J.; Fréour, S. Hygroscopic multi-scale behavior of polypropylene matrix reinforced with flax fibers. *Ind. Crops Prod.* **2019**, *140*, 111634. [CrossRef]
9. Esleman, E.A.; Önal, G. Effect of saltwater on the mechanical properties of basalt/carbon/glass-epoxy hybrid composites. *J. Compos. Mater.* **2022**, *56*, 3783–3799. [CrossRef]
10. Hu, Y.; Lang, A.W.; Li, X.; Nutt, S.R. Hygrothermal aging effects on fatigue of glass fiber/polydicyclopentadiene composites. *Polym. Degrad. Stab.* **2014**, *110*, 464–472. [CrossRef]
11. Xian, G.; Li, H.; Su, X. Effects of immersion and sustained bending on water absorption and thermomechanical properties of ultraviolet cured glass fiber-reinforced acylate polymer composites. *J. Compos. Mater.* **2013**, *47*, 2275–2285. [CrossRef]
12. Hu, Y.; Li, X.; Lang, A.W.; Zhang, Y.; Nutt, S.R. Water immersion aging of polydicyclopentadiene resin and glass fiber composites. *Polym. Degrad. Stab.* **2016**, *124*, 35–42. [CrossRef]
13. Ekielski, A.; Żelaziński, T.; Mishra, P.K.; Skudlarski, J. Properties of Biocomposites Produced with Thermoplastic Starch and Digestate: Physicochemical and Mechanical Characteristics. *Materials* **2021**, *14*, 6092. [CrossRef] [PubMed]
14. Bangar, S.P.; Whiteside, W.S. Nano-cellulose reinforced starch bio composite films- A review on green composites. *Int. J. Biol. Macromol.* **2021**, *185*, 849–860. [CrossRef] [PubMed]
15. Arbelaiz, A.; Cantero, G.; Fernández, B.; Mondragon, I.; Gañán, P.; Kenny, J.M. Flax fiber surface modifications: Effects on fiber physico mechanical and flax/polypropylene interface properties. *Polym. Compos.* **2005**, *26*, 324–332. [CrossRef]
16. Kabir, M.M.; Wang, H.; Lau, K.T.; Cardona, F. Chemical treatments on plant-based natural fibre reinforced polymer composites: An overview. *Compos. Part B Eng.* **2012**, *43*, 2883–2892. [CrossRef]
17. Dinesh, S.; Kumaran, P.; Mohanamurugan, S.; Vijay, R.; Singaravelu, D.L.; Vinod, A.; Sanjay, M.R.; Siengchin, S.; Bhat, K.S. Influence of wood dust fillers on the mechanical, thermal, water absorption and biodegradation characteristics of jute fiber epoxy composites. *J. Polym. Res.* **2019**, *27*, 9. [CrossRef]
18. Sanjeevi, S.; Shanmugam, V.; Kumar, S.; Ganesan, V.; Sas, G.; Johnson, D.J.; Shanmugam, M.; Ayyanar, A.; Naresh, K.; Neisiany, R.E.; et al. Effects of water absorption on the mechanical properties of hybrid natural fibre/phenol formaldehyde composites. *Sci. Rep.* **2021**, *11*, 13385. [CrossRef]
19. Alsubari, S.; Zuhri, M.Y.M.; Sapuan, S.M.; Ishak, M.R.; Ilyas, R.A.; Asyraf, M.R.M. Potential of Natural Fiber Reinforced Polymer Composites in Sandwich Structures: A Review on Its Mechanical Properties. *Polymers* **2021**, *13*, 423. [CrossRef]
20. Chow, C.P.L.; Xing, X.S.; Li, R.K.Y. Moisture absorption studies of sisal fibre reinforced polypropylene composites. *Compos. Sci. Technol.* **2007**, *67*, 306–313. [CrossRef]
21. Chen, H.; Miao, M.; Ding, X. Influence of moisture absorption on the interfacial strength of bamboo/vinyl ester composites. *Compos. Part A Appl. Sci. Manuf.* **2009**, *40*, 2013–2019. [CrossRef]
22. Dhakal, H.N.; Zhang, Z.Y.; Richardson, M.O.W. Effect of water absorption on the mechanical properties of hemp fibre reinforced unsaturated polyester composites. *Compos. Sci. Technol.* **2007**, *67*, 1674–1683. [CrossRef]
23. Bollino, F.; Giannella, V.; Armentani, E.; Sepe, R. Mechanical behavior of chemically-treated hemp fibers reinforced composites subjected to moisture absorption. *J. Mater. Res. Technol.* **2023**, *22*, 762–775. [CrossRef]
24. Akil, H.M.; Cheng, L.W.; Mohd Ishak, Z.A.; Abu Bakar, A.; Abd Rahman, M.A. Water absorption study on pultruded jute fibre reinforced unsaturated polyester composites. *Compos. Sci. Technol.* **2009**, *69*, 1942–1948. [CrossRef]
25. Sanjay, M.R.; Yogesha, B. Studies on Natural/Glass Fiber Reinforced Polymer Hybrid Composites: An Evolution. *Mater. Today Proc.* **2017**, *4*, 2739–2747. [CrossRef]

26. Ismail, M.; Rejab, M.R.M.; Siregar, J.P.; Mohamad, Z.; Quanjin, M.; Mohammed, A.A. Mechanical properties of hybrid glass fiber/rice husk reinforced polymer composite. *Mater. Today Proc.* **2020**, *27*, 1749–1755. [CrossRef]
27. Sujon, M.A.S.; Habib, M.A.; Abedin, M.Z. Experimental investigation of the mechanical and water absorption properties on fiber stacking sequence and orientation of jute/carbon epoxy hybrid composites. *J. Mater. Res. Technol.* **2020**, *9*, 10970–10981. [CrossRef]
28. Singh, H.; Inder Preet Singh, J.; Singh, S.; Dhawan, V.; Kumar Tiwari, S. A Brief Review of Jute Fibre and Its Composites. *Mater. Today Proc.* **2018**, *5*, 28427–28437. [CrossRef]
29. Kretsis, G. A review of the tensile, compressive, flexural and shear properties of hybrid fibre-reinforced plastics. *Composites* **1987**, *18*, 13–23. [CrossRef]
30. Velmurugan, R.; Manikandan, V. Mechanical properties of palmyra/glass fiber hybrid composites. *Compos. Part A Appl. Sci. Manuf.* **2007**, *38*, 2216–2226. [CrossRef]
31. Swolfs, Y.; Gorbatikh, L.; Verpoest, I. Fibre hybridisation in polymer composites: A review. *Compos. Part A Appl. Sci. Manuf.* **2014**, *67*, 181–200. [CrossRef]
32. Venkateshwaran, N.; ElayaPerumal, A.; Alavudeen, A.; Thiruchitrambalam, M. Mechanical and water absorption behaviour of banana/sisal reinforced hybrid composites. *Mater. Des.* **2011**, *32*, 4017–4021. [CrossRef]
33. John, M.J.; Thomas, S. Biofibres and biocomposites. *Carbohydr. Polym.* **2008**, *71*, 343–364. [CrossRef]
34. Yu, H.; Longana, M.L.; Jalalvand, M.; Wisnom, M.R.; Potter, K.D. Pseudo-ductility in intermingled carbon/glass hybrid composites with highly aligned discontinuous fibres. *Compos. Part A Appl. Sci. Manuf.* **2015**, *73*, 35–44. [CrossRef]
35. Swolfs, Y.; Verpoest, I.; Gorbatikh, L. Recent advances in fibre-hybrid composites: Materials selection, opportunities and applications. *Int. Mater. Rev.* **2019**, *64*, 181–215. [CrossRef]
36. Rajpurohit, A.; Joannès, S.; Singery, V.; Sanial, P.; Laiarinandrasana, L. Hybrid Effect in In-Plane Loading of Carbon/Glass Fibre Based Inter- and Intraply Hybrid Composites. *J. Compos. Sci.* **2020**, *4*, 6. [CrossRef]
37. Marom, G.; Fischer, S.; Tuler, F.R.; Wagner, H.D. Hybrid effects in composites: Conditions for positive or negative effects versus rule-of-mixtures behaviour. *J. Mater. Sci.* **1978**, *13*, 1419–1426. [CrossRef]
38. Hayashi, T. On the improvement of mechanical properties of composites by hybrid composition. In Proceedings of the 8th International Reinforced Plastics Conference, London, UK, 10–12 October 1972; pp. 149–152.
39. Bunsell, A.R.; Harris, B. Hybrid carbon and glass fibre composites. *Composites* **1974**, *5*, 157–164. [CrossRef]
40. Zweben, C. Tensile strength of hybrid composites. *J. Mater. Sci.* **1977**, *12*, 1325–1337. [CrossRef]
41. Summerscales, J.; Short, D. Carbon fibre and glass fibre hybrid reinforced plastics. *Composites* **1978**, *9*, 157–166. [CrossRef]
42. Manders, P.W.; Bader, M.G. The strength of hybrid glass/carbon fibre composites. *J. Mater. Sci.* **1981**, *16*, 2233–2245. [CrossRef]
43. Kim, H.; Park, M.; Hsieh, K. Fatigue fracture of embedded copper conductors in multifunctional composite structures. *Compos. Sci. Technol.* **2006**, *66*, 1010–1021. [CrossRef]
44. Yao, H.; Sui, X.; Zhao, Z.; Xu, Z.; Chen, L.; Deng, H.; Liu, Y.; Qian, X. Optimization of interfacial microstructure and mechanical properties of carbon fiber/epoxy composites via carbon nanotube sizing. *Appl. Surf. Sci.* **2015**, *347*, 583–590. [CrossRef]
45. Paran, S.M.R.; Abdorahimi, M.; Shekarabi, A.; Khonakdar, H.A.; Jafari, S.H.; Saeb, M.R. Modeling and analysis of nonlinear elastoplastic behavior of compatibilized polyolefin/polyester/clay nanocomposites with emphasis on interfacial interaction exploration. *Compos. Sci. Technol.* **2018**, *154*, 92–103. [CrossRef]
46. Ahmed, K.S.; Vijayarangan, S. Tensile, flexural and interlaminar shear properties of woven jute and jute-glass fabric reinforced polyester composites. *J. Mater. Process. Technol.* **2008**, *207*, 330–335. [CrossRef]
47. Joseph, K.; Thomas, S.; Pavithran, C. Effect of chemical treatment on the tensile properties of short sisal fibre-reinforced polyethylene composites. *Polymer* **1996**, *37*, 5139–5149. [CrossRef]
48. Devireddy, S.B.R.; Biswas, S. Physical and thermal properties of unidirectional banana–jute hybrid fiber-reinforced epoxy composites. *J. Reinf. Plast. Compos.* **2016**, *35*, 1157–1172. [CrossRef]
49. Latha, P.S.; Rao, M.V.; Kumar, V.K.; Raghavendra, G.; Ojha, S.; Inala, R. Evaluation of mechanical and tribological properties of bamboo–glass hybrid fiber reinforced polymer composite. *J. Ind. Text.* **2016**, *46*, 3–18. [CrossRef]
50. Sriranga, B.K.; Kirthan, L.J.; G, A. The mechanical properties of hybrid laminates composites on epoxy resin with natural jute fiber and S-glass fibers. *Mater. Today Proc.* **2021**, *46*, 8927–8933. [CrossRef]
51. Mishra, C.; Ranjan Deo, C.; Baskey, S. Influence of moisture absorption on mechanical properties of kenaf/glass reinforced polyester hybrid composite. *Mater. Today Proc.* **2021**, *38*, 2596–2600. [CrossRef]
52. Braga, R.A.; Magalhaes, P.A.A. Analysis of the mechanical and thermal properties of jute and glass fiber as reinforcement epoxy hybrid composites. *Mater. Sci. Eng. C* **2015**, *56*, 269–273. [CrossRef] [PubMed]
53. Kaddami, H.; Arrakhiz, F.-e.; Hafs, O.; Assimi, T.E.; Boulafrouh, L.; Ablouh, E.-H.; Mansori, M.; Banouni, H.; Bouzit, S.; Erchiqui, F.; et al. Implementation and Characterization of a Laminate Hybrid Composite Based on Palm Tree and Glass Fibers. *Polymers* **2021**, *13*, 3444. [CrossRef] [PubMed]
54. Samanta, S.; Muralidhar, M.; Singh, T.J.; Sarkar, S. Characterization of Mechanical Properties of Hybrid Bamboo/GFRP and Jute/GFRP Composites. *Mater. Today Proc.* **2015**, *2*, 1398–1405. [CrossRef]
55. Kong, K.; Hejda, M.; Young, R.J.; Eichhorn, S.J. Deformation micromechanics of a model cellulose/glass fibre hybrid composite. *Compos. Sci. Technol.* **2009**, *69*, 2218–2224. [CrossRef]

56. Karthi, N.; Kumaresan, K.; Sathish, S.; Prabhu, L.; Gokulkumar, S.; Balaji, D.; Vigneshkumar, N.; Rohinth, S.; Rafiq, S.; Muniyaraj, S.; et al. Effect of weight fraction on the mechanical properties of flax and jute fibers reinforced epoxy hybrid composites. *Mater. Today Proc.* **2021**, *45*, 8006–8010. [CrossRef]
57. Tripathy, C.; Ranjan Deo, C.; Kumar Das, S. Performance studies of polyester-based hybrid composites reinforced with palmyra-palm leaf stalk and glass fibers. *Mater. Today Proc.* **2021**, *38*, 2671–2674. [CrossRef]
58. Komai, K.; Minoshima, K.; Shiroshita, S. Hygrothermal degradation and fracture process of advanced fibre-reinforced plastics. *Mater. Sci. Eng. A* **1991**, *143*, 155–166. [CrossRef]
59. Kim, H.J.; Seo, D.W. Effect of water absorption fatigue on mechanical properties of sisal textile-reinforced composites. *Int. J. Fatigue* **2006**, *28*, 1307–1314. [CrossRef]
60. Rezgani, L.; Madani, K.; Feaugas, X.; Touzain, S.; Cohendoz, S.; Valette, J. Influence of water ingress onto the crack propagation rate in a AA2024-T3 plate repaired by a carbon/epoxy patch. *Aerosp. Sci. Technol.* **2016**, *55*, 359–365. [CrossRef]
61. Aranha, R.; Filho, M.A.A.; de Lima Santos, C.; Fonseca, V.M.; Rivera, J.L.V.; de Lima, A.G.B.; de Amorim, W.F.; Carvalho, L.H. Water Sorption in Hybrid Polyester/Glass/Jute Composites Processed via Compression Molding and Vacuum-Assisted Resin Transfer Molding. *Polymers* **2023**, *15*, 4438. [CrossRef]
62. *ASTM D3039*; Standard Test Method for Tensile Properties of Polymer Matrix Composite Materials. American Society for Testing Materials (ASTM): West Conshohocken, PA, USA, 2022.
63. Zhang, Y.; Li, Y.; Ma, H.; Yu, T. Tensile and interfacial properties of unidirectional flax/glass fiber reinforced hybrid composites. *Compos. Sci. Technol.* **2013**, *88*, 172–177. [CrossRef]
64. Khatri, S.C.; Koczak, M.J. Thick-section AS4-graphite/E-glass/PPS hybrid composites: Part I. Tensile behavior. *Compos. Sci. Technol.* **1996**, *56*, 181–192. [CrossRef]
65. Harish, S.; Michael, D.P.; Bensely, A.; Lal, D.M.; Rajadurai, A. Mechanical property evaluation of natural fiber coir composite. *Mater. Charact.* **2009**, *60*, 44–49. [CrossRef]
66. Muñoz, E.; García-Manrique, J.A. Water Absorption Behaviour and Its Effect on the Mechanical Properties of Flax Fibre Reinforced Bioepoxy Composites. *Int. J. Polym. Sci.* **2015**, *2015*, 390275. [CrossRef]
67. Dalbehera, S.; Acharya, S. Effect of cenosphere addition on the mechanical properties of jute-glass fiber hybrid epoxy composites. *J. Ind. Text.* **2016**, *46*, 177–188. [CrossRef]

Disclaimer/Publisher's Note: The statements, opinions and data contained in all publications are solely those of the individual author(s) and contributor(s) and not of MDPI and/or the editor(s). MDPI and/or the editor(s) disclaim responsibility for any injury to people or property resulting from any ideas, methods, instructions or products referred to in the content.

Article

Mechanochemical Encapsulation of Caffeine in UiO-66 and UiO-66-NH₂ to Obtain Polymeric Composites by Extrusion with Recycled Polyamide 6 or Polylactic Acid Biopolymer

Cristina Pina-Vidal [1,2], Víctor Berned-Samatán [1,2], Elena Piera [3], Miguel Ángel Caballero [3] and Carlos Téllez [1,2,*]

[1] Instituto de Nanociencia y Materiales de Aragón (INMA), CSIC-Universidad de Zaragoza, 50009 Zaragoza, Spain; crispina@unizar.es (C.P.-V.); victorberned@unizar.es (V.B.-S.)
[2] Chemical and Environmental Engineering Department, Universidad de Zaragoza, 50018 Zaragoza, Spain
[3] Research and Development Department, Nurel S.A., Ctra. Barcelona km 329, 50016 Zaragoza, Spain; epiera@samca.com (E.P.); acaballero@samca.com (M.Á.C.)
* Correspondence: ctellez@unizar.es

Citation: Pina-Vidal, C.; Berned-Samatán, V.; Piera, E.; Caballero, M.Á.; Téllez, C. Mechanochemical Encapsulation of Caffeine in UiO-66 and UiO-66-NH₂ to Obtain Polymeric Composites by Extrusion with Recycled Polyamide 6 or Polylactic Acid Biopolymer. *Polymers* **2024**, *16*, 637. https://doi.org/10.3390/polym16050637

Academic Editors: Jesús-María García-Martínez and Emilia P. Collar

Received: 31 January 2024
Revised: 19 February 2024
Accepted: 21 February 2024
Published: 27 February 2024

Copyright: © 2024 by the authors. Licensee MDPI, Basel, Switzerland. This article is an open access article distributed under the terms and conditions of the Creative Commons Attribution (CC BY) license (https:// creativecommons.org/licenses/by/ 4.0/).

Abstract: The development of capsules with additives that can be added to polymers during extrusion processing can lead to advances in the manufacturing of textile fabrics with improved and durable properties. In this work, caffeine (CAF), which has anti-cellulite properties, has been encapsulated by liquid-assisted milling in zirconium-based metal–organic frameworks (MOFs) with different textural properties and chemical functionalization: commercial UiO-66, UiO-66 synthesized without solvents, and UiO-66-NH₂ synthesized in ethanol. The CAF@MOF capsules obtained through the grinding procedure have been added during the extrusion process to recycled polyamide 6 (PA6) and to a biopolymer based on polylactic acid (PLA) to obtain a load of approximately 2.5 wt% of caffeine. The materials have been characterized by various techniques (XRD, NMR, TGA, FTIR, nitrogen sorption, UV–vis, SEM, and TEM) that confirm the caffeine encapsulation, the preservation of caffeine during the extrusion process, and the good contact between the polymer and the MOF. Studies of the capsules and PA6 polymer+capsules composites have shown that release is slower when caffeine is encapsulated than when it is free, and the textural properties of UiO-66 influence the release more prominently than the NH₂ group. However, an interaction is established between the biopolymer PLA and caffeine that delays the release of the additive.

Keywords: metal organic framework; UiO-66; UiO-66-NH₂; caffeine; microencapsulation; textile composite; polyamide; polylactic

1. Introduction

In recent years, there has been a surge in industry interest in using microencapsulation processes to develop innovative materials. Microencapsulation has various benefits and advantages since active substances, such as vitamins, scents, essential oils, biocides, and drugs, may be encapsulated to preserve them from environmental factors, increase their temperature resistance, control the release rate of the additive encapsulated, or make liquids compatible with solids [1–4]. In the textile sector, microencapsulation offers many opportunities to improve the properties of textiles and provide new functionalities. Examples of functional textiles incorporating capsules include color-changing textiles [5], insect-repellent textiles [6], textiles for thermal control [7], antimicrobial [8,9] and medical textiles [10], and cosmetotextiles [11–13], among others. A wide range of methods and procedures have been developed to create capsules with desired materials and target properties: chemical methods (interfacial polymerization, in situ polymerization), physico-chemical methods (complex coacervation, molecular inclusion), and physical procedures (spray drying, solvent evaporation) [14]. The capsules obtained through these procedures usually consist of a polymeric layer that covers the additive. An alternative to these capsules is to encapsulate additives in porous materials, both inorganic (e.g., zeolites [15]) and organic–inorganic materials (e.g., MOF [16]). The former allow processing at high temperatures while the latter preserve this property and are more compatible with textiles due to their organic–inorganic nature.

Metal–organic frameworks (MOFs) are porous crystalline hybrid materials made by linking a metal ion or cluster with organic linkers creating crystal lattices. These materials have several characteristics and properties that are of interest in many fields. They exhibit high porosity, high specific surface areas, excellent chemical and thermal stability, and the possibility of varying the pore size, shape, or chemical functionality by modifying the metal cluster and the nature of the linker. This makes them attractive for numerous applications, such as encapsulation [16], the development of selective membranes for molecular separation [17,18], adsorption and storage of gases [19], catalysis [20,21], and in biomedicine [22]. Among various MOF designs, those featuring zirconium-based structures incorporating $[Zr_6O_4(OH)_4]^{12+}$ cluster nodes and carboxylate linkers are highly appealing due to their outstanding chemical and thermal stability. In particular, UiO-66 stands out in this category. Comprising hexanuclear oxozirconium clusters and organic terephthalate ligands, it adopts a face-centered cubic (fcu) topology with a hydrophilic surface. Since its discovery in 2008, UiO-66 has become one of the most widely utilized MOFs [23]. Its popularity is attributed to its distinctive properties, such as ease of functionalization and tunability arising from its isoreticular structure [24]. This structural flexibility enables the creation of similar materials with the same UiO-66 topology by modifying either the metal cluster or the ligand. Consequently, a diverse range of materials with varying functionalities and pore sizes can be obtained. Notable among these isoreticular structures is the UiO-66-R, where R represents the functional group. Depending on the ligand used, the synthesized MOF can be denoted as UiO-66-NH_2, UiO-66-NO_2, or UiO-66-Br, among others. The pore aperture size of UiO-66 is theoretically estimated to be around 6.0 Å [23]. The remarkable thermal stability of UiO-66 is explained by the robust Zr-O bonds and the ability of the inner Zr_6 cluster to undergo reversible rearrangement following the dihydroxylation or rehydration of the μ_3-OH groups [25].

However, the preparation process of these materials mostly involves organic solvents, such as N,N-dimethylformamide (DMF) or N,N-dimethylacetamide (DMAC), which are highly toxic and hazardous to the environment and human health. Therefore, it is necessary to develop greener synthesis methods [26] in order to eliminate solvents entirely from the synthetic route or replace harsh organic solvents with water or less harmful solvents to reduce environmental costs. Some researchers have reported a green scalable modulated hydrothermal synthesis of UiO-66 in quite mild conditions using acetic acid as a modulator [27,28]. Other researchers have developed mechanochemical and solvent-free synthesis methods for UiO-66 and UiO-66-NH_2 [29]. Mechanochemistry minimizes the use of organic solvents. This technique includes a process known as liquid-assisted grinding (LAG), in which a small amount of liquid is added [30,31].

Caffeine, known as 1,3,7-trimethylxanthine, is a versatile drug with amphiphilic properties that has gained significant attention in the cosmetic industry due to its anti-cellulite properties. There are studies in the literature in which caffeine is encapsulated in the MOFs [16,32–34] by immersing the previously synthesized MOF in a caffeine solution, which entails a subsequent step of filtration and drying. To improve this process, various advances have been made to limit the steps, such as performing the in situ synthesis of the MOF with caffeine [35,36] or performing pressure encapsulation with the already synthesized MOF [37]. Additionally, a solid-state mechanochemical strategy for additive encapsulation in MOF nanoparticles via a ball milling method has been reported [38]. The encapsulation occurs using a liquid-assisted grinding (LAG) method. This is an environmentally friendly process compared to the classic liquid phase encapsulation processes, since the use of solvent is minimized and the additive is fully used, reducing waste.

The incorporation of caffeine into polymer composites has been studied in diverse fields. Labay et al. analyzed the effect on caffeine release in plasma-treated polyamide (PA) samples [39,40] that had been impregnated with a caffeine solution. Sta et al. obtained caffeine-loaded poly(N-vinylcaprolactam-co-acrylic acid) composites by electrospinning which were evaluated as drug delivery systems [41]. Li et al. manufactured polyvinyl-alcohol (PVA) nanocomposites incorporating caffeine and riboflavin in the preparation by electrospinning for evaluating a fast-dissolving delivery system [42]. Similar to this study, caffeine and paracetamol were mixed with a polyvinylpyrrolidone (PVP) solution and nanocomposites were prepared by electrospinning [43]. Li et al. investigated the effects of caffeine on the degradation of the PLA matrix in caffeine delivery systems [44]. For cosmetic purposes, Tipduangta et al. developed cellulose acetate/PVP composites for caffeine delivery [45]. Recently, caffeine has been encapsulated in porous materials (Zeolite Y, ZIF-8, and MIL-53) and introduced during the spinning of polyamide 6 fibers for textile applications [16].

In the present work, caffeine-loaded UiO-66 capsules were developed using a ball mill-encapsulation method which minimizes the use of water and reduces waste. For this encapsulation process, two materials with different textural properties were compared: commercial UiO-66 and one synthesized by a solvent-free method. In addition, UiO-66 capsules were compared with the functionalized form, UiO-66-NH$_2$, which was synthesized by a solvothermal synthesis in ethanol, avoiding the use of harmful solvents, such as DMF. Furthermore, the present study examined the use of more environmentally friendly processes and polymers by using recycled polyamide 6 (PA6) and polylactic acid (PLA), which is a biodegradable polymer used as an alternative to petroleum [46]. Then, the developed capsules were incorporated into PA6 and PLA composites during the extrusion process (see Figure 1). Therefore, the capsules remained embedded inside the composites improving caffeine durability and resistance, and consequently, the useful life of the final product. Traditionally, the addition of capsules to polymers for textile use has been performed in the last step and only superficially [47], so their durability is quite limited. The high temperatures used in the extrusion process must be taken into account, and UiO-66 capsules are suitable for resisting these temperatures.

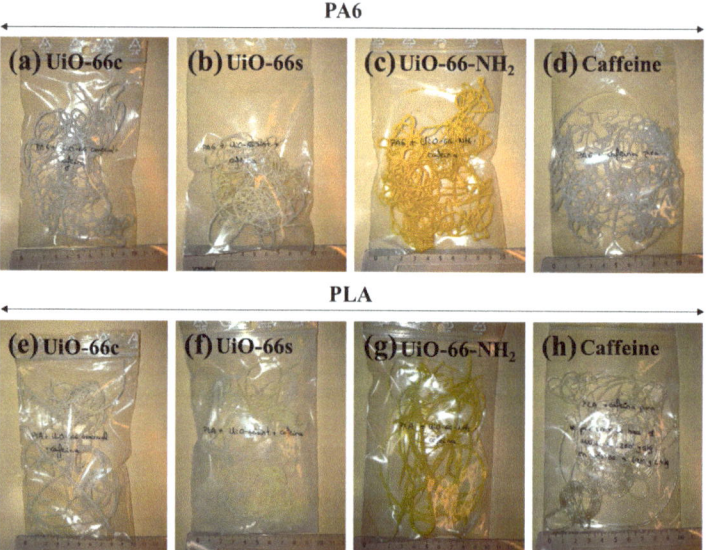

Figure 1. Photographs of the prepared polymer composites: (**a**) PA6 + CAF@UiO-66c, (**b**) PA6 + CAF@UiO-66s, (**c**) PA6 + UiO-66-NH$_2$, (**d**) PA6 + caffeine, (**e**) PLA + CAF@UiO-66c, (**f**) PLA + CAF@UiO-66s, (**g**) PLA + CAF@UiO-66-NH$_2$, (**h**) PLA + caffeine.

2. Materials and Methods

2.1. Materials

Commercial UiO-66 (Strem Chemicals Inc., Newburyport, MA, USA) was purchased from CymitQuimica (Barcelona, Spain) and activated in an oven at 300 °C for 2 h. Zirconium (IV) oxychloride octahydrate ($ZrOCl_2 \cdot 8H_2O$, Glentham Life Sciences, Corsham, UK, 98%) and tetraethylammonium bromide (TEABr, TCI chemicals, Tokyo, Japan, >98%) were also supplied by CymitQuimica. 1,4-benzenedicarboxylic acid (BDC, Aldrich, San Luis, MO, USA, 98%) and caffeine (1,3,7-trimethylxanthine, ReagentPlus, Aldrich) were purchased from Sigma-Aldrich (San Luis, MO, USA). Zirconium (IV) chloride ($ZrCl_4$, Alfa Aesar, Haverhill, MA, USA, +99.5%), 2-aminoterephthalic acid (NH_2-BDC, ThermoScientific, Waltham, MA, USA, 99%), and formic acid (HCOOH, FisherScientific, Waltham, MA, USA, 90%) were purchased from ThermoFisher Scientific (Waltham, MA, USA). As solvents, absolute ethanol (EtOH, 99.5%) and deionized water were provided by Productos Gilca S.C. (Zaragoza, Spain). Recycled Polyamide 6 (Recomyde® B30 P4) and Biopolymer PLA based (Inzea® F19) were provided by NUREL S.A. (Zaragoza, Spain).

2.2. MOFs Synthesis

An efficient solvent-free method was used for the UiO-66 preparation following a procedure described in the literature, with slight modifications [48]. The synthetic procedure was as follows: A mixture of BDC (1 mmol), $ZrOCl_2 \cdot 8H_2O$ (1 mmol), and TEABr (1.42 mmol) was ground in a mortar for around 10 min until the mixture was homogeneous. It was then transferred to an autoclave and left at 190 °C for 24 h. After this time, the autoclave was left to cool down and the resultant solid was washed with warm EtOH (three or four times). Next, the solid was dried in an oven at 80 °C overnight. Finally, for MOF activation it was placed in a vacuum oven at 200 °C for 12 h. This method is easily scalable due to the absence of solvent, so, maintaining the proportions, several grams can be produced. This MOF is labeled UiO-66s to differentiate it from the commercial MOF labeled UiO-66c.

The UiO-66-NH_2 synthesis was carried out using a solvothermal method [49]. In this synthesis, the Zr metal was provided by $ZrCl_4$ and the ligand in the amino-functionalized form (NH_2-BDC). The procedure consisted of dissolving $ZrCl_4$ (2 mmol) and NH_2-BDC (1 mmol) in 20 mL of ethanol. Next, 25 mL of 30 wt% formic acid in water was added as an acid modulator. The mixed solution was placed in an ultrasound system for 10 min. Then, the solution was transferred to an autoclave and placed in an oven at 110 °C for 12 h. The solid was washed three to five times with absolute ethanol and water. Finally, the UiO-66-NH_2 sample was activated in a vacuum oven at 200 °C for 12 h. It should be noted that the UiO-66-NH_2 sample is yellow while the UiO-66 samples are white (Figure A1).

2.3. Mechanochemical Encapsulation Method

The encapsulation process was performed following a ball milling method, in particular through liquid-assisted grinding (LAG). In this technique, the reagents, previously physically mixed, were added into the grinding jar. Additionally, a small quantity of solvent, known as assisting liquid, was added to the solids. This may facilitate mechanochemical processes [31], help to obtain a better distribution of the solids, and enhance the diffusion, and consequently, the caffeine encapsulation. A Vibratory Micro Mill PULVERISETTE 0 (Fritsch, Idar-Oberstein, Germany) equipped with a mortar and a grinding ball, both made of stainless steel, was used. The operation conditions were optimized in terms of operation time, weight of solids, and amount of solvent. The procedure was the following: UiO-66 or UiO-66-NH_2 (750 mg) and caffeine (250 mg) were physically mixed and added to the mortar. Then, 1 mL of distilled water was added as the assisting liquid, since caffeine presents a moderate water solubility (20 g/L at room temperature) and, thus, the presence of water will facilitate its diffusion within the pores of the UiO-66. After 2 h of ball mill operation, the loaded-caffeine MOF capsules, named CAF@MOF, were dried in an oven at 100 °C for 2 h to eliminate the water present in the samples. Figure A1 shows that the samples did not change color when the caffeine was encapsulated.

2.4. Preparation of Polymer Composites with UiO-66 Capsules with Caffeine

The CAF@MOF capsules were incorporated into the textile composites during the extrusion process. Two kinds of polymer were studied, polyamide (PA6) and polylactic acid (PLA). In each composite, the weight percentage of the caffeine-loaded capsules was 10 wt%, and, taking into account that the caffeine loading in the capsules was 25 wt%, the theoretical value of caffeine in the composites was 2.5 wt%. In order to determine the efficacy of the capsules, a composite was prepared with pure caffeine, without encapsulation in UiO-66. In this case, the amount of caffeine added in the extrusion process was proportional to the caffeine encapsulated in the capsules. Therefore, all the composites had the same quantity of theoretical caffeine. These composites were prepared using an Instron®'s Melt Flow Tester equipment model CEAST MF20 (Barcelona, Spain). The capsules were previously mixed with the polymer and this mixture was added to the equipment and heated. The PA6 composites were prepared at 260 °C while the PLA composites, given their lower melting temperature, were prepared at 190 °C. A weight (5 kg for PA6 composites and 2.16 kg for PLA composites) was placed on a piston which was located above the heated mixture in order for the melted polymer to flow through a hot nozzle, cooling at the outside of the oven and creating the extruded composites. Figure 1 shows images of the polymer composites with the UiO-66 capsules and caffeine. The samples have a whitish color except for the samples prepared with CAF@UiO-66-NH_2, which are yellowish, consistent with the color of the synthesized MOFs (see Figure A1).

2.5. Characterization

The formation of UiO-66 was studied by X-ray diffraction (XRD) analysis using an Empyrean PANanlytical diffractometer (Malvern Panalytical, Malvern, UK) with a copper anode and a graphite monochromator to select CuKα radiation (λ = 1.5406 Å). Data were collected in the 2.5–40° 2θ range with a scanning rate of 0.01 °/s.

The textural properties, including BET surface areas with pore volumes, were measured using a Micromeritics TriStar 3000 (Micromeritics Instrument Corporation, Norcross, GA, USA) with previous degasification at 200 °C (for UiO-66) and at 175 °C (for UiO-66-NH_2) for 10 h with a heating rate of 10 °C/min in a Micromeritics VacPrep™ 061 (Micromeritics Instrument Corporation, Norcross, GA, USA).

Thermogravimetric analysis (TGA) was performed using Mettler Toledo TGA/SDTA 854e equipment (Mettler Toledo, Columbus, OH, USA). Samples placed in 70 μL alumina pans were heated up to 700 °C with a heating rate of 10 °C/min.

Attenuated total reflection-Fourier transformed infrared spectroscopy (ATR-FTIR) was carried out in a Bruker Vertex 70 FTIR spectrometer (Bruker, Billerica, MA, USA) with a DTGS detector and a Golden Gate diamond ATR accessory. The spectra were recorded by averaging 40 scans in the 600–4000 cm^{-1} wavenumber range with a resolution of 4 cm^{-1}.

The surface morphology of the samples and their particle sizes were studied by SEM. Moreover, SEM-EDX analysis was performed on the polymer composites in order to determine the atomic percentage at different points and to confirm the presence of UiO-66. The samples were cut under a liquid nitrogen atmosphere and their cross sections were viewed through electron microscopy using an FEI-Inspect F50 microscope at a voltage of 10 kV. The samples were previously coated with palladium under vacuum conditions.

The powder samples were also observed by Transmission Electron Microscopy (TEM) using a FEI Tecnai T20 microscope (FEI Company, Hillsboro, OR, USA) operated at 200 kV accelerating voltage. The samples were dispersed in ethanol and drops of the solution were placed onto a carbon-coated grid.

Caffeine release from the CAF@UiO-66 samples was evaluated at 25 °C using distilled water. Suspensions of 10 mg CAF@MOF samples in 100 mL of water were prepared and the concentration of caffeine in each 3 mL aliquot was determined using a calibration curve.

Each release experiment was performed twice. In order to achieve a gradual release, these experiments were carried out without stirring. Additionally, this solid–liquid extraction was performed with a physical mixture of MOF and caffeine in the same ratio as the ball mill samples (25 wt% of caffeine). Caffeine release from the polymer composites was carried out similarly. Around 500 mg of composites were cut into similar pieces and added to 100 mL of distilled water. Periodically, a liquid sample (3 mL) was taken from the release media. The initial temperature was 25 °C, which was increased to 50 °C, and then to 80 °C in order to achieve a gradual release over time. The amount of caffeine in the liquids obtained from solid–liquid extraction was quantified by ultraviolet spectroscopy (UV–Vis) using a JASCO V-670 spectrophotometer (JASCO International, Tokyo, Japan) using water as a solvent at a wavelength of 272 nm.

Solid-state cross-polarization magic angle spinning (CP MAS) ^{13}C-Nuclear Magnetic Resonance (NMR) spectra were acquired with a Bruker Avance III 400 MHz Wide Bore spectrometer (Bruker, Billerica, MA, USA) equipped with a 4 mm CP MAS 1H-BB probe. Approximately 100 mg of the product was packed in a 4 mm zirconium rotor and sealed with a Kel-F cap. The ^{13}C CP spectra were acquired with a MAS rate of 10 kHz, a ramp-CP contact time of 3 ms, and a 7 s recycle delay. For the ^{1}H NMR measurement, the liquid samples taken from the release experiments were evaporated and dispersed in deuterated water. NMR spectra were acquired with a Bruker Neo 500 MHz spectrometer equipped with a 5 mm iProbe 1H-BB at 25 °C using TMS as the internal standard. The ^{1}H and ^{1}H presaturation experiments were conducted with 64 scans.

During the preparation of the polymer composites (previously explained in Section 2.4), the equipment used also recorded their fluidity values in accordance with the ISO 1133 standard (reference [50]) test method. The equipment measured the fluidity at regular intervals, obtaining an average value at the end of the process. The results of the measurements, known as the melt volume rate (MVRs), are expressed in units of cm^3/10 min.

3. Results and Discussion

3.1. MOF Characterization

To evaluate the encapsulation efficiency in porous materials, various MOF samples were used in this work. Two types of UiO-66 with different textural properties, and the amino-functionalized form UiO-66-NH$_2$, were tested. Therefore, first of all, the MOF samples were characterized by nitrogen adsorption–desorption (Figure 2 and Table 1).

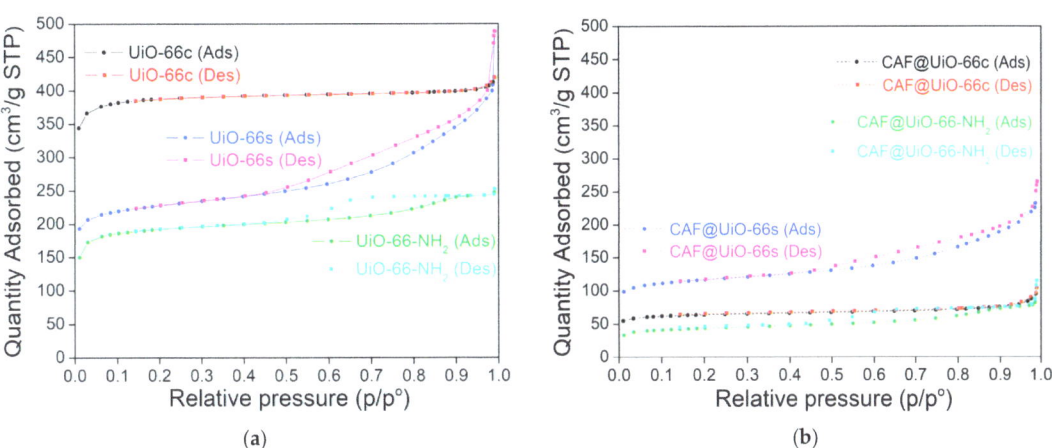

Figure 2. Nitrogen adsorption–desorption (Ads-Des) isotherms of UiO-66c, UiO-66s, and UiO-66-NH$_2$ before (**a**) and after (**b**) caffeine encapsulation.

Table 1. Textural properties of UiO-66c, UiO-66s, and UiO-66-NH$_2$ before and after caffeine encapsulation.

	BET Surface Area (m^2/g)	Total Pore Volume [1] (cm^3/g)	Micropore Volume [2] (cm^3/g)
UiO-66c	1298 ± 24	0.626	0.544
UiO-66s	771 ± 12	0.600	0.271
UiO-66-NH$_2$	650 ± 11	0.380	0.240
CAF@UiO-66c	216 ± 3	0.130	0.078
CAF@UiO-66s	396 ± 6	0.340	0.129
CAF@UiO-66-NH$_2$	150 ± 2	0.120	0.044

[1] At P/P$_0$ = 0.97. [2] Using t-plot.

The different textural properties of UiO-66c and UiO-66s can be seen from the sorption isotherms represented in Figure 2. For UiO-66c, the isotherm is a type-I according to the IUPAC classification, indicating a microporous structure of this material (see also micropore volume in Table 1). For the synthesized MOFs (UiO-66 and UiO-66-NH$_2$), the isotherms are a combination of type I and type IV, indicating the presence of micropores and mesopores, which can influence caffeine encapsulation and its posterior gradual release. Furthermore, the synthesized MOFs that have a smaller size, as seen below, show an increase in nitrogen adsorption at values close to P/P$_0$ ≈ 1, which is related to the condensation between nanoparticles. In the literature, the reported BET surface areas for UiO-66 range from ~600 to 1800 m^2/g [51]. The surface area values found here for UiO-66 are in that order. The lower BET surface values of UiO-66s (771 m^2/g) compared to UiO-66c (1298 m^2/g) are in agreement, as will be seen below, with the XRD pattern of this material, which was not as crystalline as UiO-66c. The BET surface area (650 m^2/g) and the micro-pore volume (0.240 cm^3/g) of UiO-66-NH$_2$ were slightly lower than those obtained for UiO-66s (771 m^2/g and 0.271 cm^3/g). This slight decrease could be related to the presence of the amino group that reduces the space in the pores. In any case, the surface area for UiO-66-NH$_2$ is within the range of values reported in the literature, ~630–996 m^2/g [25,30].

Figure 3 shows the XRD spectra obtained after the activation of UiO-66 synthesized using a green solventless method, UiO-66-NH$_2$ synthesized using a solvothermal method, UiO-66c, and simulated UiO-66. The XRD pattern of the synthesized UiO-66 is in line with the simulated pattern. The main peaks at approximately 2·θ = 7.4, 8.5, and 25.7° are related to the crystallographic planes (111), (002), and (224) [52], and affirm the successful synthesis of the desired MOF structure. The agreement of peaks in the XRD shows that UiO-66-NH$_2$ was also properly synthesized by the solvothermal method. It was noted that UiO-66 and UiO-66-NH$_2$ displayed comparable diffraction patterns, suggesting that ligand functionalization had no impact on the MOF crystal structure [53]. In the case of the UiO-66s, the peaks are broader than those of the UiO-66c, which may indicate a lower degree of crystallinity and also a smaller crystal size, as explained in the microscopy study below.

The average crystal size (D) was calculated by Scherrer's equation (Equation (1)) using the peaks at 7.4, 8.5, and 25.7° for the three MOF samples.

$$D = \frac{K \cdot \lambda}{\beta \cdot \cos\theta} \quad (1)$$

The values obtained with the highest intensity peak at 7.4° were 49.5 nm, 13.4 nm, and 40.2 nm for the UiO-66c, UiO-66s, and UiO-66-NH$_2$, respectively. The calculated values for the other peaks are similar to these previous values (see Table A1). In the subsequent sections, the trend of these values will be corroborated by observing SEM images. The highest particle size was obtained for UiO-66c and the lowest for UiO-66s.

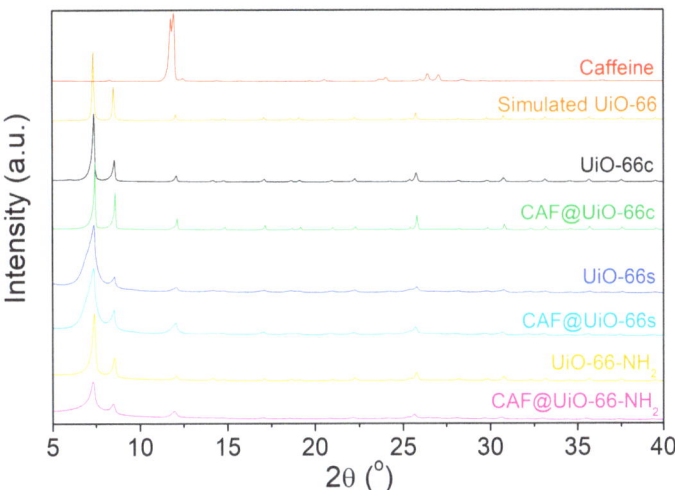

Figure 3. XRD spectra of UiO-66s, UiO-66c, UiO-66-NH$_2$, and simulated UiO-66 using Mercury 3.8 software [54] and CCDC 1018045 [55] from Cambridge Crystallographic Data Centre. The XRD of the samples after the caffeine encapsulation process are also shown.

The FTIR spectra (Figure 4) of the UiO-66c and the synthesized MOFs are compared to confirm the MOF synthesis and identify the main molecular groups. As can be seen, the synthesized UiO-66 presents a similar pattern to that of the UiO-66c. In particular, the band at 1578 cm^{-1} is assigned to the C=O carbonyl bond stretching vibration of the carboxyl group. The weak band at 1507 cm^{-1} is related to the C=C vibration of the benzene ring. The strong band at 1391 cm^{-1} is ascribed to the C-O bond of the C-OH group of the carboxylic acid [56]. Next to this peak appears a small peak at 1424 cm^{-1} related to the C-C of the benzene ring. Other typical bands for UiO-66 are at 743 and 664 cm^{-1} for the symmetric and asymmetric stretching of the O-Zr-O bonds, respectively [57]. In the case of UiO-66-NH$_2$, the bands that appear relate to the functional groups corresponding to the amino group. There is a broad absorption band at around 3343 cm^{-1} that corresponds with the symmetric and asymmetric vibration bands of –NH$_2$. The small peak at 1662 cm^{-1} corresponds to the N-H bending vibration [58]. The band at 1562 cm^{-1} is attributed to the C=O bond in the carboxyl group. The band at 1424 cm^{-1}, characteristic of C-C in the benzene ring, is more pronounced than in the UiO-66 sample. Additionally, two bands are observed at 1254 and 1378 cm^{-1} corresponding to the band of C-N bonded between aromatic carbon and nitrogen [49].

SEM images of the nanoparticles are shown in Figure 5. In this figure, it can be seen that the nanoparticles are relatively homogeneous, and the geometry is not fully developed, presenting an appearance between cubic and spherical. The particle size of these samples was measured using the ImageJ software [59], obtaining mean values of 77 ± 12 nm, 38 ± 5 nm, and 59 ± 7 nm for UiO-66c, UiO-66s and UiO-66-NH$_2$, respectively. The size distribution of the MOF samples was calculated from these SEM images (Figure A2). As can be observed, the MOF samples present a uniform particle size distribution, having a maximum of around 70 nm, 30 nm, and 60 nm for UiO-66c, UiO-66s, and UiO-66-NH$_2$, respectively, which are very close to the mean values.

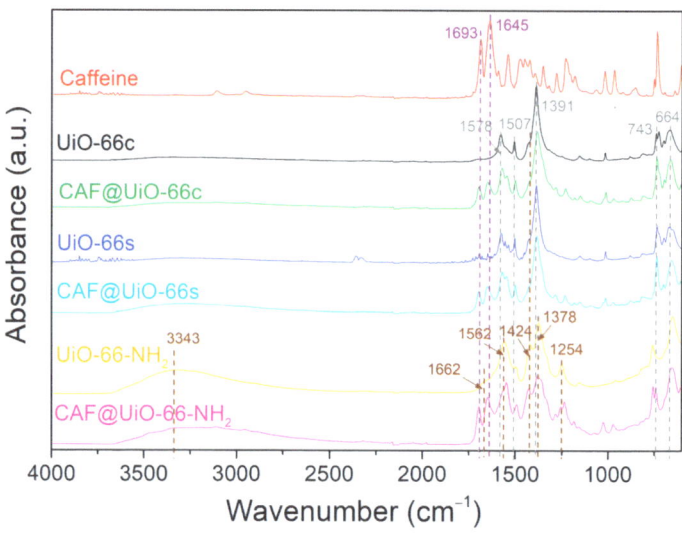

Figure 4. FTIR spectra of UiO-66c, UiO-66s, and UiO-66-NH$_2$ before and after caffeine encapsulation.

Figure 5. SEM images of (**a**) UiO-66c, (**b**) UiO-66s, and (**c**) UiO-66-NH$_2$. SEM images of UiO-66 with encapsulated caffeine (**d**) CAF@UiO-66c, (**e**) CAF@UiO-66s, and (**f**) CAF@UiO-66-NH$_2$.

The smaller particle size observed for the synthesized UiO-66 is in agreement with the lower crystallinity observed by XRD. In addition, the particle sizes observed by SEM analysis are similar to the values obtained by Scherrer's equation (see Table A1). Using the highest intensity XRD peak, for UiO-66c the difference between values from the SEM and Scherrer's equation is from 77 to 54 nm, for UiO-66s it is from 13.4 to 38 nm, and for UiO-66-NH$_2$ from 59 to 43 nm. The difference is more pronounced in the case of the UiO-66s. This may be because the peak at 7.4° for the Scherrer's equation calculation is more amorphous in the synthesized MOF sample than in the other samples (UiO-66c and

UiO-66-NH$_2$). In any case, the trends in particle size are predicted both by SEM and by the calculations made with Scherrer's equation for any of the peaks used.

In order to examine these MOF samples in more detail, they were also characterized by TEM. Figure A3 shows that the nanoparticles in the UiO-66-NH$_2$ sample were better dispersed than in the UiO-66 samples, which appeared more agglomerated, so the particle sizes in the latter were not as clear. The particle sizes obtained from this technique were similar to the values determined by SEM analysis.

The thermal stability of the samples was tested by thermogravimetric analysis (TGA). Figure 6 shows the weight loss of the samples under the flow of nitrogen. The initial weight loss (until 100 °C) of the MOF samples corresponds to the removal of moisture from the samples. UiO-66c and UiO-66s presented high thermal stability (up to 500 °C), as reported in the literature [34]. This weight loss is related to ligand decomposition under pyrolytic conditions and the subsequent amorphization of the structure and pore collapse. The synthesized sample showed a higher content of moisture. The thermal stability of the UiO-66-NH$_2$ was lower than that of the pristine UiO-66, because of the presence of an additional functional group [25], showing a weight loss from 300 to nearly 600 °C, which was more gradual than that of the UiO-66. After TGA under N$_2$ flow, the residue originated from the carbonization of the ligand and the formation of ZrO$_2$ coming from Zr and O elements in the UiO-66 [60].

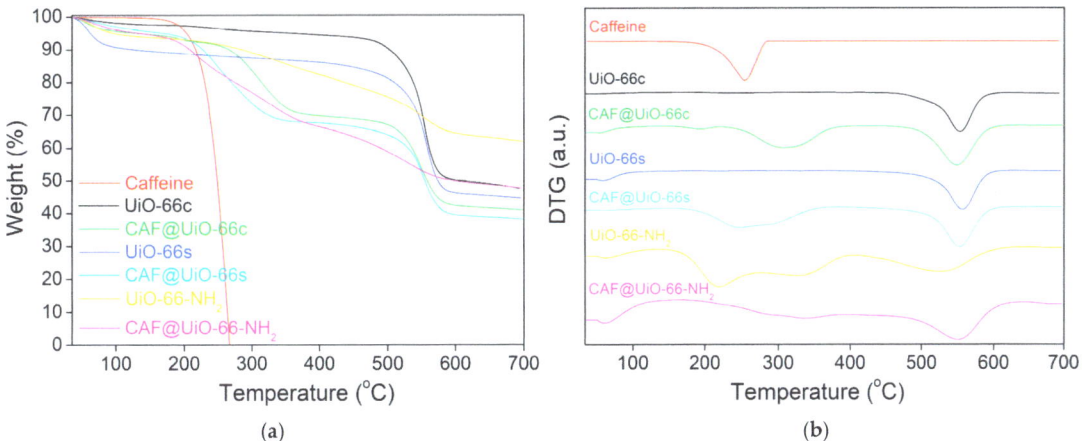

Figure 6. TGA (**a**) and DTG (**b**) analysis of UiO-66c, UiO-66s, and UiO-66-NH$_2$ samples before and after caffeine encapsulation. TGA was carried out under nitrogen flow.

3.2. Caffeine-Loaded Capsules

The XRD patterns of the UiO-66 samples before and after caffeine encapsulation are shown in Figure 3. Peaks corresponding to caffeine do not appear in the CAF@MOF samples, which indicates that there were no caffeine crystals in the samples, suggesting the encapsulation of the additive in the porous framework of UiO-66 and UiO-66-NH$_2$. Slight shifts in the peaks can be noted after caffeine encapsulation. Particularly in CAF@UiO-66-NH$_2$, the XRD peaks were displaced to a lower angle. This shift observed may involve an increase in the d-spacing or the distance between crystal planes in the UiO-66 structure. In the bibliography, the introduction of ionic liquids into the micropores of UiO-66 resulted in a slight expansion of the crystal lattice [61]. Nevertheless, the preservation of the XRD peaks indicates that the crystal structure remains after caffeine loading.

The presence of caffeine in the encapsulated samples was affirmed by using FTIR spectroscopy, as shown in Figure 4. In the spectra of caffeine, some typical bands were identified at 1693 and 1645 cm^{-1}, corresponding to the carbonyl bond C=O. Slight shifts in carbonyl groups were detected in the spectra of CAF@UiO-66c (from 1693 and 1645 cm^{-1}

to 1701 and 1653 cm^{-1}), in the spectra of CAF@UiO-66s (to 1701 and 1654 cm^{-1}), and in the spectra of CAF@UiO-66-NH$_2$ (to 1698 and 1652 cm^{-1}). These shifts suggest interactions between caffeine molecules and UiO-66 groups [34]. In the FTIR spectra, the main functional groups in the MOF are shown to have been preserved in the CAF@UiO-66 samples.

The thermal stability of the capsules was measured by using TGA. Additionally, with this technique, the caffeine encapsulation was estimated and was later corroborated by the release experiments and analyzed by UV–Vis. Therefore, the efficiency of additive encapsulation can be observed in terms of the initial degradation temperature of caffeine in the encapsulated samples. Figure 6 represents the TGA and DTG (derivative thermogravimetry) for the caffeine and UiO-66 samples before and after encapsulation.

For the CAF@UiO-66c sample, the biggest weight loss related to caffeine appears delayed (maximum weight loss at 310 °C and degradation between 220 and 380 °C) compared to the curve of pure caffeine (maximum weight loss at 250 °C and degradation between 180 and 290 °C). This delay in the caffeine degradation may be explained by its encapsulation in the UiO-66 framework. The percentage of encapsulated caffeine measured for this peak is 24.2 wt%. There is also a small weight loss (approximately 1.5 wt%) between 180 and 220 °C that may be related to free caffeine or caffeine superficially adsorbed on the MOF. The total value of caffeine present in this sample is close to the theoretical value added in the encapsulation process, which is 25 wt%. The MOF decomposition occurs at the same temperature as it occurs in the crystalline MOF before encapsulation; therefore, the UiO-66 structure remains unaffected after the grinding encapsulation process.

In the case of CAF@UiO-66s, the delay related to caffeine degradation is lower than for CAF@UiO-66c. In this case, the degradation of caffeine takes place between 180 °C and 370 °C with two maximums, one at approximately 240 °C and the other at 290 °C. This fact can be explained by the lower surface area of UiO-66s, which is not as crystalline as UiO-66c, and, therefore, the caffeine encapsulated inside the micropores will be less, and a part of the caffeine is adsorbed in mesopores or superficially adsorbed in the MOF since its external area is high given its small particle size. The caffeine load of this sample was 26.4 wt%, again, close to the theoretical value of 25 wt%.

In the case of CAF@UiO-66-NH$_2$, the degradation of caffeine takes place between 180 °C and 380 °C with two maximums, one at approximately 230 °C and the other at 320 °C, which again can be related to two types of caffeine encapsulated in the micropores and the mesopores/superficial. Analyzing the weight loss corresponding to caffeine for the UiO-66-NH$_2$ sample the additive load was 25.8 wt%.

The sorption isotherms before and after encapsulation have similar shapes for each MOF, although the quantity adsorbed in the encapsulated samples is much lower (Figure 2). The BET analysis was performed after the caffeine encapsulation (Table 1) in order to verify the decrease in the surface area resulting from the additive encapsulation. The specific surface area of UiO-66c decreased by 83% (from 1298 m^2/g to 216 m^2/g) while the decrease in UiO-66s was slightly lower at 49% (from 771 m^2/g to 396 m^2/g). Following the same trend, UiO-66-NH$_2$ decreased its area by 77% (from 650 m^2/g to 150 m^2/g).

The ^{13}C CP MAS NMR spectra (Figure 7) were achieved for pure caffeine, both UiO-66, UiO-66-NH$_2$, and the caffeine encapsulated samples (CAF@UiO-66c, CAF@UiO-66s and CAF@UiO-66-NH$_2$). The observed chemical shifts (in ppm) for caffeine correspond to 27.8 (C10), 29.61 (C12), 33.82 (C14), 104.56 (C5), 141.40 (C8), 146.55 (C4), 149.65 (C2), and 153.01 (C6) [62]. The NMR spectra of UiO-66 contain three peaks located at isotropic chemical shifts of 127.52 ppm (C2, C3, C5, C6), 135.53 ppm (C1, C4), and 169.78 ppm (C7, C8). The peak at the high chemical shift corresponds to carbon atoms from a carboxylic group. The peak at 127.52 ppm is assigned to carbon atoms bonded to a proton. The peak at 135.53 ppm corresponds to quaternary aromatic carbon atoms [63]. Peaks corresponding to pure caffeine are clearly seen in the encapsulated samples, in C5 105.50, 105.24, and 105.79 ppm for CAF@UiO-66c, CAF@UiO-66s, and CAF@UiO-66-NH$_2$, respectively, with a slight change in chemical shift as compared with the caffeine spectra (104.56 ppm), suggesting the encapsulation of the additive. The intensity of these peaks is lower for encapsulated samples. In addition, the peaks of UiO-66 were observed in the ball mill samples, indicating the preservation of the UiO-66 structure.

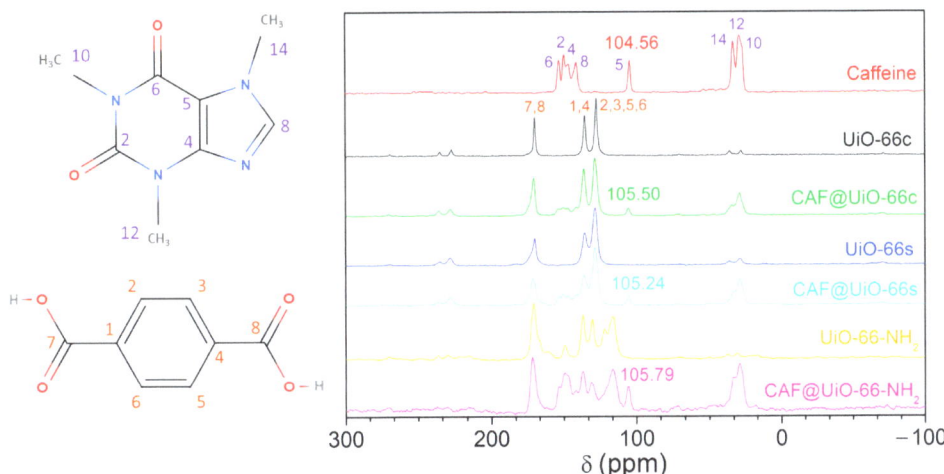

Figure 7. ^{13}C CP MAS NMR spectra of caffeine, UiO-66c, UiO-66s, and UiO-66-NH$_2$, and the corresponding encapsulated samples (CAF@UiO-66c, CAF@UiO-66s, and CAF@UiO-66-NH$_2$).

SEM analysis was used to observe the physical morphology of the samples after caffeine encapsulation (Figure 5). As can be observed, the UiO-66 morphology of all the samples was preserved after ball mill encapsulation; therefore, the crystalline framework and, by extension, most of the UiO-66 properties were maintained after the process. A pure caffeine sample was also seen using SEM (Figure A4). Caffeine crystals have an elongated shape with a particle size of approximately 7.8 ± 3.3 μm. This caffeine crystals do not appear in the encapsulated samples.

To summarize the characterization of the capsules with caffeine encapsulated in MOF by ball milling, the following indications ensure the correct encapsulation of caffeine: the XRD patterns do not show peaks corresponding to pure caffeine; the TGA analysis shows a delayed degradation of caffeine compared to pure caffeine and whose weight loss approximately coincides with the expected load of 25 wt% of additive; the FTIR and NMR analyses show the characteristic peaks of caffeine with slight shifts; the specific surface and pore volume of the capsules decrease significantly compared to the pure MOF; and the SEM images preserve the shape of the MOF without observing caffeine crystals. These indications are corroborated by the slower release of caffeine from the capsules than caffeine without encapsulating, as described in the caffeine-release section below.

3.3. Characterization of Textile Composites

Figure 8 represents the XRD spectra of the PA6 and PLA composites with capsules and with pure caffeine. The pure polyamide shows a wide peak around 2θ = 21.4°, characteristic of γ-PA6 [64], which indicates its semi-crystalline character. With the presence of capsules or caffeine, three peaks are related to PA6, the one just indicated above at 2θ = 21.4°, and two new peaks at 2θ = 20.4° and 23.7° related to α-PA6 [64,65]. This would indicate a certain degree of change in the crystalline character of PA6 with the presence of the filler. The pure PLA shows a practically amorphous character with a very wide peak centered at 2θ = 16° that seems to vary with the addition of capsules based on caffeine or pure caffeine, which would suggest a certain degree of interaction of these materials with the PLA polymer. According to the XRD results, some characteristic peaks of the UiO-66 are present in the sample composites with capsules, indicating the presence of the capsules and providing evidence that the crystalline structure of the MOF was maintained after the extrusion process. For polymer composites with synthesized UiO-66, the observed peaks are broader, as occurred in the XRD of the solid samples (see Figure 3). In the case of the PA6 and PLA composites with pure caffeine, there are no visible peaks corresponding to this additive.

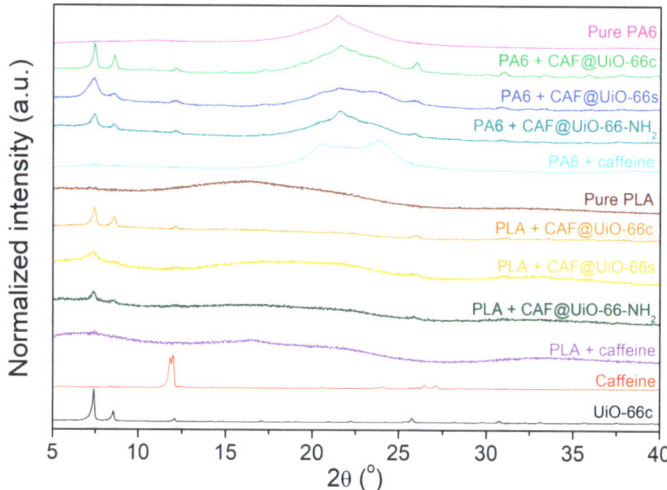

Figure 8. XRD of PA6 and PLA composites of UiO-66 encapsulated samples and pure caffeine.

In addition, the presence of caffeine and UiO-66 in the polymer composites was studied by FTIR analysis. Figure 9 shows the spectra of the PA6 and PLA composites. In this figure, signals appear corresponding to the caffeine-loaded capsules and to each polymer. For the pure PA6 composite, the main bands of this polymer can be observed [66]. The bands located at 3300 and 1538 cm^{-1} were assigned to the stretching and bending vibration of the hydrogen bonds in the amide II. These same bonds for amide I present a band at 1636 cm^{-1}. The carbonyl group (C=O) is represented by the band at 1636 cm^{-1}. Additionally, the signal at 3080 cm^{-1} corresponds to the intramolecular bond between amide and carbonyl groups (-CONH$_2$). The O=C-N bond related to the ring opening is confirmed by the presence of the band at 1262 cm^{-1} which corresponds to the C$_{aliphatic}$-N bond. The bands observed at 2932 and 2861 cm^{-1} were assigned to the vibration of symmetrical and asymmetrical stretching of –CH$_2$ and –CH$_3$ groups, respectively. Additionally, the presence of groups was confirmed due to the presence of the bands at 1462 and 1418 cm^{-1} for –CH$_2$, and the band at 1371 cm^{-1} corresponding to the –CH$_3$ group. Finally, the γ-PA6 phase is denoted by the presence of a band at 977 cm^{-1} [64]. For the PA6 composites with capsules or caffeine, a band appears at 928 cm^1 related to the α-PA6 phase. In Figure 9, the most important bands of UiO-66 at 1578 cm^{-1} (C=O carbonyl bond), 1507 cm^{-1} (C=C phenyl ring), 1391 cm^{-1} (C-O of carboxylic group), and 743 and 664 cm^{-1} (O-Zr-O) are reflected in the PA6 composites. These bands are marked by dark dash lines in Figure 9. Likewise, there are some bands that could correspond to caffeine, such as 1700 cm^{-1}, and, although it coincides with a UiO-66 band, that of 743 cm^{-1}.

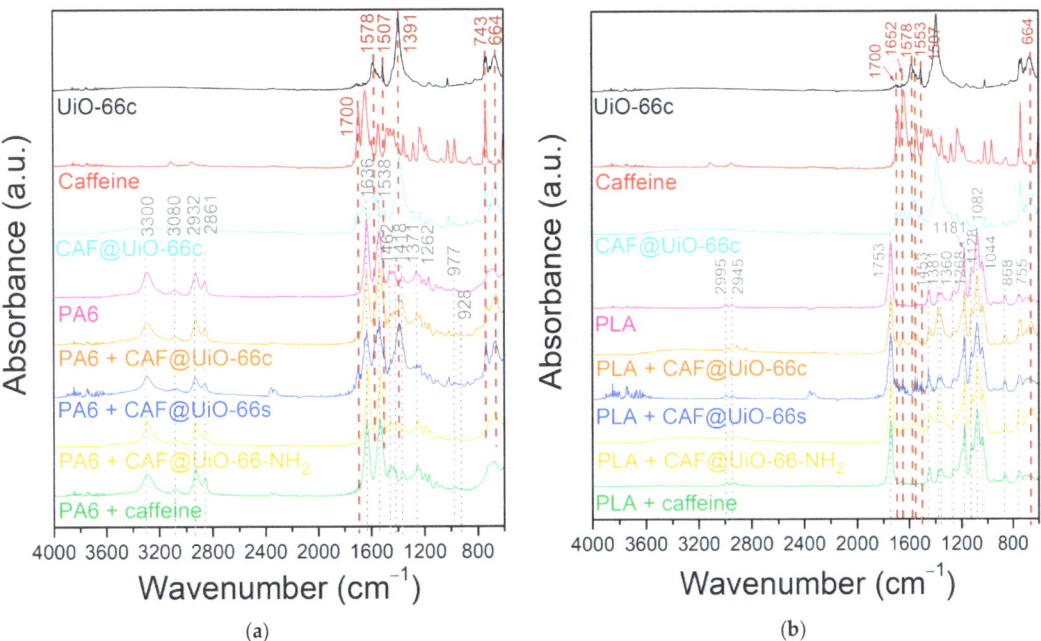

Figure 9. FTIR spectra of polymer composites of UiO-66 encapsulated samples and pure caffeine: (**a**) PA6. (**b**) PLA.

The FTIR spectra of the pure PLA composites show the main bands for this polymer (Figure 9b) [67]. The bands located at 2995 and 2945 cm^{-1} correspond to the vibration of the symmetrical and asymmetrical stretching of C-CH$_3$. The stretch of the carbonyl bond (C=O) is represented by the band at 1753 cm^{-1}. The band located at 1453 cm^{-1} is due to the asymmetrical bending of the CH$_3$ group. Additionally, the bands observed at 1381 and 1360 cm^{-1} represent the asymmetrical and symmetrical bending of CH. The bending of C=O is denoted by the presence of a band at 1268 cm^{-1}. The stretching symmetrical vibration of the C-O-C bond corresponds to the bands at 1181, 1128, and 1082 cm^{-1}. The band at 1044 cm^{-1} was assigned to the C-CH$_3$ stretching. The C-C stretching is denoted by the band at 868 cm^{-1}. In the case of the PLA samples with capsules (CAF@UiO-66c), their corresponding bands also appear in the composite's spectra, especially the bands corresponding to the MOF, the caffeine bands being not very noticeable. The bands corresponding to the UiO-66 present in the composite samples are at 1578 cm^{-1} (C=O) and 1507 cm^{-1} (phenyl ring C=C), and the band at 664 cm^{-1} is due to the (O-Zr-O) bond. Moreover, there are slight bands that could correspond to caffeine, specifically at 1700 and 1652 cm^{-1}, which denote the presence of the carbonyl (C=O) bond. The results are in agreement with what was observed by XRD in terms of the preservation of the structure and functional groups of the MOF. For polymer composites with pure caffeine, bands corresponding to this additive are not visible in the spectra.

The prepared PA6 and PLA composites were characterized by TGA (Figure 10). The TGA curves of the pure PA6 composites (Figure 10a) show differences between the pure polymer and PA6 with capsules. In that of PA6, there is a small continuous degradation (<5 wt%) between 100 and 250 °C that may be related to unreacted reagents and small amounts of water. The large mass loss related to the degradation of the polyamide begins at around 350 °C with a maximum in DTG at 445 °C (Figure 10b). In the PA6 with capsules, the degradation of the sample occurs earlier than in the pure PA6, with a maximum in DTG at 432 °C, 420 °C, and 430 °C for UiO-66c, UiO-66s, and UiO-NH$_2$, respectively. This faster degradation may be due to a slight catalytic effect of the MOF present in the capsules since the degradation of PA6 with pure caffeine has a maximum DTG (446 °C) at the same temperature as the pure PA6. In the polymer composites with capsules and pure caffeine, there is a detectable weight loss in the DTG before the degradation of the polymer, which could be related to the presence of caffeine as well as stabilization since it occurs at higher temperatures than pure and encapsulated caffeine. Furthermore, in the polymer composites with capsules, a clear weight loss is shown after the degradation of the polymer, which may be due to the MOF. This represents an acceleration in the degradation of the MOF, probably due to the caloric effects of the degradation of the polymer, which indicates good integration of the capsules in the polymer. It should be noted that, in these samples, the final residue of the TGA carried out in the air is ZrO_2.

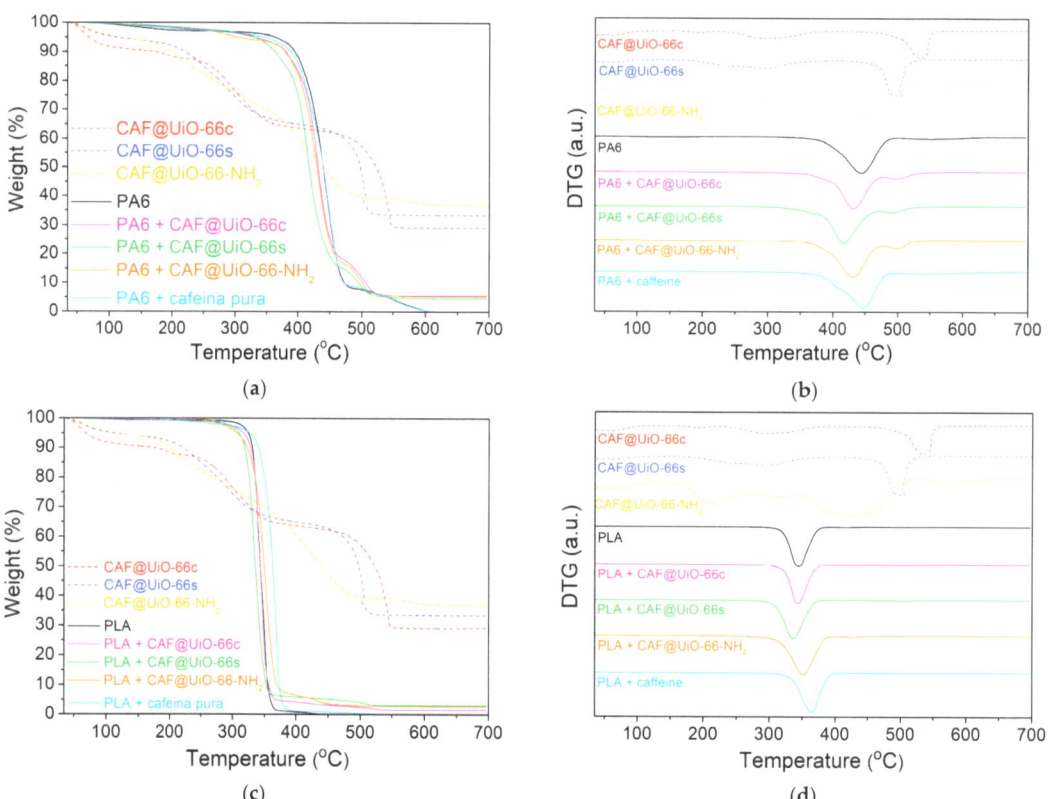

Figure 10. TGA (**a**) and DTG (**b**) characterization of PA6 composites. TGA (**c**) and DTG (**d**) characterization of PLA composites. TGA was carried out under air flow.

Regarding the residual weight loss after each analysis, the amount of zirconium dioxide (zirconia) in the sample can be calculated. Therefore, the experimental percentage of capsules inside the composites can be determined. For the PA6 composites, the final percentage of ZrO_2 was 5.21 wt%, 4.57 wt%, and 5.40 wt% for UiO-66c, UiO-66s, and UiO-66-NH_2, respectively. Taking into account the previous values and UiO-66 as the molecular formula $C_{48}H_{28}O_{32}Zr_6$ and UiO-66-NH_2 as $C_{48}H_{28}N_6O_{32}Zr_6$, the experimental percentage of capsules in the composites has been calculated, as shown in Table 2. The theoretical percentage of capsules was 10 wt%. The difference between this value and the previous ones may be due to the small quantity of sample used in the TGA analysis, which may not be very reproducible from the total sample.

Table 2. Experimental percentage of capsules inside the PA6 and PLA composites. The theoretical value is 10 wt% in all cases.

Polymer	Capsule	wt% Capsules (Experimental) [1]
PA6	CAF@UiO-66c	9.4 ± 5.5
	CAF@UiO-66s	9.0 ± 6.7
	CAF@UiO-66-NH$_2$	11.4 ± 8.0
PLA	CAF@UiO-66c	5.7 ± 1.5
	CAF@UiO-66s	11 ± 6.0
	CAF@UiO-66-NH$_2$	8.9 ± 0.4

[1] The value has been calculated by TGA with at least two samples.

Figure 10c,d shows little variation in the degradation of the PLA composites with and without UiO-66 capsules. In the sample with pure caffeine, there is a certain delay in the degradation of the polymer. In the samples with capsules, there is a very slight advance in the degradation of the polymer. In any case, a residual weight was observed after each analysis, reaching values of 1.56 wt% (composite with UiO-66c), 3.11 wt% (composite with UiO-66s), and 2.75 wt% (composite with UiO-66-NH$_2$). With these data, as with the PA6 composites, the percentage of capsules in the PLA composite was determined (Table 2). The values, although they are of the expected order, do not agree with the theoretical percentage, which is related to the small amount of sample used in the analysis.

To confirm the presence of caffeine in the polymer composites, a ^{13}C NMR analysis (Figure A5) was also performed of the PA6 composites with caffeine and with CAF@UiO-66c, the sample which gave the best release results, as will be shown below. The ^{13}C NMR spectra of PA6 can present different forms depending on the polyamide structure [68]. The PA6 composite spectra obtained with caffeine and capsules are similar to the spectra of pure PA6. Neither caffeine nor UiO-66 were detected inside the composites.

Figures 11 and 12 show the SEM images of the PA6 and PLA composites. As expected, for the pure PA6 composite no particle was observed (Figure 11a). In the composites with CAF@MOF, UiO-66 particles were seen in the PA6 composite cross-section (Figure 11b–d) with a shape very similar to the particles before being added to the composite and showing good contact between the capsules and the PA6. This would be related to the organic character of the MOF that makes it compatible with the polymer. In general, regardless of the type of capsule, both isolated particles (Figure 11b) and particle aggregates (Figure 11c,d) can be found in the samples, although in the sample with UiO-66c, the particles are less aggregated. On the samples with capsules (Figure 11b–d), EDX analysis was performed at the indicated points. The Zr atomic percentage values of the EDX analysis of these composites are shown in Figure 11f. In the PA6 + CAF@UiO-66c sample, the Zr element was detected by EDX analysis at different points of the solid particle. Another zone away from this particle was also analyzed (point 4), where the Zr percentage was zero. In the composite sample with synthesized UiO-66, zirconium was also detected in an aggregate of particles reaching values around 2.5 at%. If a larger aggregate of particles is analyzed, such as the one in the composite of UiO-66-NH$_2$ capsules, zirconium is detected in greater quantities (Figure 11f). In any case, with the EDX analysis and the preservation of its shape, it is possible to confirm the presence of the MOF after the composite manufacturing process. It should be noted that the composite manufacturing process occurs at 260 °C, a temperature which, as has been observed in the TGA, the MOF capsules with caffeine can withstand. Figure 11e clearly shows the caffeine crystals inside the PA6 composite, which indicates that the pure caffeine added to the composite is also capable of withstanding the processing conditions, as was already intuited in the TGAs of the composites.

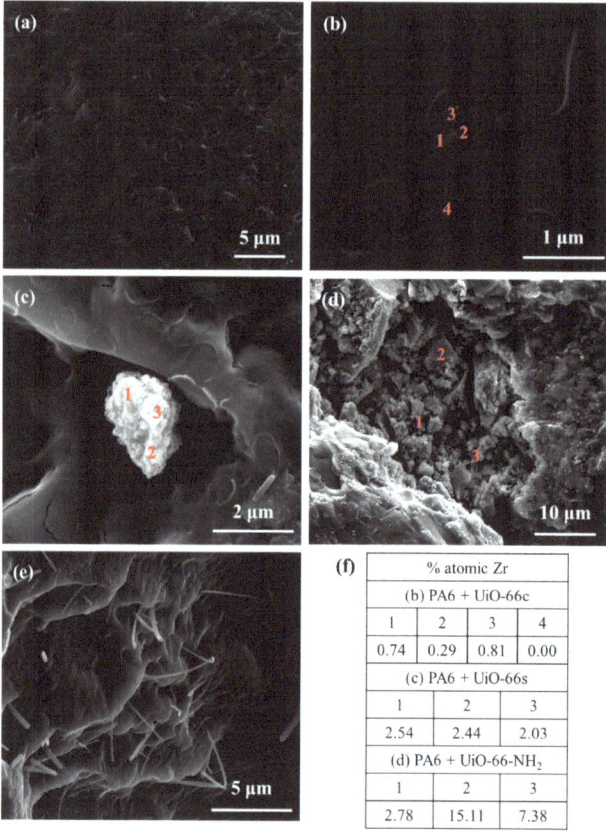

Figure 11. SEM images of PA6 composites (**a**) pure PA6, (**b**) PA6 + CAF@UiO-66c, (**c**) PA6 + CAF@UiO-66s, (**d**) PA6 + CAF@UiO-66-NH$_2$, (**e**) PA6 + caffeine, (**f**) Atomic percentages of Zr in cross-section SEM images obtained from EDX analysis.

In the PLA composites with capsules (Figure 12b–d), MOF particles were also observed preserving their appearance, showing good contact with the polymer, and sometimes forming aggregates. In addition, the presence of the Zr element was confirmed by EDX analysis, reaching values between 2–4 at%. It should be noted that the processing of PLA composites was at 190 °C, which is slightly lower than the processing of PA6 (260 °C) in which, as mentioned above, the capsules were not degraded. Unlike the PA6 + caffeine composite, for the PLA + caffeine sample, no caffeine crystals were observed in the cross-section. This may be due to the reaction that takes place between PLA and caffeine, which is explained in more detail in the caffeine delivery section.

The melt volume rate values of the PA6 and PLA composites were measured during composite preparation, and the values are represented in Table 3. Composites with the capsules, especially PLA, presented lower fluency rates in comparison with the pure polymers. This decrease is not very pronounced, so it does not affect the composites' processing or their mechanical properties. The values for the PLA composites are particularly low with respect to the PA6 composites due to the lower fluidity of the pure polymer.

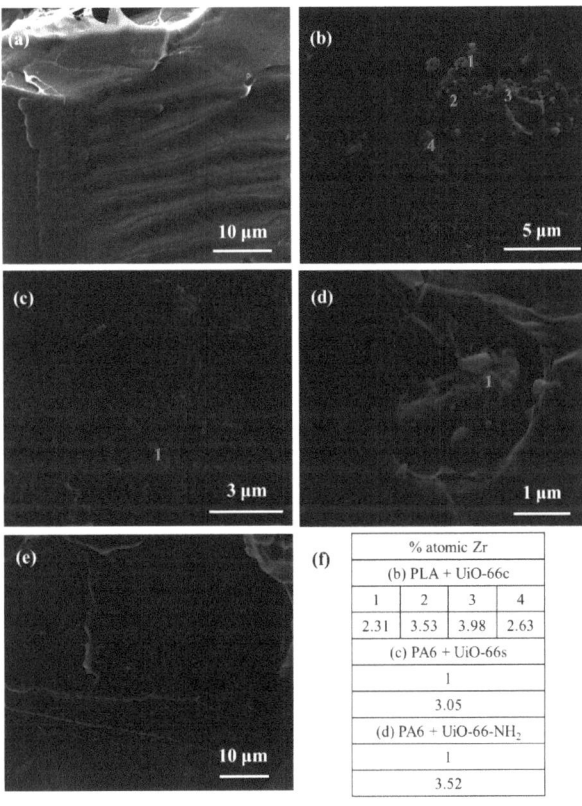

Figure 12. SEM images of PLA composites (**a**) pure PLA, (**b**) PLA + CAF@UiO-66c, (**c**) PLA + CAF@UiO-66s, (**d**) PLA + CAF@UiO-66-NH$_2$, (**e**) PLA + caffeine, (**f**) atomic percentages of Zr in cross-section SEM images obtained from EDX analysis.

Table 3. Melt volume rate values (MVR) of PA6 and PLA composites.

Polymer	Capsule	MVR (cm^3/10 min)
PA6	Pure	78.78 ± 4.50
	CAF@UiO-66c	73.42 ± 6.37
	CAF@UiO-66s	60.32 ± 14.17
	CAF@UiO-66-NH$_2$	63.88 ± 6.91
	Caffeine	77.01 ± 4.65
PLA	Pure	5.75 ± 0.45
	CAF@UiO-66c	2.40 ± 0.88
	CAF@UiO-66s	2.96 ± 1.45
	CAF@UiO-66-NH$_2$	1.25 ± 0.36
	Caffeine	3.66 ± 0.19

In summary, it can be stated that the incorporation of the capsules with caffeine in the PA6 and PLA composites has been successful given the following evidence obtained in the characterization of the composite materials: the XRD patterns show the preservation of the structure of the capsules; FTIR analysis denotes the presence of the main functional groups of the capsules; TGA reveals a certain influence in the degradation of the composite due to the presence of the capsules observing a final residue related to them; the SEM images

3.4. Caffeine-Release
3.4.1. CAF@MOF

Figure 13 shows the evolution with time of caffeine release in the CAF@MOF samples. As shown in the figure, the best encapsulation result was achieved for the CAF@UiO-66c sample since the release is more gradual than for the other samples. With this sample, the maximum caffeine percentage was reached at seven days until a value of 20 wt%. For the CAF@UiO-66s sample, the initial release is similar to the CAF@UiO-66c, but at shorter times the release is faster, achieving a value of around 20 wt% at 3 h. Then, for longer times, the value stabilizes at four days (23 wt% of caffeine in the capsules). This difference can be attributed to the textural properties of UiO-66. On the one hand, due to the smaller particle size of the UiO-66s than the UiO-66c, the external surface in contact with water is higher, so the release would be faster. On the other hand, the specific surface area of the UiO-66s is lower and with a smaller volume of micropores, so its capacity is more limited than the UiO-66c. Although UiO-66s has mesopores, these appear to release caffeine more quickly. Furthermore, as seen by TGA, there appear to be two types of caffeine in this sample and in general less interaction with the MOF than in the CAF@UiO-66c sample. In these release experiments, the temperature was increased to 80 °C for two days (from day 8 to day 10), in order to know the maximum caffeine content in the capsules, and reached values of 23 wt% and 27 wt% for UiO-66c and UiO-66s, respectively. In both cases, the percentages were similar to those calculated by TGA and, therefore, also to the theoretical value of encapsulated caffeine (25 wt%). For the physical mixture, the release was faster in all cases, since at 10 min the caffeine release was around 18 wt%. In both CAF@UiO-66, there was an improvement in the rate of caffeine release related to the physical mixture. This difference was more significant for UiO-66c, as already mentioned.

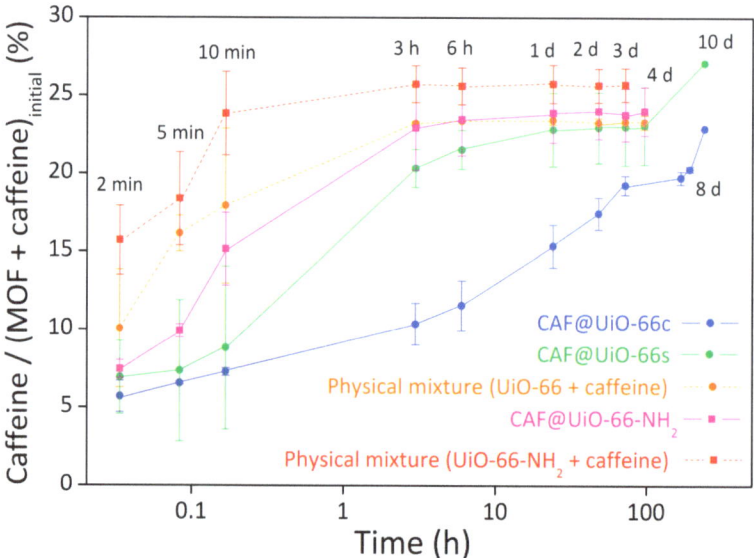

Figure 13. Release experiments of solid samples, CAF@UiO-66c, CAF@UiO-66s, CAF@UiO-66-NH$_2$, and a physical mixture of both MOF with caffeine, determined by UV–Vis absorption. The error bars come from the measurement of two samples prepared in different batches.

The results for UiO-66-NH$_2$ showed an improvement in the caffeine release for the sample encapsulated in the ball mill (CAF@UiO-66-NH$_2$) compared to the physical mixture (caffeine + UiO-66-NH$_2$). Comparing the ball mill samples, the synthesized MOFs (UiO-66s and UiO-66-NH$_2$) presented similar results. These release results were in accordance with the surface area and TGA results, which are very similar. At low times, the release of caffeine is higher for UiO-66-NH$_2$, which may be due to the amino group hindering caffeine encapsulation, so the caffeine was not well encapsulated. These results are in agreement with observations in the literature that the −NH$_2$ group causes competitive adsorption of the water used as a medium for impregnation, which worsens the encapsulation of caffeine [69].

3.4.2. Polymer Composites

The caffeine delivery from the polymer composites was also studied. Samples were put in contact with distilled water for a period of time (until 14 days) and the temperature was increased from 25 °C to 80 °C. The polymer composites were maintained for several days at this final temperature, until reaching a constant value of caffeine in the solution. Figure 14 shows a comparison of the release results of both kinds of polymers with all the caffeinated composites. Since the mixing method used for composite processing did not produce a homogeneous distribution of the capsules and the polymer, as observed by SEM, the results obtained, given the small amounts of composites taken in the release, will depend on the pieces of composite selected to carry out the release. The values in Figure 14 have been normalized taking into account the maximum value obtained in each composite sample, which in most cases corresponds to the final value (14 days) already stabilized over time.

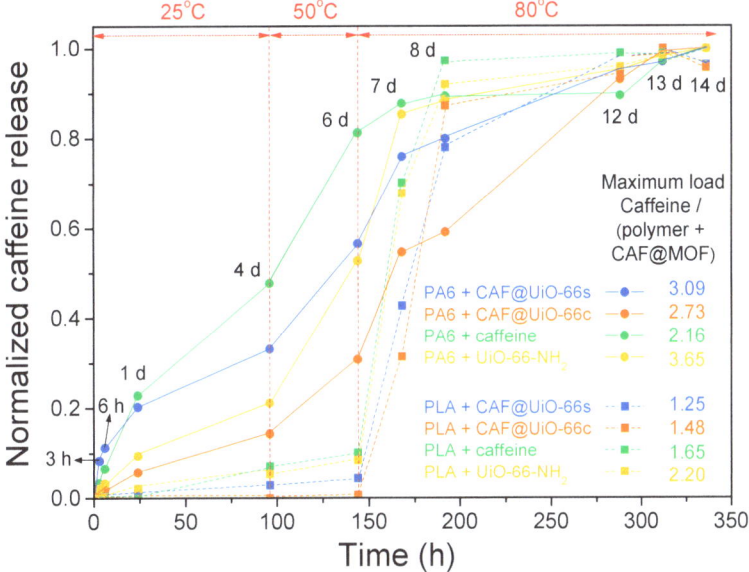

Figure 14. Normalized values of caffeine release in PA6 and PLA composites showing the evolution with time and temperature. The amount of polymer composites introduced was 0.5 g for all samples.

For PA6, the release is faster in the polymer composite containing caffeine without encapsulation. Therefore, the encapsulation process in the UiO-66 makes the release of caffeine more gradual over time, extending the service life of the textile product. Comparing the PA6 polymer composites with the capsules, the ones prepared with UiO-66c exhibit more gradual caffeine release, as occurred with the capsules prepared with this MOF (see

Figure 13). As indicated, the sample with UiO-66c, which has the largest surface area, microporosity, and particle size, and in the TGA analysis delays the degradation of caffeine indicating a deeper encapsulation, shows the most gradual release. In the case of the sample with UiO-66-NH$_2$, the release of caffeine seems slower than for UiO-66s in the PA6 composite but much faster than UiO-66c. In any case, it seems that the textural properties have a more prominent role in the release than the presence of the amino group.

In the case of the PLA composites, the release was very slow at initial times when the temperature was relatively low and experienced a sharp increase when the temperature was increased up to 80 °C. Studies in the literature have analyzed the effect of basic drugs, such as caffeine [44], on PLA degradation. They have concluded that basic drugs neutralize the carboxyl end groups and minimize the autocatalytic effect of acidic chain ends in polymer degradation and, therefore, the drug release rate decreases [70]. Furthermore, the release of caffeine is delayed when it is encapsulated, in this case being evident at high temperatures. In any case, it seems that an interaction is established between PLA and caffeine, which means that even the caffeine sample without encapsulation shows a remarkably slow release. This interaction between caffeine and PLA could occur during extrusion.

To verify the reproducibility of the trends obtained, an experiment similar to the previous one was repeated with some representative samples but with a greater amount of composite, between 7.4–9.4 g (Figure A6). In this experiment, it can be observed again that in the PA6 polymer, the encapsulation of caffeine delays its release, with the CAF@UiO-66c capsules showing the slowest release. In the PLA polymer, it is observed that there is a differential delay in the release of caffeine with respect to the PA6 polymer. In PLA at low temperatures, it does not seem that the encapsulation of caffeine favors its retention, and high temperatures are needed to clearly show that the encapsulation in UiO-66c delays its release. These results suggest that an interaction is established between caffeine and PLA that could occur during extrusion and that undoubtedly delays its release. In any case, for the PA6 composites, the encapsulation in UiO-66 is essential in order to achieve a more gradual release.

To confirm the presence of caffeine, the liquids obtained from the solid–liquid extractions shown in Figure 14 (at 12 days) were analyzed by ^{1}H NMR (Figure A7). The ^{1}H NMR spectrum of caffeine consists of the sole ring ^{1}H signal (7.58 ppm) and the signals from three methyl groups (3.37, 3.55, and 4.01 ppm) [71]. In Figure A7a (PA6) and A7b (PLA), peaks can be observed corresponding to caffeine in the ^{1}H NMR spectra of the polymer composites. For the PA6 composites, there are some little peaks apart from these, which may be due to polyamide decomposition. In the case of the PLA composites, peaks other than caffeine appear with higher intensity, due to the fact that polylactic acid at 80 °C decomposes more than polyamide. These peaks appear in other regions than the caffeine ones, therefore they are not influenced by the presence of PLA or PA6.

3.5. Benchmark

The use of UiO-66 in encapsulation has been widely reported by other authors due to its great adsorption properties and high thermal stability. The encapsulated compound was caffeine in several studies, among other additives. Sarker et al. [34] reported the storage and controlled release of caffeine from UiO-66 and the functionalized form UiO-66-COOH, achieving caffeine loadings of 21.4 and 17.5 wt%, respectively. The release of caffeine from pristine UiO-66 studied in a buffer solution (PBS, 0.01 M, pH: 7.4), was fast and delivered 85% of the encapsulated caffeine in 12 h. Cunha et al. [32] achieved a caffeine loading of 22.4 wt% and 13.2 wt% for UiO-66 and UiO-66-NH$_2$, respectively. In a later work, they analyzed the caffeine delivery of these MOFs in aqueous media at 37 °C. The experiments achieved 75% and 60% caffeine release in 24 h for the UiO-66 and amino-functionalized form, respectively [33]. Zhou et al. reported the encapsulation of caffeic acid (CA) in UiO-66, improving its stability. CA@UiO-66 showed the highest CA loading rate of 56% and a release ability of CA of 83% until 100 h [72]. Caffeine was also encapsulated in UiO-66 by solvent-free encapsulation under high-pressure contact of the additive and the MOF. The encapsulation value achieved with UiO-66 was 15 wt% [37]. Moreover, by a simple impregnation method, caffeine was encapsulated in UiO-66 and UiO-66-NH$_2$ by Devautour-Vinot et al., obtaining encapsulation values of 22.4 wt% and 13.2 wt%, respectively [69]. In our study, 20–25 wt% of caffeine was encapsulated in the capsule (0.333 mg of caffeine/g MOF), which is slightly above the values mentioned in the literature. Although very dependent on the operational conditions, with the best material used in this work (UiO-66c) approximately 75% of the encapsulated caffeine was released in 24 h at room temperature, which is in line with the data discussed above.

In the production of PA6 fibers through extrusion, Pérez et al. [15] achieved the incorporation of α-tocopheryl acetate into porous materials, specifically zeolite Y. The resulting PA6 fibers contained 0.075 wt% of the additive. In a similar process, Zornoza et al. [16] encapsulated 0.03%, 0.06%, and 0.17 wt% of caffeine in zeolite Y, MIL-53, and ZIF-8, respectively. In our work with the UiO-66 MOF incorporated in PA6, a higher load of up to 2.5 wt% of caffeine was achieved. However, it must be taken into account that in the cited work the addition was performed during the spinning process to obtain fibers with diameters of 15–20 μm, while in our work melt extrusion was carried out to obtain composites of about 1 mm in diameter.

4. Conclusions

In the present work, composite materials of polymers and metal–organic frameworks based on zirconium (UiO-66) with encapsulated caffeine were successfully developed by extrusion. In particular, the following conclusions can be listed:

- As indicated by the different characterization techniques (XRD, FTIR, SEM, TGA, NMR, and nitrogen adsorption), UiO-66 has been prepared through a solvent-free synthesis with textural properties (specific surface area and pore size distribution) and a particle size that are different from those of commercially purchased UiO-66.
- With the same characterization techniques indicated in the previous point, it has been demonstrated that UiO-66-NH$_2$ has been prepared through a synthesis with ethanol, avoiding the use of toxic solvents, such as DMF.
- The encapsulation of 25 wt% of caffeine by milling assisted by a small amount of water in the capsules (MOF@CAF) has been carried out correctly with the available UiO-66 and also with UiO-66-NH$_2$. This assertion is supported by the comparison of MOF and caffeine with these capsules using various techniques, such as XRD, FTIR, SEM, TGA, NMR, and nitrogen adsorption.
- The caffeine release study of the MOF@CAF samples shows a much slower release compared to the physical mixture of caffeine with the MOF. The release is especially slow in sample UiO-66c.
- The PA6 + MOF@CAF and PLA + MOF@CAF composites have been prepared by extrusion with a theoretical caffeine load of 2.5 wt% and, despite the high processing temperatures (190 °C and 260 °C for PLA and PA6, respectively), the capsules retain their fundamental characteristics, as indicated by their characterization (XRD, FTIR, SEM, EDX, and TGA).
- The caffeine release study of the PA6 + MOF@CAF composites shows a much slower release compared to incorporating pure caffeine into the polymer. Therefore, the encapsulation of caffeine has a determining effect on delaying its release in PA6 polymers.
- Among the MOFs, UiO-66c, with its greater surface area, microporosity, and larger particle size, shows the slowest caffeine release in the polymeric composites, which is also in accordance with the release in the capsules.
- The amino group in the UiO-66 does not appear to play a prominent role in the release of caffeine.
- Both in the CAF@MOF capsules and in pure caffeine, the PLA polymer establishes an interaction with caffeine that delays its release markedly compared to the PA6 polymer. In the PLA composites, the effect of encapsulation is observed at high temperatures.

Finally, it should be mentioned that procedures respecting sustainability have been developed, so that the syntheses of the zirconium-based MOFs have been carried out avoiding the use of toxic solvents. Likewise, the encapsulation of caffeine in these synthesized MOFs has been carried out by liquid-assisted milling, considerably reducing the amount of solvents, reducing the process steps, and avoiding residues. Following the same trend, the MOF@CAF capsules have been embedded during the extrusion process in recycled polyamide 6 (PA6) and in a biopolymer based on polylactic acid (PLA) that is biodegradable. These developed composite materials are the basis for the improvement of sustainable functionalized textile fibers with embedded durable capsules that can dose additives, such as caffeine, slowly.

Author Contributions: Conceptualization, C.P.-V., V.B.-S., E.P., M.Á.C. and C.T.; methodology, C.P.-V., V.B.-S., E.P., M.Á.C. and C.T.; software, C.P.-V.; validation, C.P.-V. and V.B.-S.; formal analysis, C.P.-V.; investigation, C.P.-V. and V.B.-S.; resources, E.P., M.Á.C. and C.T.; writing—original draft preparation, C.P.-V., V.B.-S., E.P., M.Á.C. and C.T.; writing—review and editing, C.P.-V., V.B.-S., E.P., M.Á.C. and C.T.; visualization, C.P.-V., V.B.-S., E.P., M.Á.C. and C.T.; supervision, E.P., M.Á.C. and C.T.; project administration, E.P., M.Á.C. and C.T.; funding acquisition, E.P., M.Á.C. and C.T. All authors have read and agreed to the published version of the manuscript.

Funding: This research was funded by the Government of Aragon, grant number LMP53_21, and the APC was funded by the Government of Aragon. The LMP53_21 project also includes a financial contribution from Nurel S.A. Thanks are also due to the T68-23R project financed by the Government of Aragón.

Institutional Review Board Statement: Not applicable.

Data Availability Statement: Complementary data are available as an appendix and upon request to the authors.

Acknowledgments: Department of Science, University and Knowledge Society of the Government of Aragon for the LMP53_21 project is gratefully acknowledged. The authors would like to thank the University of Zaragoza for the use of the Servicio General de Apoyo a la Investigación-SAI and the use of instrumentation as well as the technical advice provided by the National Facility ELECMI ICTS, node "Laboratorio de Microscopias Avanzadas", at the Universidad de Zaragoza. The Government of Aragon is also gratefully acknowledged for financing the T68_23R group. Thanks to Clara Marquina from INMA for the use of the ball mill.

Conflicts of Interest: The authors declare that this research work has been carried out within the LMP53_21 project that was funded by the Government of Aragon in collaboration with the company Nurel S.A. Elena Piera and Miguel Ángel Caballero were employed by the company Nurel. The remaining authors declare that the research was conducted in the absence of any commercial or financial relationships that could be construed as a potential conflict of interest.

Appendix A

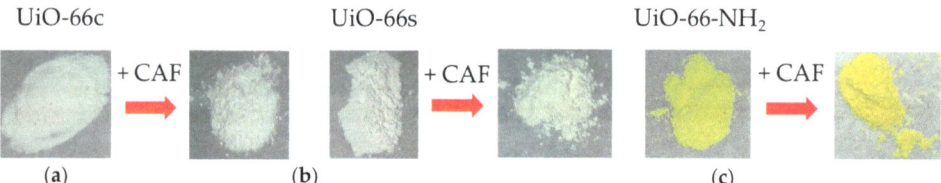

Figure A1. Images of (**a**) UiO-66c, (**b**) UiO-66s, and (**c**) UiO-66-NH$_2$, before and after caffeine encapsulation.

Table A1. Particle size values (nm) obtained from Scherrer's equation calculated at different diffraction peaks (2·θ) for UiO-66c, UiO-66s, and UiO-66-NH$_2$.

	Particle Size (nm)		
2·θ	7.4°	8.5°	25.7°
UiO-66c	49.5	56.9	53.6
UiO-66s	13.4	21.9	23.5
UiO-66-NH$_2$	40.2	42.4	43.6

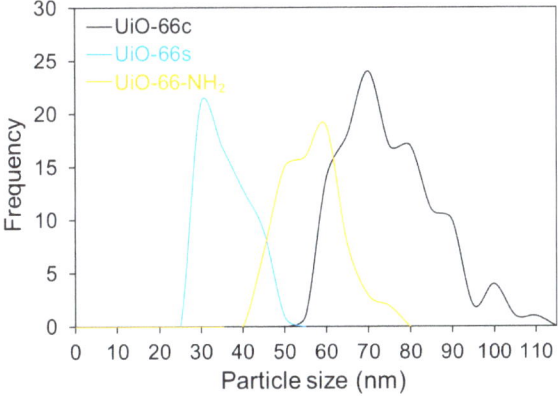

Figure A2. Particle size distribution of UiO-66c, UiO-66s, and UiO-66-NH$_2$.

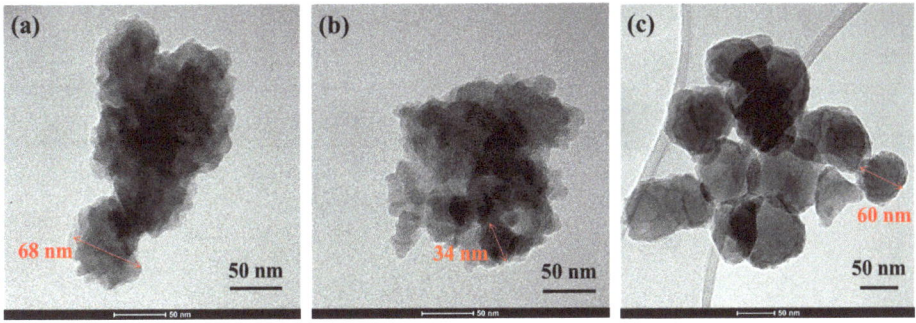

Figure A3. TEM images of (a) UiO-66c, (b) UiO-66s, and (c) UiO-66-NH$_2$.

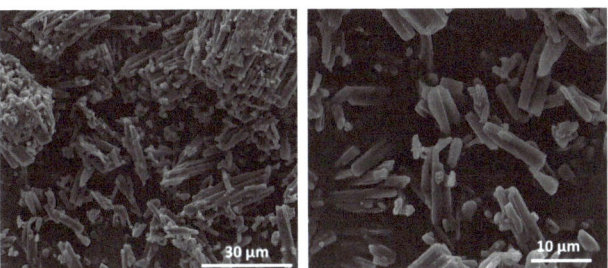

Figure A4. SEM images of caffeine crystals.

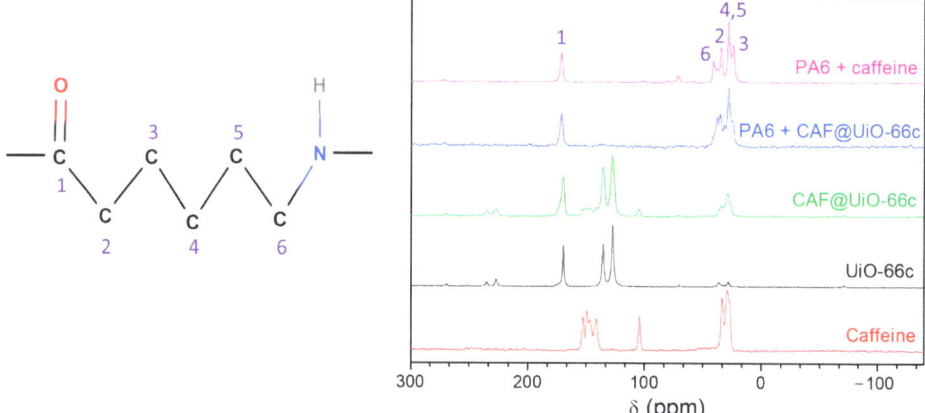

Figure A5. ^{13}C NMR spectra of PA6 composites compared with the solid sample's spectra.

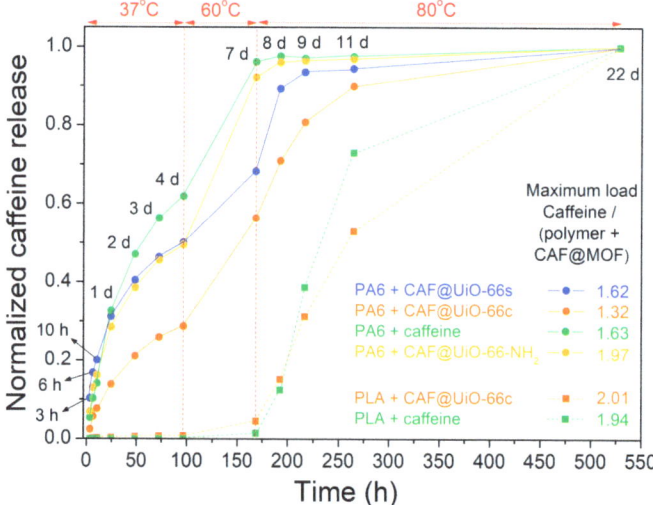

Figure A6. Normalized values of caffeine release in PA6 and PLA composites showing the evolution with time and temperature. The amount of composites introduced was 7.956 g, 8.414 g, 7.398 g, and 7.981 g for PA6 composites (UiO-66c, UiO-66s, caffeine, and UiO-66-NH$_2$) and 9.380 g and 9.4261 g for PLA composites (UiO-66c and caffeine).

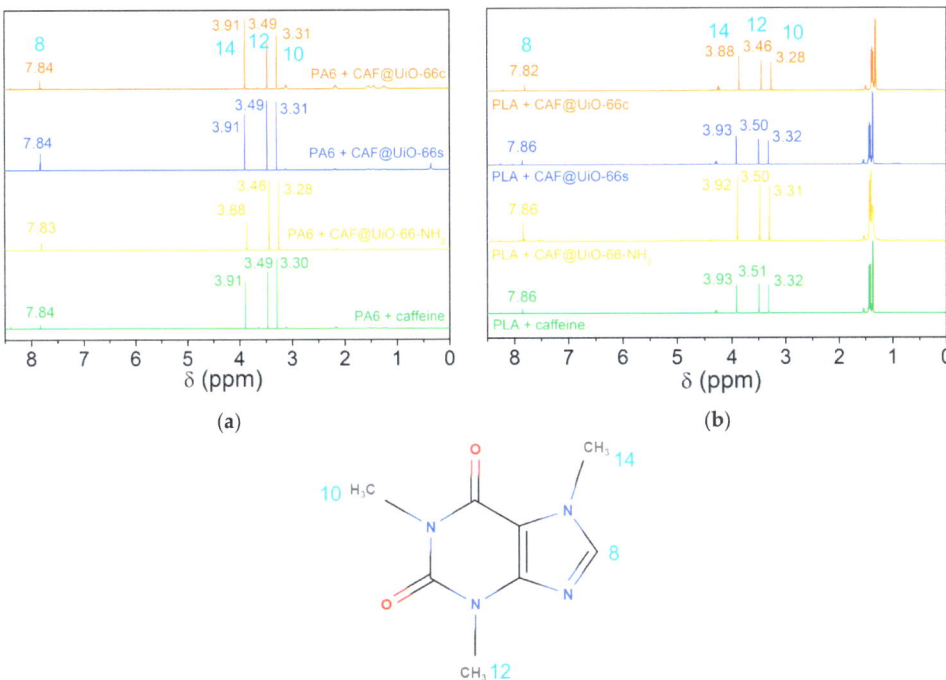

Figure A7. ^{1}H NMR spectra of liquids obtained after liquid–solid extraction at different temperatures (25–50–80 °C) at 12 days (80 °C): (**a**) PA6 composites. (**b**) PLA composites.

References

1. Peng, X.; Umer, M.; Pervez, M.N.; Hasan, K.M.F.; Habib, M.A.; Islam, M.S.; Lin, L.; Xiong, X.; Naddeo, V.; Cai, Y. Biopolymers-based microencapsulation technology for sustainable textiles development: A short review. *Case Stud. Chem. Environ. Eng.* **2023**, *7*, 100349. [CrossRef]
2. García, A.; Ramos, M.; Sanahuja, A.; Garrigós, M. Recent Trends in Microencapsulation for Smart and Active Innovative Textile Products. *Curr. Org. Chem.* **2018**, *22*, 1237–1248.
3. Timilsena, Y.P.; Akanbi, T.O.; Khalid, N.; Adhikari, B.; Barrow, C.J. Complex coacervation: Principles, mechanisms and applications in microencapsulation. *Int. J. Biol. Macromol.* **2019**, *121*, 1276–1286. [CrossRef]
4. Paseta, L.; Simón-Gaudó, E.; Gracia-Gorría, F.; Coronas, J. Encapsulation of essential oils in porous silica and MOFs for trichloroisocyanuric acid tablets used for water treatment in swimming pools. *Chem. Eng. J.* **2016**, *292*, 28–34. [CrossRef]
5. Shi, M.; Lu, B.; Li, X.; Jin, Y.; Ge, M. Thermochromic luminescent fiber based on yellow thermochromic microcapsules: Preparation, properties, and potential application areas. *Cellulose* **2021**, *28*, 5005–5018. [CrossRef]
6. Tariq, Z.; Izhar, F.; Malik, M.H.; Oneeb, M.; Anwar, F.; Abbas, M.; Khan, A. Development of functional textile via microencapsulation of peppermint oils: A novel approach in textile finishing. *Res. J. Text. Apparel* **2022**, ahead-of-print.
7. García-Viñuales, S.; Rubio, C.; Martínez-Izquierdo, L.; Zornoza, B.; Piera, E.; Caballero, M.Á.; Téllez, C. Study of Melamine-Formaldehyde/Phase Change Material Microcapsules for the Preparation of Polymer Films by Extrusion. *Membranes* **2022**, *12*, 266. [CrossRef]
8. Julaeha, E.; Puspita, S.; Eddy, D.R.; Wahyudi, T.; Nurzaman, M.; Nugraha, J.; Herlina, T.; Al Anshori, J. Microencapsulation of lime (*Citrus aurantifolia*) oil for antibacterial finishing of cotton fabric. *RSC Adv.* **2021**, *11*, 1743–1749. [CrossRef]
9. Yaman Turan, N.; Turker, E.; Insaatci, Ö. Microparticles loaded with propolis to make antibacterial cotton. *Cellulose* **2021**, *28*, 4469–4483. [CrossRef]
10. Abdelhameed, R.M.; Rehan, M.; Emam, H.E. Figuration of Zr-based MOF@cotton fabric composite for potential kidney application. *Carbohydr. Polym.* **2018**, *195*, 460–467. [CrossRef] [PubMed]
11. Azizi, N.; Chevalier, Y.; Majdoub, M. Isosorbide-based microcapsules for cosmeto-textiles. *Ind. Crops Prod.* **2014**, *52*, 150–157. [CrossRef]
12. Ghaheh, F.S.; Khoddami, A.; Alihosseini, F.; Jing, S.; Ribeiro, A.; Cavaco-Paulo, A.; Silva, C. Antioxidant cosmetotextiles: Cotton coating with nanoparticles containing vitamin E. *Process Biochem.* **2017**, *59*, 46–51. [CrossRef]

13. Ma, J.; Fan, J.; Xia, Y.; Kou, X.; Ke, Q.; Zhao, Y. Preparation of aromatic β-cyclodextrin nano/microcapsules and corresponding aromatic textiles: A review. *Carbohydr. Polym.* **2023**, *308*, 120661. [CrossRef] [PubMed]
14. Boh Podgornik, B.; Šandrić, S.; Kert, M. Microencapsulation for Functional Textile Coatings with Emphasis on Biodegradability—A Systematic Review. *Coatings* **2021**, *11*, 1371. [CrossRef]
15. Pérez, E.; Martin, L.; Rubio, C.; Urieta, J.; Piera, E.; Caballero, M.; Téllez, C.; Coronas, J.n. Encapsulation of α-Tocopheryl Acetate into Zeolite Y for Textile Application. *Ind. Eng. Chem. Res.* **2010**, *49*, 8495–8500. [CrossRef]
16. Zornoza, B.; Rubio, C.; Piera, E.; Caballero, M.A.; Julve, D.; Pérez, J.; Téllez, C.; Coronas, J. Caffeine Encapsulation in Metal Organic Framework MIL-53(Al) at Pilot Plant Scale for Preparation of Polyamide Textile Fibers with Cosmetic Properties. *ACS Appl. Mater. Interfaces* **2022**, *14*, 22476–22488. [CrossRef]
17. Liu, X.; Demir, N.K.; Wu, Z.; Li, K. Highly Water-Stable Zirconium Metal–Organic Framework UiO-66 Membranes Supported on Alumina Hollow Fibers for Desalination. *J. Am. Chem. Soc.* **2015**, *137*, 6999–7002. [CrossRef]
18. Ma, D.; Han, G.; Gao, Z.F.; Chen, S.B. Continuous UiO-66-Type Metal–Organic Framework Thin Film on Polymeric Support for Organic Solvent Nanofiltration. *ACS Appl. Mater. Interfaces* **2019**, *11*, 45290–45300. [CrossRef]
19. Trickett, C.A.; Helal, A.; Al-Maythalony, B.A.; Yamani, Z.H.; Cordova, K.E.; Yaghi, O.M. The chemistry of metal–organic frameworks for CO_2 capture, regeneration and conversion. *Nat. Rev. Mater.* **2017**, *2*, 17045. [CrossRef]
20. Cirujano, F.G.; Llabrés I Xamena, F.X. Tuning the Catalytic Properties of UiO-66 Metal–Organic Frameworks: From Lewis to Defect-Induced Brønsted Acidity. *J. Phys. Chem. Lett.* **2020**, *11*, 4879–4890. [CrossRef] [PubMed]
21. Zhang, X.; Wang, X.; Li, C.; Hu, T.; Fan, L. Nanoporous {Co_3}-Organic framework for efficiently separating gases and catalyzing cycloaddition of epoxides with CO_2 and Knoevenagel condensation. *J. Colloid Interface Sci.* **2024**, *656*, 127–136. [CrossRef] [PubMed]
22. Pourmadadi, M.; Eshaghi, M.M.; Ostovar, S.; Shamsabadipour, A.; Safakhah, S.; Mousavi, M.S.; Rahdar, A.; Pandey, S. UiO-66 metal-organic framework nanoparticles as gifted MOFs to the biomedical application: A comprehensive review. *J. Drug Delivery Sci. Technol.* **2022**, *76*, 103758. [CrossRef]
23. Cavka, J.H.; Jakobsen, S.; Olsbye, U.; Guillou, N.; Lamberti, C.; Bordiga, S.; Lillerud, K.P. A New Zirconium Inorganic Building Brick Forming Metal Organic Frameworks with Exceptional Stability. *J. Am. Chem. Soc.* **2008**, *130*, 13850–13851. [CrossRef] [PubMed]
24. Chen, Z.; Wang, X.; Noh, H.; Ayoub, G.; Peterson, G.W.; Buru, C.T.; Islamoglu, T.; Farha, O.K. Scalable, room temperature, and water-based synthesis of functionalized zirconium-based metal–organic frameworks for toxic chemical removal. *CrystEngComm* **2019**, *21*, 2409–2415. [CrossRef]
25. Kandiah, M.; Nilsen, M.H.; Usseglio, S.; Jakobsen, S.; Olsbye, U.; Tilset, M.; Larabi, C.; Quadrelli, E.A.; Bonino, F.; Lillerud, K.P. Synthesis and Stability of Tagged UiO-66 Zr-MOFs. *Chem. Mater.* **2010**, *22*, 6632–6640. [CrossRef]
26. Julien, P.A.; Mottillo, C.; Friščić, T. Metal–organic frameworks meet scalable and sustainable synthesis. *Green Chem.* **2017**, *19*, 2729–2747. [CrossRef]
27. Hu, Z.; Peng, Y.; Kang, Z.; Qian, Y.; Zhao, D. A Modulated Hydrothermal (MHT) Approach for the Facile Synthesis of UiO-66-Type MOFs. *Inorg. Chem.* **2015**, *54*, 4862–4868. [CrossRef]
28. Pakamorė, I.; Rousseau, J.; Rousseau, C.; Monflier, E.; Szilágyi, P.Á. An ambient-temperature aqueous synthesis of zirconium-based metal–organic frameworks. *Green Chem.* **2018**, *20*, 5292–5298. [CrossRef]
29. Užarević, K.; Wang, T.C.; Moon, S.-Y.; Fidelli, A.M.; Hupp, J.T.; Farha, O.K.; Friščić, T. Mechanochemical and solvent-free assembly of zirconium-based metal–organic frameworks. *Chem. Commun.* **2016**, *52*, 2133–2136. [CrossRef]
30. D'Amato, R.; Bondi, R.; Moghdad, I.; Marmottini, F.; Mcpherson, M.J.; Naïli, H.; Taddei, M.; Costantino, F. "Shake 'n Bake" Route to Functionalized Zr-UiO-66 Metal–Organic Frameworks. *Inorg. Chem* **2021**, *60*, 14294–14301. [CrossRef]
31. Huang, Y.-H.; Lo, W.-S.; Kuo, Y.-W.; Chen, W.-J.; Lin, C.-H.; Shieh, F.-K. Green and rapid synthesis of zirconium metal–organic frameworks via mechanochemistry: UiO-66 analog nanocrystals obtained in one hundred seconds. *Chem. Commun.* **2017**, *53*, 5818–5821. [CrossRef]
32. Cunha, D.; Gaudin, C.; Colinet, I.; Horcajada, P.; Maurin, G.; Serre, C. Rationalization of the entrapping of bioactive molecules into a series of functionalized porous zirconium terephthalate MOFs. *J. Mater. Chem. B* **2013**, *1*, 1101–1108. [CrossRef]
33. Cunha, D.; Ben Yahia, M.; Hall, S.; Miller, S.R.; Chevreau, H.; Elkaïm, E.; Maurin, G.; Horcajada, P.; Serre, C. Rationale of Drug Encapsulation and Release from Biocompatible Porous Metal–Organic Frameworks. *Chem. Mater.* **2013**, *25*, 2767–2776. [CrossRef]
34. Sarker, M.; Jhung, S.H. Zr-MOF with free carboxylic acid for storage and controlled release of caffeine. *J. Mol. Liq.* **2019**, *296*, 112060. [CrossRef]
35. Liédana, N.; Galve, A.; Rubio, C.; Téllez, C.; Coronas, J. CAF@ZIF-8: One-Step Encapsulation of Caffeine in MOF. *ACS Appl. Mater. Interfaces* **2012**, *4*, 5016–5021. [CrossRef] [PubMed]
36. Liédana, N.; Lozano, P.; Galve, A.; Téllez, C.; Coronas, J. The template role of caffeine in its one-step encapsulation in MOF NH_2-MIL-88B(Fe). *J. Mater. Chem. B* **2014**, *2*, 1144–1151. [CrossRef] [PubMed]
37. Monteagudo-Olivan, R.; Paseta, L.; Potier, G.; López-Ram-de-Viu, P.; Coronas, J. Solvent-Free Encapsulation at High Pressure with Carboxylate-Based MOFs. *Eur. J. Inorg. Chem.* **2019**, *2019*, 29–36. [CrossRef]
38. Wei, T.-H.; Wu, S.-H.; Huang, Y.-D.; Lo, W.-S.; Williams, B.P.; Chen, S.-Y.; Yang, H.-C.; Hsu, Y.-S.; Lin, Z.-Y.; Chen, X.-H.; et al. Rapid mechanochemical encapsulation of biocatalysts into robust metal–organic frameworks. *Nat. Commun.* **2019**, *10*, 5002. [CrossRef]
39. Labay, C.; Canal, J.M.; Navarro, A.; Canal, C. Corona plasma modification of polyamide 66 for the design of textile delivery systems for cosmetic therapy. *Appl. Surf. Sci.* **2014**, *316*, 251–258. [CrossRef]

40. Labay, C.; Canal, J.M.; Canal, C. Relevance of Surface Modification of Polyamide 6.6 Fibers by Air Plasma Treatment on the Release of Caffeine. *Plasma Processes Polym.* **2012**, *9*, 165–173. [CrossRef]
41. Sta, M.; Tada, D.B.; Medeiros, S.F.; Santos, A.M.; Demarquette, N.R. Hydrophilic drug release from electrospun membranes made out of thermo and pH-sensitive polymers. *J. Drug Delivery Sci. Technol.* **2022**, *71*, 103284. [CrossRef]
42. Li, X.; Kanjwal, M.A.; Lin, L.; Chronakis, I.S. Electrospun polyvinyl-alcohol nanofibers as oral fast-dissolving delivery system of caffeine and riboflavin. *Colloids Surf. B* **2013**, *103*, 182–188. [CrossRef] [PubMed]
43. Illangakoon, U.E.; Gill, H.; Shearman, G.C.; Parhizkar, M.; Mahalingam, S.; Chatterton, N.P.; Williams, G.R. Fast dissolving paracetamol/caffeine nanofibers prepared by electrospinning. *Int. J. Pharm.* **2014**, *477*, 369–379. [CrossRef]
44. Li, S.; Girod-Holland, S.; Vert, M. Hydrolytic degradation of poly(dl-lactic acid) in the presence of caffeine base. *J. Control. Release* **1996**, *40*, 41–53. [CrossRef]
45. Tipduangta, P.; Watcharathirawongs, W.; Waritdecha, P.; Sirithunyalug, B.; Leelapornpisid, P.; Chaiyana, W.; Goh, C.F. Electrospun cellulose acetate/polyvinylpyrrolidone fiber mats as potential cosmetic under-eye masks for caffeine delivery. *J. Drug Delivery Sci. Technol.* **2023**, *86*, 104732. [CrossRef]
46. Rezvani Ghomi, E.R.; Khosravi, F.; Saedi Ardahaei, A.S.; Dai, Y.; Neisiany, R.E.; Foroughi, F.; Wu, M.; Das, O.; Ramakrishna, S. The Life Cycle Assessment for Polylactic Acid (PLA) to Make It a Low-Carbon Material. *Polymers* **2021**, *13*, 1854. [CrossRef]
47. Arias, M.J.L.; López, A.; Vilaseca, M.; Vallès, B.; Prieto, R.; Simó, M.; Valle, J.A.B.; Valle, R.D.C.S.C.; Bezerra, F.M.; Bellalta, J.P. Influence of Chitosan Characteristics in the Microencapsulation of Essential Oils. *J. Biomed. Sci. Eng.* **2021**, *14*, 119–129. [CrossRef]
48. Zou, C.; Vagin, S.; Kronast, A.; Rieger, B. Template mediated and solvent-free route to a variety of UiO-66 metal–organic frameworks. *RSC Adv.* **2016**, *6*, 102968–102971. [CrossRef]
49. Zhang, X.; Liu, M.; Han, R. Adsorption of phosphate on UiO-66-NH_2 prepared by a green synthesis method. *J. Environ. Chem. Eng.* **2021**, *9*, 106672. [CrossRef]
50. UNE-EN ISO 1133; Plásticos. Determinación del Índice de Fluidez de Materiales Termoplásticos en Masa (IFM) y en Volumen (IFV). AENOR: Madrid, Spain, 2002.
51. Decker, G.E.; Stillman, Z.; Attia, L.; Fromen, C.A.; Bloch, E.D. Controlling Size, Defectiveness, and Fluorescence in Nanoparticle UiO-66 through Water and Ligand Modulation. *Chem. Mater.* **2019**, *31*, 4831–4839. [CrossRef]
52. Liao, X.; Wang, X.; Wang, F.; Yao, Y.; Lu, S. Ligand Modified Metal Organic Framework UiO-66: A Highly Efficient and Stable Catalyst for Oxidative Desulfurization. *J. Inorg. Organomet. Polym. Mater.* **2021**, *31*, 756–762. [CrossRef]
53. Jiang, X.; Li, S.; He, S.; Bai, Y.; Shao, L. Interface manipulation of CO_2-philic composite membranes containing designed UiO-66 derivatives towards highly efficient CO_2 capture. *J. Mater. Chem. A* **2018**, *6*, 15064–15073. [CrossRef]
54. Macrae, C.F.; Sovago, I.; Cottrell, S.J.; Galek, P.T.A.; Mccabe, P.; Pidcock, E.; Platings, M.; Shields, G.P.; Stevens, J.S.; Towler, M.; et al. Mercury 4.0: From visualization to analysis, design and prediction. *J. Appl. Crystallogr.* **2020**, *53*, 226–235. [CrossRef]
55. Øien, S.; Wragg, D.; Reinsch, H.; Svelle, S.; Bordiga, S.; Lamberti, C.; Lillerud, K.P. Detailed Structure Analysis of Atomic Positions and Defects in Zirconium Metal–Organic Frameworks. *Cryst. Growth Des.* **2014**, *14*, 5370–5372. [CrossRef]
56. Abid, H.R.; Tian, H.; Ang, H.-M.; Tade, M.O.; Buckley, C.E.; Wang, S. Nanosize Zr-metal organic framework (UiO-66) for hydrogen and carbon dioxide storage. *Chem. Eng. J.* **2012**, *187*, 415–420. [CrossRef]
57. Rodrigues, M.A.; Ribeiro, J.d.S.; Costa, E.d.S.; Miranda, J.L.d.; Ferraz, H.C. Nanostructured membranes containing UiO-66 (Zr) and MIL-101 (Cr) for O_2/N_2 and CO_2/N_2 separation. *Sep. Purif. Technol.* **2018**, *192*, 491–500. [CrossRef]
58. Wang, Y.L.; Zhang, S.; Zhao, Y.F.; Bedia, J.; Rodriguez, J.J.; Belver, C. UiO-66-based metal organic frameworks for the photodegradation of acetaminophen under simulated solar irradiation. *J. Environ. Chem. Eng.* **2021**, *9*, 106087. [CrossRef]
59. Rasband, W.S. ImageJ, U.S. National Institutes of Health, Bethesda, Maryland, USA, 1997–2018. Available online: https://imagej.net/ij/ (accessed on 18 February 2024).
60. Dong, C.; Wei, F.; Li, J.; Lu, Q.; Han, X. Uniform octahedral ZrO_2@C from carbonized UiO-66 for electrocatalytic nitrogen reduction. *Mater. Today Energy* **2021**, *22*, 100884. [CrossRef]
61. Majid, M.F.; Mohd Zaid, H.F.; Abd Shukur, M.F.; Ahmad, A.; Jumbri, K. Physicochemical properties and density functional theory calculation of octahedral UiO-66 with Bis(Trifluoromethanesulfonyl)imide ionic liquids. *Heliyon* **2023**, *9*, e20743. [CrossRef]
62. Enright, G.D.; Terskikh, V.V.; Brouwer, D.H.; Ripmeester, J.A. The Structure of Two Anhydrous Polymorphs of Caffeine from Single-Crystal Diffraction and Ultrahigh-Field Solid-State ^{13}C NMR Spectroscopy. *Cryst. Growth Des.* **2007**, *7*, 1406–1410. [CrossRef]
63. Devautour-Vinot, S.; Maurin, G.; Serre, C.; Horcajada, P.; Paula Da Cunha, D.; Guillerm, V.; De Souza Costa, E.; Taulelle, F.; Martineau, C. Structure and Dynamics of the Functionalized MOF Type UiO-66(Zr): NMR and Dielectric Relaxation Spectroscopies Coupled with DFT Calculations. *Chem. Mater.* **2012**, *24*, 2168–2177. [CrossRef]
64. Vasanthan, N.; Salem, D.R. FTIR spectroscopic characterization of structural changes in polyamide-6 fibers during annealing and drawing. *J. Polym. Sci. Part B Polym. Phys.* **2001**, *39*, 536–547. [CrossRef]
65. Yebra-Rodríguez, A.; Alvarez-Lloret, P.; Rodríguez-Navarro, A.B.; Martín-Ramos, J.D.; Cardell, C. Thermo-XRD and differential scanning calorimetry to trace epitaxial crystallization in PA6/montmorillonite nanocomposites. *Mater. Lett.* **2009**, *63*, 1159–1161. [CrossRef]
66. Farias-Aguilar, J.C.; Ramírez-Moreno, M.J.; Téllez-Jurado, L.; Balmori-Ramírez, H. Low pressure and low temperature synthesis of polyamide-6 (PA6) using Na0 as catalyst. *Mater. Lett.* **2014**, *136*, 388–392. [CrossRef]
67. Riba, J.R.; Cailloux, J.; Cantero, R.; Canals, T.; Maspoch, M.L. Multivariable methods applied to FTIR: A powerful technique to highlight architectural changes in poly(lactic acid). *Polym. Test.* **2018**, *65*, 264–269. [CrossRef]

68. Vanderhart, D.L.; Asano, A.; Gilman, J.W. Solid-State NMR Investigation of Paramagnetic Nylon-6 Clay Nanocomposites. 1. Crystallinity, Morphology, and the Direct Influence of Fe^{3+} on Nuclear Spins. *Chem. Mater.* **2001**, *13*, 3781–3795. [CrossRef]
69. Devautour-Vinot, S.; Martineau, C.; Diaby, S.; Ben-Yahia, M.; Miller, S.; Serre, C.; Horcajada, P.; Cunha, D.; Taulelle, F.; Maurin, G. Caffeine Confinement into a Series of Functionalized Porous Zirconium MOFs: A Joint Experimental/Modeling Exploration. *J. Phys. Chem. C* **2013**, *117*, 11694–11704. [CrossRef]
70. Proikakis, C.S.; Tarantili, P.A.; Andreopoulos, A.G. The role of polymer/drug interactions on the sustained release from poly(dl-lactic acid) tablets. *Eur. Polym. J.* **2006**, *42*, 3269–3276. [CrossRef]
71. Sitkowski, J.; Stefaniak, L.; Nicol, L.; Martin, M.L.; Martin, G.; Webb, G.A. Complete assignments of the ^1H, ^{13}C and ^{15}N NMR spectra of caffeine. *Spectrochim. Acta Part A* **1995**, *51*, 839–841. [CrossRef]
72. Zhou, J.; Guo, M.; Wu, D.; Shen, M.; Liu, D.; Ding, T. Synthesis of UiO-66 loaded-caffeic acid and study of its antibacterial mechanism. *Food Chem.* **2023**, *402*, 134248. [CrossRef]

Disclaimer/Publisher's Note: The statements, opinions and data contained in all publications are solely those of the individual author(s) and contributor(s) and not of MDPI and/or the editor(s). MDPI and/or the editor(s) disclaim responsibility for any injury to people or property resulting from any ideas, methods, instructions or products referred to in the content.

Article

Composites of Poly(3-hydroxybutyrate) and Mesoporous SBA-15 Silica: Crystalline Characteristics, Confinement and Final Properties

Tamara M. Díez-Rodríguez, Enrique Blázquez-Blázquez, Ernesto Pérez and María L. Cerrada *

Instituto de Ciencia y Tecnología de Polímeros (ICTP-CSIC), Juan de la Cierva 3, 28006 Madrid, Spain; t.diez@ictp.csic.es (T.M.D.-R.); enrique.blazquez@ictp.csic.es (E.B.-B.); ernestop@ictp.csic.es (E.P.)
* Correspondence: mlcerrada@ictp.csic.es; Tel.: +34-91-258-7474

Abstract: Several composites based on poly(3-hydroxybutyrate) (PHB) and mesoporous SBA-15 silica were prepared by solvent-casting followed by a further stage of compression molding. The thermal stability, phase transitions and crystalline details of these composites were studied, paying special attention to the confinement of the PHB polymeric chains into the mesopores of the silica. For that, differential scanning calorimetry (DSC) and real-time variable-temperature X-ray scattering at small angles (SAXS) were performed. Confinement was stated first by the existence of a small endotherm at temperatures around 20 °C below the main melting or crystallization peak, being later confirmed by a notable discontinuity in the intensity of the main (100) diffraction from the mesoporous silica observed through SAXS experiments, which is related to the change in the scattering contrast before and after the crystallization or melting of the polymer chains. Furthermore, the usual α modification of PHB was developed in all samples. Finally, a preliminary investigation of mechanical and relaxation parameters was carried out through dynamic–mechanical thermal analysis (DMTA). The results show, in the temperature interval analyzed, two relaxations, named α and β (the latest related to the glass transition) in order of decreasing temperatures, in all specimens. The role of silica as a filler is mainly observed at temperatures higher than the glass transition. In such cases, stiffness is dependent on SBA-15 content.

Keywords: poly(3-hydroxybutyrate) (PHB); mesoporous SBA-15 silica; composites; confinement; thermal properties; synchrotron SAXS; DMTA

Citation: Díez-Rodríguez, T.M.; Blázquez-Blázquez, E.; Pérez, E.; Cerrada, M.L. Composites of Poly(3-hydroxybutyrate) and Mesoporous SBA-15 Silica: Crystalline Characteristics, Confinement and Final Properties. *Polymers* **2024**, *16*, 1037. https://doi.org/10.3390/polym16081037

Academic Editor: Alexander Malkin

Received: 28 February 2024
Revised: 21 March 2024
Accepted: 8 April 2024
Published: 10 April 2024

Copyright: © 2024 by the authors. Licensee MDPI, Basel, Switzerland. This article is an open access article distributed under the terms and conditions of the Creative Commons Attribution (CC BY) license (https:// creativecommons.org/licenses/by/ 4.0/).

1. Introduction

There is, nowadays, a big concern about sustainability and preservation of the environment, both by demands of society and by the increasingly restrictive directives on environmental issues. For these reasons, the use of commodity plastics, widespread for decades, is being reduced especially for single-use applications: packaging, catering or agriculture, among others. Moreover, an efficient recycling of these actual polymer wastes is mandatory.

A suitable alternative for commodities is the use of biodegradable polymers, but unfortunately, most of them have significant disadvantages, precluding their massive usage in areas demanding great consumption.

Polyhydroxyalkanoates (PHA), which are biopolyesters of bacterial origin, are a very remarkable biodegradable and biobased polymer family. Among them, poly(3-hydroxybutyrate) (PHB) and their copolymers with 3-hydroxyvalerate (PHBHV) are the best-known systems. In fact, PHB displays certain performances that are similar to those in some commodities, showing, however, some practical deficiencies [1–4] that prevent its full attractiveness. Thus, high manufacture prices stand out together with a low toughness arising from its high crystallinity, embrittlement effects with time owing to secondary crystallization, and also a narrow processing window because of its prompt thermal degradation at relatively low temperatures [5].

In order to overcome these drawbacks, several strategies have been proposed. These approaches include the addition of nucleating or plasticizing agents (or both), a mixture with other polymers, and the use of adequate fillers [6,7]. Concerning the latter, mesoporous silicas are especially versatile among the wide variety of materials used as fillers, with the added advantage of involving nanodimensions. These silicas were first disclosed by Exxon Mobil in 1992 [8]. The Mobil Crystalline Materials No. 41, labeled commonly as MCM-41, is characterized by showing a hexagonal arrangement with parallel one-dimensional pores of a diameter of circa 3 nm. Years later, alternative mesoporous silicas were synthesized in 1998 at the Santa Barbara University of California. Among them, the best-known member is the Santa Barbara Amorphous SBA-15, displaying also a hexagonal array similar to that exhibited by MCM-41, but now the pore diameters range from around 5 to 10 nm [9]. These silicas are, therefore, nanostructurally ordered. Moreover, they display rather interesting properties, including their easy functionalization, so that they are being used in different fields, like catalysis, coatings, cosmetics, diagnostics, drug delivery, gas and bio-separation, optics and nanotechnology, among others.

Another very remarkable aspect of these materials is the fact that polymeric chains might go inside the nanometric channels existing in the silica mesostructure, thus leading to their confinement. This confinement has been observed in several systems [10–13], initially detected by Differential Scanning Calorimetry (DSC). Later on, experiments of Small Angle X-ray Scattering with synchrotron radiation (SAXS) were described as a very useful and conclusive tool. Thus, confinement, which is dependent on polymer molecular weight [14] and on mesoporous pore size [15], is characterized in real-time variable-temperature SAXS measurements through a notable discontinuity in the intensity of the primary (100) diffraction of the mesoporous silica, which is related to the change in the scattering contrast before and after the melting of the polymer chains [16,17].

The evaluation of confinement of organic solids in glasses with controlled pores has been an interesting issue for a long time [18,19]. The development of those crystals with restricted dimensions can be initially described by Gibbs theories [20], which were further adapted to small crystals in confined geometries, leading to the well-known Gibbs–Thomson equation [21,22].

The advent of nanotechnology has led to a renewed interest in the study of confinement effects, with the final purpose of analyzing the influence of such confined crystals on the final bulk properties, which is especially interesting in polymeric systems [12]. In addition, the macromolecular chains that may be partially inserted into the pores of the mesoporous structures can be of capital importance for obtaining polymeric nanocomposites, which in turn can influence the final performance of those polymeric systems.

The aim of the present study is the preparation and characterization of composites based on PHB and different contents of SBA-15 silica, in order to analyze their crystalline characteristics, the existence (or absence) of confinement of PHB chains into the pores of the silica, and, preliminarily, their stiffness as well as relaxations characteristics. For that, differential scanning calorimetry (DSC) and real-time variable-temperature X-ray scattering at both small angles (SAXS) and wide angles (WAXS) were performed, with the SAXS experiments especially suited for verifying the existence of confinement in the composite materials. Finally, an initial estimation of values for storage modulus, as a mechanical parameter, and of the different relaxation processes was carried out through dynamic–mechanical thermal analysis (DMTA).

2. Materials and Methods
2.1. Materials and Chemicals

A commercially available PHB (purchased from Ercros, Barcelona, Spain) with an MFI of about 30 g/10 min at 190 °C with a load of 2.16 kg, was used as polymeric matrix in this research. SBA-15 particles were purchased from Sigma-Aldrich, St. Louis, MO,

USA (specific surface area, S_{BET} = 517 m^2/g; total pore volume, V_t = 0.83 cm^3/g; average mesopore diameter, D_p = 6.25 nm) [23] and were used as received.

2.2. (Nano)composite and Film Preparation

Composites of PHB with different contents of SBA-15 particles (nominal amounts of 2, 4, 8 and 16% by weight) were prepared. They were labeled as PHBSBA2, PHBSBA4, PHBSBA8 and PHBSBA16, respectively. Drying of the components was carried out prior to their obtainment. The PHB was placed in an oven at 50 °C for 20 min followed by a drying under vacuum also at 50 °C for 20 h. The SBA-15 particles were dried at 100 °C for 24 h under vacuum. After this drying stage, the protocol followed was: a suitable content of SBA-15 silica was dispersed in chloroform at the same time that a PHB/chloroform solution (6 wt.% in PHB) was prepared. Both dispersion and solution were stirred for 18 h at room temperature. Afterward, the dispersion containing the silica particles was added to the PHB/chloroform solution. This PHB/silica/chloroform dispersion was stirred for a further 6 h at room temperature before it was poured into Petri dishes and dried at room temperature for 48 h. The resultant composite films were additionally dried in a vacuum oven at 85 °C for 2 h. A sample of neat PHB was also prepared with that same protocol.

These materials were subsequently processed by compression molding in a hot plate Collin press, New York, NY, USA. Initially, the material was maintained at a temperature of 200 °C, firstly without pressure for 2 min and, later, under a pressure of 30 bar for 4 min. Afterward, a cooling process at a relatively rapid rate of around 80 °C/min and at a pressure of 30 bar was applied to the different composites from their molten state to room temperature.

2.3. Transmission Electron Microscopy

Measurements were performed at room temperature in a 200 kV JEM-2100 JEOL microscope (Tokyo, Japan). The particles were dispersed in acetone in an ultrasonic bath for 5 min and then deposited in a holder prior to observation.

2.4. Scanning Electron Microscopy

Images of scanning electron microscopy (SEM) were attained in a Philips XL30 microscope (Amsterdam, The Netherlands). The samples were coated with a layer of 80:20 Au/Pd alloy and deposited in a holder before visualization.

2.5. Thermogravimetric Analysis

Thermogravimetric analyses (TGA) were carried out using TGA 2 equipment from METTLER TOLEDO (Columbus, OH, USA) at a heating rate of 10 °C/min under a nitrogen atmosphere. From these experiments, both the degradation behavior of the different samples and the particular SBA-15 amount incorporated into the composites during extrusion were estimated.

2.6. Differential Scanning Calorimetry

Calorimetric experiments were performed in a TA Instruments Q100 calorimeter connected (New Castle, DE, USA) to a cooling system and calibrated with different standards. The sample weights were around 3 mg. A heating–cooling–heating cycle was used, with a scanning rate of 20 °C/min in the temperature range from −50 to 200 °C. The crystallinity was estimated by considering a melting enthalpy of the 100% crystalline PHB of 146 J/g [24].

2.7. X-ray Experiments with Synchrotron Radiation

Simultaneous real-time variable-temperature SAXS/WAXS experiments were carried out with synchrotron radiation in beamline BL11-NCD-SWEET at ALBA (Cerdanyola del Vallès, Barcelona, Spain) at a fixed wavelength of 0.1 nm. A Pilatus detector was used for SAXS (off beam, at a distance of 294 cm from the sample) and a Rayonix one for WAXS (at about 12 cm from the sample, and a tilt angle of around 29 degrees). A Linkam Unit, connected to a cooling system of liquid nitrogen, was employed for the temperature control. The calibration of spacings was obtained by means of silver behenate and Cr_2O_3 standards. The initial 2D X-ray images were converted into 1D diffractograms, as a function of the inverse scattering vector, $s = 1/d = 2 \sin \theta/\lambda$, by means of pyFAI python code (ESRF), modified by ALBA beamline staff. Film samples of around 5 mm × 5 mm × 0.2 mm were used in the synchrotron analysis.

2.8. Dynamic Mechanical Thermal Analysis (DMTA)

Dynamic Mechanical Thermal Analyses were carried out in a TA Q800 Dynamic Mechanical Thermal Analyzer (New Castle, DE, USA), working in a tensile mode. The storage modulus, E', loss modulus, E'', and the loss tangent, tan δ, of PHB and its composites with SBA-15 were determined as functions of temperature over the range from −55 to 170 °C, at fixed frequencies of 0.5, 1, 3 and 10 Hz, and at a heating rate of 1.5 C/min. Strips of 2.2 mm wide and 15 mm long, cut from the compressed-molded films, were used for this analysis. Composite PHBSBA16 was not possible to analyze, since it is too rigid and fragile. The viscoelastic relaxations were determined from these studies. A smoothing function using an FFT filter with three points was applied to the loss modulus curves at different frequencies.

3. Results and Discussion

3.1. Morphological Characteristics

Figure 1a displays the TEM micrograph of the SBA-15 particles used for preparing the composites. This picture allows observation of the interior of particles, showing the existence of their ordered arrangements in a hexagonal morphology as well as the long parallel channels that constitute that particular ordering.

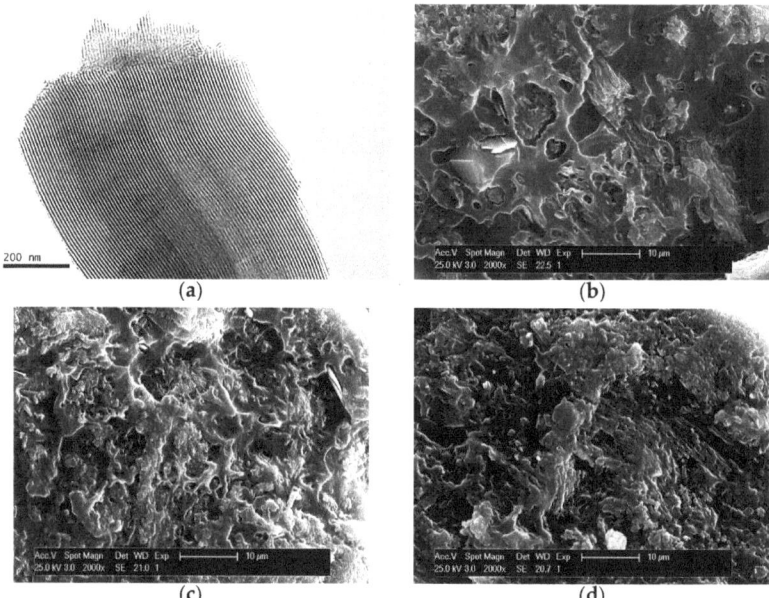

Figure 1. TEM micrographs of SBA-15 particles, dispersed in acetone and further deposited in a holder (**a**). SEM pictures at fracture surface for different composites: PHBSBA4 (**b**), PHBSBA8 (**c**), and PHBSBA16 (**d**).

SEM pictures at fracture surfaces of different composites are depicted in Figure 1b–d. The average size of an individual SBA-15 particle is around 350 nm wide and 0.9 µm long [13]. Accordingly, a suitable dispersion of SBA-15 particles was obtained within the PHB matrix since there is no observation of detectable bulky inorganic domains across the materials at the different mesoporous silica contents. These results seem to indicate that the preparation protocol followed has allowed a rather good contact at interfaces between silica particles and the PHB chains. On the other hand, a change is also noted in the PHB surface, from relatively ductile to rather fragile as the amount of SBA-15 in the composites increases. This fact is deduced from the differences in the fracture surface. This is rather continuous in the PHBSBA4 composite while it becomes coarser in the PHBSBA8 material and even more in the PHBSBA16 one. The appearance of coarseness in the fracture surface is characteristic of fragility. This fragility will be further discussed.

3.2. Thermal Stability

The TGA curves, under a nitrogen environment, for the different samples are shown in Figure 2a. These curves indicate that a single primary stage is present in the decomposition of all the specimens, with an inflection point of about 290 °C. Furthermore, they also display a final inert residue, which allows determining the actual SBA-15 content in the samples. This content is collected in Table 1, compared with the nominal values.

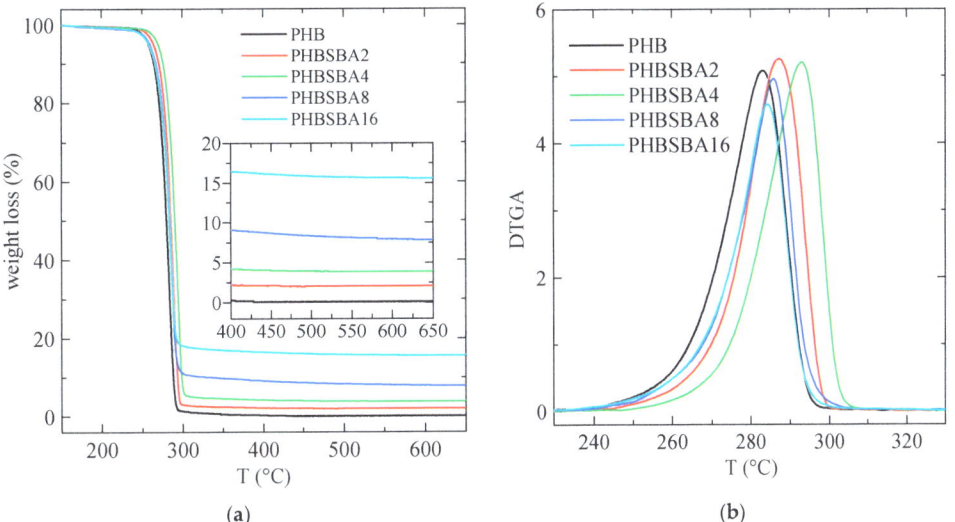

Figure 2. TGA (**a**) and DTGA (**b**) curves under nitrogen for PHB and its composites with SBA-15. The amplified inset in (**a**) reflects the content in SBA-15 of the composites.

Table 1. TGA results under nitrogen for PHB and its composites, prepared by melt extrusion: SBA-15 content (determined at 650 °C) and temperature at the maximum in the DTGA curves (T_{max}^{DTGA}).

Specimen	SBA-15 Content (wt.%)		T_{max}^{DTGA} (°C)
	Nominal	From TGA	
PHB	0	0	283.2
PHBSBA2	2	2.1	287.2
PHBSBA4	4	3.9	292.8
PHBSBA8	8	7.8	285.8
PHBSBA16	16	15.6	284.3

Moreover, the DTGA curves, depicted in Figure 2b, are characteristic of that single decomposition stage, and the corresponding values of T_{max} are deduced from them, being also collected in Table 1. These values are close to others reported before in the literature [1].

Although mesoporous silicas were also used, sometimes, as catalysts for thermal decomposition [25,26], however the results in Figure 2 and Table 1 indicate a slightly higher thermal stability in the composites. In fact, the value of T_{max}^{DTGA} increases first in composites PHBSBA2 and PHBSBA4, then decreases for the other two composites, although the values of T_{max}^{DTGA} are always higher than the ones for neat PHB. This is a behavior found in some polymer composites: a certain property may display better performance than the neat polymer at low filler contents, but when the content increases, at some point the property begins to weaken somewhat, probably due to the formation of some aggregates that are hindered at smaller filler amounts.

3.3. DSC Studies: Phase Transitions and Confinement of PHB Chains

DSC experiments were used, first, to determine the different phase transitions. Thus, Figure 3 shows the DSC heating curves obtained in PHB and its composites with SBA-15 during the first melting at 20 °C/min. Two clear temperature intervals are observed: a main melting endotherm, at around 166 °C for the neat PHB with slightly higher values for the composites, and a second interval at lower temperatures (between around 130 and 155 °C). No features are observed for neat PHB in this inferior interval, but one or two small melting endotherms are observed in the composites, involving an enthalpy that increases with the SBA-15 content, as observed in the amplified inset.

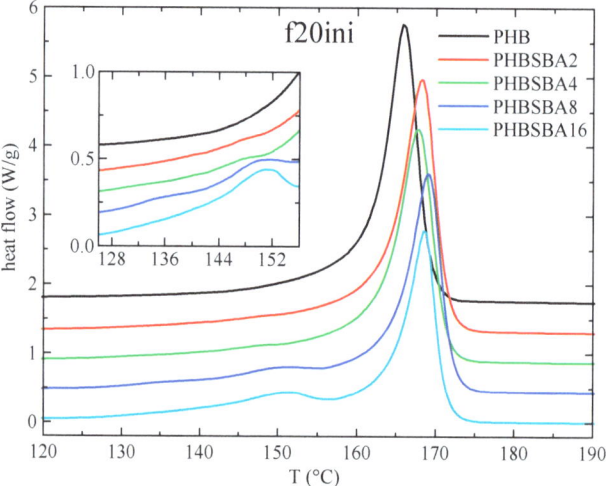

Figure 3. DSC heating curves (endo up) for PHB and its composites with SBA-15 in the first melting at 20 °C/min. The inset shows the amplification in the region of the melting of confined crystals of PHB.

These small endotherms are attributed, eventually, to the melting of the PHB crystallites that are confined inside the pores of SBA-15. Since these crystallites shall be much thinner than the ones outside of the mesoporous silica particles, their melting temperature is going to be then considerably lower than that of the PHB crystals outside the channels. The reason for that is found when considering the Gibbs–Thomson equation, mentioned above. Thus, a simplified equation can be considered [27–29] in the case of "regular" lamellar crystals, but a more general form [30,31] is required for the thinner crystallites confined in the mesoporous silica channels, which are supposed to present very low values in their lateral size, restricted by the pore diameter. Consequently, significantly depressed melting temperatures shall be exhibited by those confined crystals.

Moreover, it seems obvious that the more SBA-15 silica is present in the composites, the more channels will be available for confinement, and, therefore, the enthalpy involved in the melting of these confined crystallites will increase. Regarding the fact that two components seem to be observed in the melting of the confined crystallites, we will come back below to this issue.

Figure 4 shows the DSC curves for PHB and its composites in the subsequent cooling from the melt (at 20 °C/min). It is important to emphasize at this point that the PHB samples do not crystallize on cooling from the melt at regular DSC rates in most of the published works, most probably because the molecular weight is too high, and only in a few cases [1,32] that crystallization is readily observed, as it happens with the present PHB matrix. Thus, the main crystallization exotherm can be seen in neat PHB with a peak crystallization temperature, T_c, of around 115 °C, and for the composites (similarly to the behavior of T_{max}^{DTGA}) there is an initial increase on passing from PHB to PHBSBA2, PHB-SBA4 and PHBSBA8, decreasing then for PHBSBA16. But in all cases T_c of the composites is higher than the one for PHB, indicating a certain nucleating effect of the mesoporous silica. At lower temperatures, and similarly to the first melting, there is a region with small exotherms, whose enthalpy increases as the SBA-15 content increases. But now it appears to be a single component in these exotherms, as observed in the amplified inset of Figure 4.

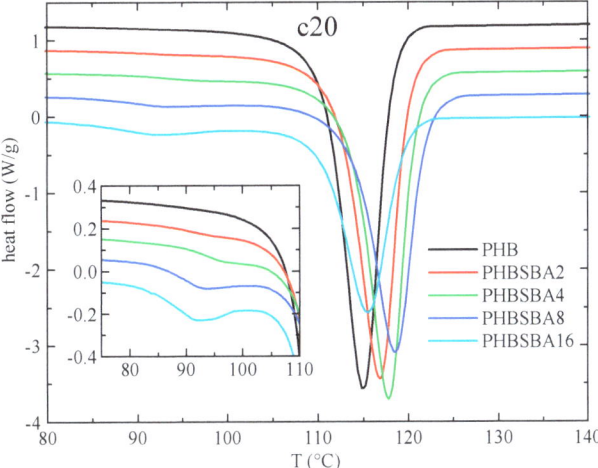

Figure 4. DSC curves (exo down) for PHB and its composites with SBA-15 in the cooling from the melt at 20 °C/min. The inset shows the amplification in the region of the crystallization of confined PHB.

The DSC heating curves in the subsequent second melting at 20 °C/min are presented in Figure 5. The behavior is rather similar to that in the first melting, with the eventual small melting endotherms arising from confined crystals showing also two components. Having taken into account that there is a single component in the confined crystallization, our tentative explanation for the two components of melting is that one of them arises from the confined crystals formed on cooling, while the second component is attributed to a confined recrystallization during melting.

Variation with the SBA-15 content of the different thermal transitions observed in Figures 3–5 is represented in Figure 6 for PHB and its composites. The differences among samples are almost not significant, although an increase in passing from PHB to the composites is usually observed, more evident at low SBA-15 amounts. More clear differences are deduced for the values of T_g (not shown in Figures 3–5) where a significant increase from neat PHB to the composites is well evident.

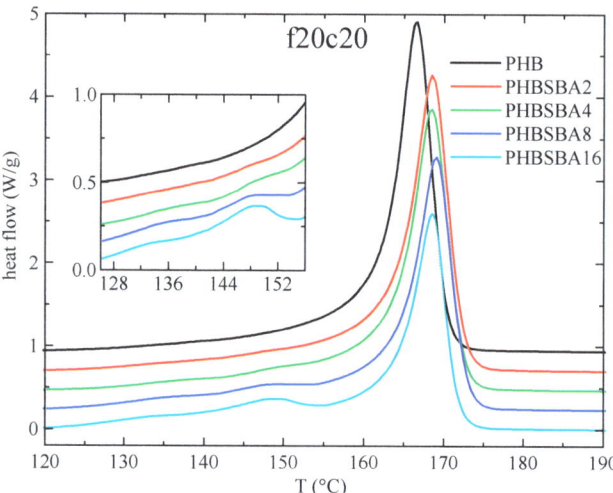

Figure 5. DSC heating curves (endo up) for PHB and its composites with SBA-15 in the second melting at 20 °C/min. The inset shows the amplification in the region of the melting of confined crystals of PHB.

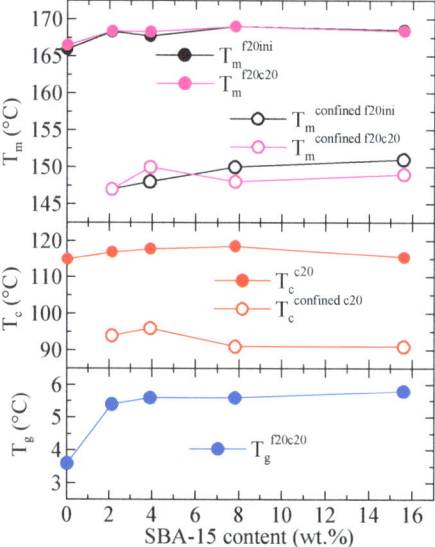

Figure 6. Variation with the SBA-15 content of different thermal transitions for PHB and its composites with this mesoporous silica: top: T_m for the first and second meltings for both the main endotherm and the confined component; middle: T_c for the cooling from the melt for both the main exotherm and the confined component; bottom: T_g for the second melting.

Regarding the values of the enthalpies involved, Figure 7 shows the variation with the SBA-15 content for different parameters in the PHB and its composites both the enthalpy for the confined component and the total enthalpy, for the different DSC ramps, and normalized to the actual PHB content in the specimen. It can be observed that the total enthalpy first increases with the SBA-15 content, and then reduces at the higher silica contents. The enthalpy for the confined components, however, increases markedly with the

SBA-15 content (top frame in Figure 7), in such a way that the percentage of that enthalpy over the total one reaches values as considerable as 12–13%. Obviously, that increase shall be a consequence of the more total volume of silica channels available for the confinement of PHB chains as SBA-15 content increases.

The values indicated in the right axis of Figure 7 represent the corresponding percentage of crystallinity, x_c, estimated by considering a melting enthalpy of the 100% crystalline PHB of 146 J/g [24]. As observed, a total crystallinity of around 58–62% is deduced, and the crystallinity involved in the confined crystal reaches values as high as around 6–8%.

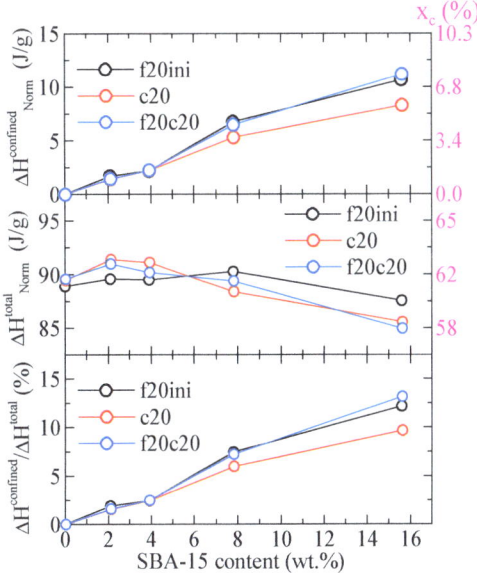

Figure 7. Variation with the SBA-15 content of different parameters for PHB and its composites with this mesoporous silica: top: enthalpy for the confined component for the indicated DSC ramps; middle: total enthalpy for the indicated DSC ramps; bottom: percentage of the enthalpy of the confined components over the total values. All enthalpies were normalized to the actual PHB content in the specimen. The corresponding degree of crystallinity is indicated in the right axes.

3.4. X-ray Experiments with Synchrotron Radiation

As mentioned above, SAXS experiments turned out a very useful and conclusive tool for analyzing the confinement in composites where mesoporous SBA-15 or MCM-41 particles were involved, independently of the nature of the polymeric matrix, since the intensity of the main (100) diffraction of the mesoporous silica shows a notable discontinuity. Therefore, synchrotron X-ray experiments were performed on selected samples. Firstly, Figure 8 shows the room-temperature WAXS diffractograms for neat PHB, displaying the profile of the initial compressed-molded specimen and that after cooling from the melt at 20 °C/min. In both cases, the usual α modification of PHB is observed [6,24,33,34], with the two main diffractions corresponding to planes (020) and (110). Moreover, the diffractions corresponding to the sample crystallized from the melt at 20 °C/min are slightly narrower than those of the initial compression-molded specimen, indicative of more perfect crystals in the former.

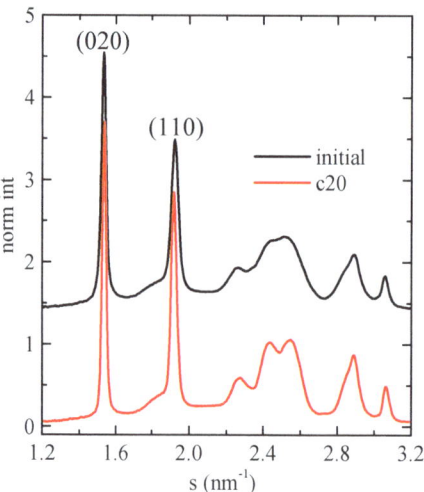

Figure 8. Room-temperature synchrotron WAXS diffractograms for neat PHB, showing the profile of the initial compressed-molded specimen and that after cooling from the melt at 20 °C/min.

Figure 9 represents the Lorentz-corrected SAXS profiles for composite PHBSBA8 during the first melting, the cooling from the melt, and the subsequent second melting. In the selected region, two features are clearly observed: first, the prominent (100) diffraction of SBA-15 [8,9,12], and second, the wide long spacing peak from the PHB crystals, which appears overlapped to the SBA-15 signal.

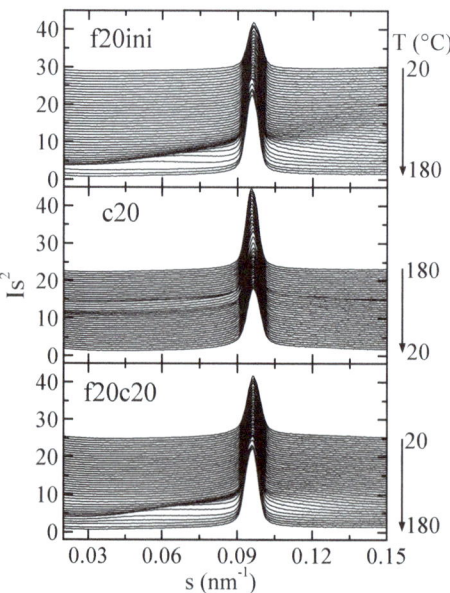

Figure 9. Lorentz-corrected SAXS profiles for composite PHBSBA8 during the first melting at 20 °C/min (**top**), the cooling from the melt at 20 °C/min (**middle**), and the subsequent second melting (**bottom**). Only one out of every two frames is plotted, for clarity.

The amplification of the SBA-15 main reflection is rather informative, as observed in Figure 10. Focusing the attention on the first melting (top frame in Figure 10), it is seen that after an initial rather constant intensity of the SBA-15 diffraction, there is a pronounced increase centered at around 144 °C (blue diffractogram), and a much smaller change at around 170 °C (red diffractogram). Compared with the DSC results, these changes correspond to the melting of the confined crystals and to the main melting, respectively, but now the rather relevant issue is that confinement is much better observed than the total melting, contrary to the case in the DSC results.

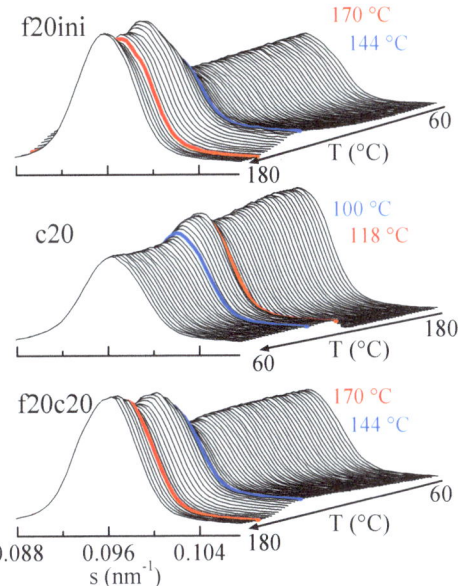

Figure 10. Lorentz-corrected SAXS profiles for composite PHBSBA8, showing the amplified region of the SBA-15 main reflection, during the first melting (**top plot**), the cooling from the melt (**middle plot**), and the subsequent second melting (**bottom plot**). The blue and red highlighted frames correspond to the events (crystallization or melting) arising from the confined component and the main one, respectively.

Similar arguments apply for the cooling from the melt (middle representation in Figure 10) and for the second melting (lower plot in Figure 10), where the decrease in the cooling and the increase in the heating are much greater for confinement than for main crystallization or main melting.

As mentioned above, these variations in the intensity of SBA-15 diffraction depend on the scattering contrast between the walls and the inside of the SBA-15 channels before and after the melting of the confined polymer chains (and also, evidently, on the amount of pore filling).

The quantification of the results in Figure 10 regarding the intensity of the main SBA-15 diffraction is represented in Figure 11, showing, in the upper frame, the variation with temperature of that intensity in composite PHBSBA8 during the first melting (f20ini), cooling (c20) and second melting (f20c20) processes, and compared with the corresponding DSC curves (lower plot). The concordance of the transition temperatures for the different events, indicated by the straight lines, is excellent among the two techniques. Moreover, it is well evident that confinement is much better observed through the variation of the intensity of the SBA-15 peak.

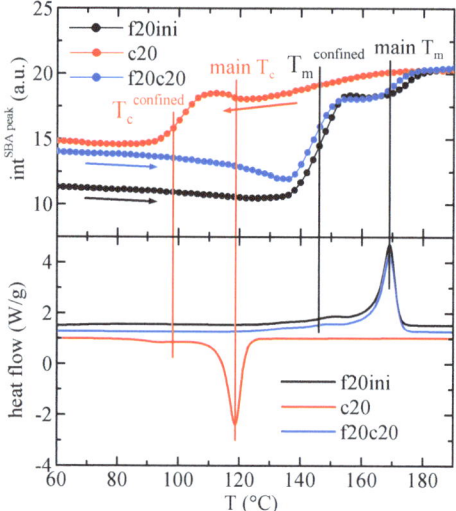

Figure 11. Variation with temperature of the intensity of the main SBA-15 diffraction in composite PHBSBA8 (**upper**) during the first melting (f20ini) (black arrow), cooling (c20) (red arrow) and second melting (f20c20) (blue arrow) processes, compared with the corresponding DSC curves (**lower**). The straight lines indicate the different thermal events.

3.5. Dynamic Mechanical Thermal Analysis (DMTA) Experiments

Figure 12 shows the variation in temperature of storage modulus, E′, loss modulus, E″, and tan δ curves (at 3 Hz) for PHB and its composites with SBA-15. Two relaxations, named α and β in order of decreasing temperatures are observed in the E″ and tan δ curves. Sample PHBSBA16 was too brittle and could not be analyzed by DMTA.

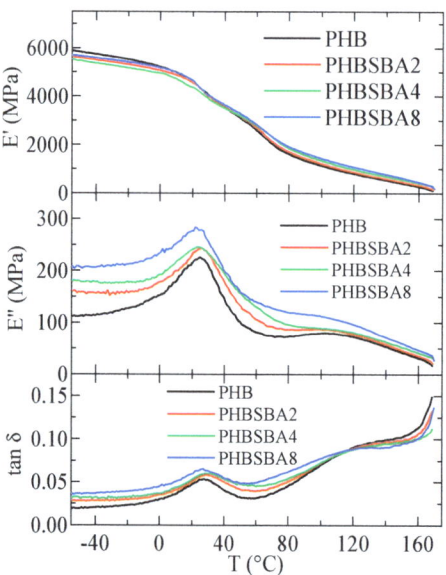

Figure 12. Variation with temperature of storage modulus, E′, loss modulus, E″, and tan δ curves (at 3 Hz) for PHB and its composites with SBA-15.

There have been different studies about the relaxation behavior in PHB. Some of them are based on DMTA measurements [1,35], and others refer to electrical analysis [36,37] or piezoelectricity [38,39]. Another relaxation was reported at temperatures lower than those analyzed here. That lowest-temperature relaxation, designated as γ, was attributed by some authors [35] to the water absorbed in PHB, but more recent works [36,37] are in favor of associating its origin to motions of the ester groups, as it happens in other polyesters.

Regarding the here named β relaxation, there is clear consensus being ascribed to the glass transition [1,35], while the one located at the highest temperature, the α relaxation, is attributed to movement within the crystalline regions as they become soft due to their partial melting.

The variation with the SBA-15 content, for PHB and its composites, of the temperature location of the α and β relaxations (E'' basis at 3 Hz) is shown in Figure 13a. It can be observed that the temperature for the α relaxation increases with the silica content, while that for the β relaxation (the glass transition) decreases. This behavior is different from the one deduced from the DSC experiments, although values reported in Figure 6 (bottom frame) correspond to the ones derived from the melting after crystallization at 20 °C/min and not to the initial films. This discrepancy could be associated with differences in the residual secondary crystallization (which is a detrimental characteristic in the PHB polymer) between the pure matrix and the distinct composites. Specimens for DMTA have remained at a certain time at room temperature before these measurements were performed. Thus, the secondary crystallization of PHB chains could take place, occurring at a higher extent in the pristine polymer than in the composites, pointing out another positive role of the presence of mesoporous SBA-15 particles. Nevertheless, determination by DSC in the second heating process does not show any change in the crystalline regions, so mobility within the amorphous phase does not allow further cold crystallization. On the other hand, the β process, ascribed to the glass transition, is located in the DMTA experiments at considerably higher temperatures than that obtained from the DSC ones, as usual [40,41].

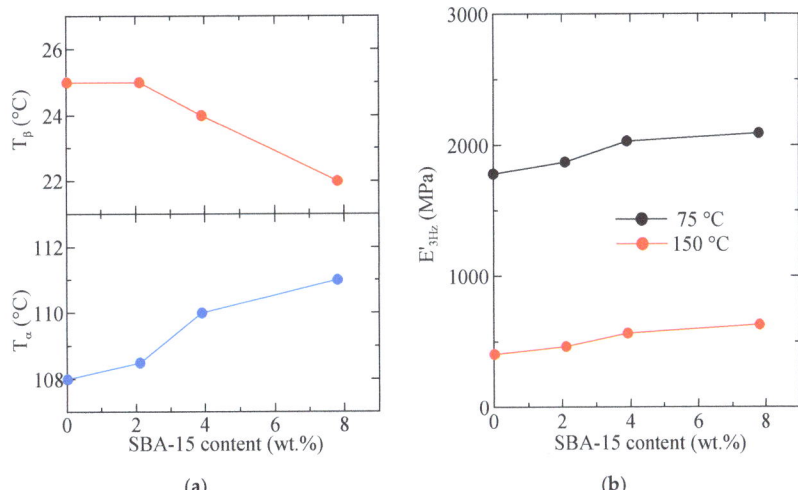

Figure 13. Variation with the SBA-15 content, for PHB and its composites, of (**a**) the temperature location of the α and β relaxations (E'' basis at 3 Hz); (**b**) the values of storage modulus at two different representative temperatures.

Figure 13b shows the variation in the silica content of the values of storage modulus at two different representative temperatures. Ideally, the best temperature to consider is 25 °C (ambient conditions) but, as observed in Figure 12, this temperature lies in the middle of the β relaxation (T_g), which is not convenient. Thus, the following representative temperatures were chosen: 75 °C: in between T_β and T_α, and 150 °C: above T_α. It can be observed that at these two temperatures (both above T_g) the rigidity increases with the SBA-15 content. The reinforcement role of silica is mainly played as the polymeric matrix is softened, and values of storage moduli increase as the SBA-15 content does.

4. Conclusions

Several composites based on poly(3-hydroxybutyrate) (PHB) and mesoporous SBA-15 silica were prepared by solvent-casting, followed by compression molding. The thermal properties and crystalline details of these composites were studied, paying special attention to the confinement of polymeric chains of PHB into the pores of the silica.

TGA measurements, under nitrogen, allow determining the actual SBA-15 content in the samples and also indicate slightly higher thermal stability in the composites. In fact, the value of T_{max}^{DTGA} increases first in composites PHBSBA2 and PHBSBA4, then decreases for the other two composites, although the values of T_{max}^{DTGA} are always higher than the ones for neat PHB.

Confinement was analyzed first by DSC and it was confirmed by SAXS experiments, which show a notable discontinuity in the intensity of the first-order (100) diffraction of the mesoporous silica. The rather relevant issue is that in the DSC results the exo- or endotherm related to confinement is much smaller than the main peaks, while in the SAXS results the variation of the intensity of the silica diffraction is considerably higher for confinement than for the total melting or crystallization.

In addition, WAXS diffractograms indicate that the usual α modification of PHB is observed in all samples, either neat PHB or its composites.

DMTA results show, in the temperature interval analyzed, two relaxations, named α and β in order of decreasing temperatures, in all specimens. Sample PHBSBA16 is too brittle and cannot be analyzed by DMTA. The temperature for the α relaxation increases with the silica content, while that for the β relaxation (the glass transition) decreases. The role of silica as a filler is mainly observed at temperatures higher than the relaxation mechanism ascribed to generalized motions within the amorphous regions, i.e., the glass transition. In such cases, stiffness is dependent on SBA-15 content.

Author Contributions: Conceptualization, T.M.D.-R., E.B.-B., M.L.C. and E.P.; methodology, T.M.D.-R., E.B.-B., M.L.C. and E.P.; software, E.P.; formal analysis, E.P. and T.M.D.-R.; investigation, T.M.D.-R., E.B.-B. and E.P.; resources, E.P. and M.L.C.; writing—original draft preparation, E.P. and M.L.C.; writing—review and editing, T.M.D.-R., E.B.-B., M.L.C. and E.P.; supervision, M.L.C. and E.P.; project administration, E.P. and M.L.C.; funding acquisition, E.P. and M.L.C. All authors have read and agreed to the published version of the manuscript.

Funding: This research was funded by MCIN/AEI/10.13039/501100011033, grant number PID2020-114930GB-I00.

Institutional Review Board Statement: Not applicable.

Data Availability Statement: Data are contained within the article.

Acknowledgments: Authors are grateful to the Characterization Service at ICTP-CSIC for SEM, TGA, and DMTA facilities as well as to its personnel for support. The synchrotron experiments were carried out in the beamline BL11-NCD-SWEET at the ALBA Synchrotron Light Facility with the collaboration of ALBA staff, to whom the authors are grateful for their support.

Conflicts of Interest: The authors declare no conflict of interest.

References

1. Corre, Y.-M.; Bruzaud, S.; Audic, J.-L.; Grohens, Y. Morphology and functional properties of commercial polyhydroxyalkanoates: A comprehensive and comparative study. *Polym. Test.* **2012**, *31*, 226–235. [CrossRef]
2. Laycock, B.; Halley, P.; Pratt, S.; Werker, A.; Lant, P. The chemomechanical properties of microbial polyhydroxyalkanoates. *Prog. Polym. Sci.* **2013**, *38*, 536–583. [CrossRef]

3. Bugnicourt, E.; Cinelli, P.; Lazzeri, A.; Alvarez, V. Polyhydroxyalkanoate (PHA): Review of synthesis, characteristics, processing and potential applications in packaging. *Express Polym. Lett.* **2014**, *8*, 791–808. [CrossRef]
4. Esposito, A.; Delpouve, N.; Causin, V.; Dhotel, A.; Delbreilh, L.; Dargent, E. From a Three-Phase Model to a Continuous Description of Molecular Mobility in Semicrystalline Poly(hydroxybutyrate-cohydroxyvalerate). *Macromolecules* **2016**, *49*, 4850–4861. [CrossRef]
5. Anjum, A.; Zuber, M.; Zia, K.M.; Noreen, A.; Anjum, M.N.; Tabasum, S. Microbial production of polyhydroxyalkanoates (PHAs) and its copolymers: A review of recent advancements. *Int. J. Biol. Macromol.* **2016**, *89*, 161–174. [CrossRef] [PubMed]
6. Pan, P.; Inoue, Y. Polymorphism and isomorphism in biodegradable polyesters. *Prog. Polym. Sci.* **2009**, *34*, 605–640. [CrossRef]
7. Saeidlou, S.; Huneault, M.A.; Li, H.; Park, C.B. Poly(lactic acid) crystallization. *Prog. Polym. Sci.* **2012**, *37*, 1657–1677. [CrossRef]
8. Beck, J.S.; Vartuli, J.C.; Roth, W.J.; Leonowicz, M.E.; Kresge, C.T.; Schmitt, K.D.; Chu, C.T.-W.; Olson, D.H.; Sheppard, E.W.; McCullen, S.B.; et al. A new family of mesoporous molecular sieves prepared with liquid crystal templates. *J. Am. Chem. Soc.* **1992**, *114*, 10834–10843. [CrossRef]
9. Zhao, D.Y.; Feng, J.L.; Huo, Q.S.; Melosh, N.; Fredrickson, G.H.; Chmelka, B.F.; Stucky, G.D. Triblock copolymer syntheses of mesoporous silica with periodic 50 to 300 angstrom pores. *Science* **1998**, *279*, 548–552. [CrossRef]
10. Cerrada, M.L.; Bento, A.; Pérez, E.; Lorenzo, V.; Lourenço, J.P.; Ribeiro, M.R. Hybrid Materials Based on Polyethylene and MCM-41 Particles Functionalized with Silanes: Catalytic Aspects of In Situ Polymerization, Crystalline Features and Mechanical Properties. *Microporous Mesoporous Mater.* **2016**, *232*, 86–96. [CrossRef]
11. Barranco-García, R.; Ferreira, A.E.; Ribeiro, M.R.; Lorenzo, V.; García-Peñas, A.; Gómez-Elvira, J.M.; Pérez, E.; Cerrada, M.L. Hybrid materials obtained by in situ polymerization based on polypropylene and mesoporous SBA-15 silica particles: Catalytic aspects, crystalline details and mechanical behavior. *Polymer* **2018**, *151*, 218–230. [CrossRef]
12. Díez-Rodríguez, T.M.; Blázquez-Blázquez, E.; Antunes, N.L.C.; Ribeiro, M.R.; Pérez, E.; Cerrada, M.L. Confinement in Extruded Nanocomposites based on PCL and Mesoporous Silicas: Effect of Pore Sizes and their Influence in Ultimate Mechanical Response. *J. Compos. Sci.* **2021**, *5*, 321. [CrossRef]
13. Díez-Rodríguez, T.M.; Blázquez-Blázquez, E.; Pérez, E.; Cerrada, M.L. Composites Based on Poly(Lactic Acid) (PLA) and SBA-15: Effect of Mesoporous Silica on Thermal Stability and on Isothermal Crystallization from Either Glass or Molten State. *Polymers* **2020**, *12*, 2743. [CrossRef] [PubMed]
14. Barranco-García, R.; Gómez-Elvira, J.M.; Ressia, J.A.; Quinzani, L.; Vallés, E.M.; Pérez, E.; Cerrada, M.L. Effect of iPP molecular weight on its confinement within mesoporous SBA-15 silica in extruded iPP-SBA-15 nanocomposites. *Microporous Mesoporous Mater.* **2020**, *294*, 109945. [CrossRef]
15. Barranco-García, R.; Gómez-Elvira, J.M.; Ressia, J.A.; Quinzani, L.; Vallés, E.M.; Pérez, E.; Cerrada, M.L. Variation of Ultimate Properties in Extruded iPP-Mesoporous Silica Nanocomposites by Effect of iPP Confinement within the Mesostructures. *Polymers* **2020**, *12*, 70. [CrossRef] [PubMed]
16. Hammond, W.; Prouzet, E.; Mahanti, S.D.; Pinnavaia, T.J. Structure factor for the periodic walls of mesoporous MCM-41 molecular sieves. *Microporous Mesoporous Mater.* **1999**, *27*, 19–25. [CrossRef]
17. Sauer, J.; Marlow, F.; Schüth, F. Simulation of powder diffraction patterns of modified ordered mesoporous materials. *Phys. Chem. Chem. Phys.* **2001**, *3*, 5579–5584. [CrossRef]
18. Haller, W. Chromatography on Glass of Controlled Pore Size. *Nature* **1965**, *206*, 693–696. [CrossRef]
19. Jackson, C.L.; McKenna, G.B. The melting behavior of organic materials confined in porous solids. *J. Chem. Phys.* **1990**, *93*, 9002–9011. [CrossRef]
20. Gibbs, J.W. On the Equilibrium of Heterogeneous Substances. In *Collected Works, Volume I: Thermodynamics*; Chapter III; Longmans: Green, NY, USA, 1928; pp. 55–353.
21. Defay, R.; Prigogine, I.; Bellemans, A.; Everett, D.H. *Surface Tension and Adsorption*; Wiley: New York, NY, USA, 1966.
22. Thomson, W. On the equilibrium of vapor at a curved surface of liquid. *Philos. Mag.* **1871**, *42*, 448–452. [CrossRef]
23. Deryło-Marczewska, A.; Zienkiewicz-Strzałka, M.; Skrzypczyńska, K.; Światkowski, A.; Kuśmierek, K. Evaluation of the SBA-15 materials ability to accumulation of 4-chlorophenol on carbon paste electrode. *Adsorption* **2016**, *22*, 801–812. [CrossRef]
24. Barham, P.J.; Keller, A.; Otun, E.L.; Holmes, P.A. Crystallization and morphology of a bacterial thermoplastic: Poly-3-hydroxybutyrate. *J. Mater. Sci.* **1984**, *19*, 2781–2794. [CrossRef]
25. Marcilla, A.; Gomez, A.; Garcia, A.N.; Olaya, M.M. Kinetic study of the catalytic decomposition of different commercial polyethylenes over an MCM-41 catalyst. *J. Anal. Appl. Pyrol.* **2002**, *64*, 85–101. [CrossRef]
26. Campos, J.; Lourenço, J.P.; Perez, E.; Cerrada, M.L.; Ribeiro, M.D.R. Self-Reinforced Hybrid Polyethylene/MCM-41 Nanocomposites: In-Situ Polymerisation and Effect of MCM-41 Content on Rigidity. *J. Nanosci. Nanotechnol.* **2009**, *9*, 3966–3974. [CrossRef] [PubMed]
27. Wunderlich, B. *Macromolecular Physics*; Academic Press: New York, NY, USA, 1980; Volume 3.
28. Darras, O.; Séguéla, R. Surface free energy of the chain-folding crystal faces of ethylene-butene random copolymers. *Polymer* **1993**, *34*, 2946–2950. [CrossRef]
29. Lu, L.; Alamo, R.G.; Mandelkern, L. Lamellar thickness distribution in linear polyethylene and ethylene copolymers. *Macromolecules* **1994**, *27*, 6571–6576. [CrossRef]
30. Gedde, U.W. *Polymer Physics*; Chapman & Hall: London, UK, 1995.

31. Shin, K.; Woo, E.; Jeong, Y.G.; Kim, C.; Huh, J.; Kim, K.-W. Crystalline structures, melting, and crystallization of linear polyethylene in cylindrical nanopores. *Macromolecules* **2007**, *40*, 6617–6623. [CrossRef]
32. Wang, S.; Capoen, L.; D'hooge, D.R.; Cardon, L. Can the melt flow index be used to predict the success of fused deposition modelling of commercial poly(lactic acid) filaments into 3D printed materials? *Plast. Rubber Compos.* **2018**, *47*, 9–16. [CrossRef]
33. Yokouchi, M.; Chatani, Y.; Tadokoro, H.; Teranish, K.; Tani, H. Structural studies of polyesters. 5. Molecular and crystal structures of optically-active and racemic poly(β-hydroxybutyrate). *Polymer* **1973**, *14*, 267–272. [CrossRef]
34. Cornibert, J.; Marchessault, R.H. Physical properties of poly-β-hydroxybutyrate. IV. Conformational analysis and crystalline structure. *J. Mol. Biol.* **1972**, *71*, 735–756.
35. Scandola, M.; Pizzofi, M.; Ceccorulli, G.; Cesàro, A.; Paoletti, S.; Navarini, L. Viscoelastic and thermal properties of bacterial poly(D-(-)-β-hydroxybutyrate). *Int. J. Biol. Macromol.* **1988**, *10*, 373–377. [CrossRef]
36. Pratt, G.J.; Smith, M.J.A. Dielectric relaxation spectroscopy of a poly-β-hydroxybutyrate homopolymer. *Eur. Polym. J.* **1997**, *33*, 857–861. [CrossRef]
37. Šics, I.; Tupureina, V.; Kalniņš, M.; Ezquerra, T.A.; Baltá-Calleja, F.J. Dielectric relaxation of poly-(β-hydroxybutyrate) relating to microstructure. *J. Macromol. Sci.-Phys.* **1998**, *37*, 851–862. [CrossRef]
38. Ando, Y.; Fukada, E. Piezoelectric Properties and Molecular Motion of Poly(β-Hydroxybutyrate) Films. *J. Polym. Sci. Polym. Phys.* **1984**, *22*, 1821. [CrossRef]
39. Fukada, E.; Ando, Y. Piezoelectric properties of poly-β-hydroxybutyrate and copolymers of β-hydroxybutyrate and β-hydroxyvalerate. *Int. J. Biol. Macromol.* **1986**, *8*, 361–366. [CrossRef]
40. Cerrada, M.L.; Pereña, J.M.; Benavente, R.; Pérez, E. Viscoelastic processes in vinyl alcohol-ethylene copolymers. Influence of composition and thermal treatment. *Polymer* **2000**, *41*, 6655–6661. [CrossRef]
41. Ward, I.M.; Sweeney, J. *Mechanical Properties of Solids Polymers*, 3rd ed.; Wiley: Chichester, UK, 2012.

Disclaimer/Publisher's Note: The statements, opinions and data contained in all publications are solely those of the individual author(s) and contributor(s) and not of MDPI and/or the editor(s). MDPI and/or the editor(s) disclaim responsibility for any injury to people or property resulting from any ideas, methods, instructions or products referred to in the content.

Article

Bioactivity and Antibacterial Analysis of Plasticized PLA Electrospun Fibers Reinforced with MgO and Mg(OH)$_2$ Nanoparticles

Adrián Leonés [1], Valentina Salaris [1], Laura Peponi [1,*], Marcela Lieblich [2], Alexandra Muñoz-Bonilla [1], Marta Fernández-García [1] and Daniel López [1,*]

1 Instituto de Ciencia y Tecnología de Polímeros (ICTP-CSIC), C/Juan de la Cierva 3, 28006 Madrid, Spain; aleones@ictp.csic.es (A.L.); v.salaris@ictp.csic.es (V.S.)
2 Centro Nacional de Investigaciones Metalúrgicas (CENIM-CSIC), 28040 Madrid, Spain
* Correspondence: lpeponi@ictp.csic.es (L.P.); daniel.l.g@csic.es (D.L.)

Citation: Leonés, A.; Salaris, V.; Peponi, L.; Lieblich, M.; Muñoz-Bonilla, A.; Fernández-García, M.; López, D. Bioactivity and Antibacterial Analysis of Plasticized PLA Electrospun Fibers Reinforced with MgO and Mg(OH)$_2$ Nanoparticles. *Polymers* **2024**, *16*, 1727. https://doi.org/10.3390/polym16121727

Academic Editor: Alexey L. Iordanskii

Received: 19 April 2024
Revised: 7 June 2024
Accepted: 12 June 2024
Published: 18 June 2024

Copyright: © 2024 by the authors. Licensee MDPI, Basel, Switzerland. This article is an open access article distributed under the terms and conditions of the Creative Commons Attribution (CC BY) license (https://creativecommons.org/licenses/by/4.0/).

Abstract: In this work, we focused on the bioactivity and antibacterial behavior of PLA-based electrospun fibers, efibers, reinforced with both MgO and Mg(OH)$_2$ nanoparticles, NPs. The evolution of PLA-based efibers was followed in terms of morphology, FTIR, XRD, and visual appearance. The bioactivity was discussed in terms of hydroxyapatite growth after 28 days, considered as T28, of immersion in simulated body fluid, SBF. In particular, the biomineralization process evidenced after immersion in SBF started at T14 in both systems. The number of precipitated crystals increased by increasing the amount of both NPs. The chemical composition of the precipitated crystals was also characterized in terms of the Ca/P molar ratio after T28 of immersion in SBF, indicating the presence of hydroxyapatite on the surface of both reinforced efibers. Moreover, a reduction in the average diameter of the PLA-based efibers was observed, reaching a maximum reduction of 46 and 60% in the average diameter of neat PLA and PLA:OLA efibers, respectively, after 28 days of immersion in SBF. The antibacterial behavior of the MgO and Mg(OH)$_2$ NPs in the PLA-based electrospun fibers was tested against *Escherichia coli*, *E. coli*, as the Gram-negative bacteria, and *Staphylococcus aureus*, *S. aureus*, as the Gram-positive bacteria, obtaining the best antibacterial activity against the Gram-negative bacteria *E. coli* of 21 ± 2% and 34 ± 6% for the highest concentration of MgO and Mg(OH)$_2$ NPs, respectively.

Keywords: electrospinning; polylactic acid; antibacterial properties; bioactivity; MgO NPs; Mg(OH)$_2$ NPs

1. Introduction

For the regeneration of tissues, biomaterials are required to treat damaged body areas. Ideally, biomaterials for tissue regeneration should display three main properties. The first is the ability to degrade under physiological conditions while remaining safe for human bodies [1]. The second is to resemble the mechanical properties associated with human tissue in terms of elastic modulus, tensile strength, and elongation at break [2]. Finally, biomaterials should display bioactivity, i.e., the ability to properly adapt to the biological environment, designed to stimulate appropriate cellular and tissue responses [3]. This could be achieved by the development of materials based on a polymeric nanocomposite [4]. In particular, combinations of biodegradable polymers and functional nanoparticles, NPs, offer numerous opportunities for improving mechanical properties and bioactivity.

Nanocomposite biomaterials can be processed by numerous techniques, such as solvent casting [5], 3D printing [6], or electrospinning [7]. In particular, the electrospinning technique obtains electrospun fibers, efibers, with a high surface area ratio, which properly recreates the extracellular matrix, enhancing the bioactivity of biomaterials. Among the biodegradable polymers used in electrospinning, poly(lactic acid), PLA, has been widely investigated for tissue engineering due to its biodegradability into non-toxic products under physiological conditions [8]. Additionally, the mechanical properties of PLA efibers can be enhanced by the addition of both organic and inorganic NPs, improving their elastic modulus, tensile strength, or elongation at break [9,10]. However, among the inorganic NPs, few studies have investigated the use of Mg-based NPs to enhance the bioactivity of electrospun biomaterials [1].

Magnesium, Mg, is a biocompatible metal that is already present in the human body in a concentration of 0.4 g of Mg·kg^{-1} [11]. Beyond their cooperative role with hydroxyapatite, HA, in maintaining bone health [12], Mg ions play an important role in mediating the functions of all cells in the body. In particular, Mg ions participate in cell functions such as attachment, proliferation, and migration [11,12]. Thus, the use of Mg in tissue engineering could potentially improve the bioactivity of nanocomposite biomaterials.

Among the Mg-based NPs available, magnesium oxide, MgO, and magnesium hydroxide, Mg(OH)$_2$, NPs were used in our work because of their biocompatibility [13,14]. Furthermore, both MgO and Mg(OH)$_2$ NPs show antimicrobial activity against bacteria, which can potentially reduce infections [15,16]. Many inorganic oxide NPs such as zinc oxide, ZnO, or titanium oxide, TiO$_2$, NPs have shown antimicrobial properties against a broad spectrum of microorganisms [17]. However, these NPs cause significant concerns considering their toxicity due to the risks associated with heavy metal elements and their accumulation in the body [18].

In contrast, MgO and Mg(OH)$_2$ NPs are considered an alternative to heavy metal oxide NPs because they can be efficiently degraded and metabolized inside the body, with Mg^{2+} and OH$^-$ ions as final degradation products [13]. In addition, MgO and Mg(OH)$_2$ NPs can enhance the mechanical properties of polymer efibers in terms of elastic modulus, tensile strength, or elongation at break [19]. Thus, some studies have recently been conducted investigating the addition of Mg-based NPs into polymer efibers for use in different fields such as environmental applications [20], energetic devices [21], or biomedical applications [22]. However, very few works involve the use of both MgO and Mg(OH)$_2$ NPs in PLA efibers. In particular, on the one hand, Canales et al. [23]. added MgO nanoparticles at 10 and 20 wt%, comparing them with bioglass nanoparticles and a mix of both NPs, resulting in a 10 and 20 wt% of their final concentration with respect to the neat PLA matrix. They focused their attention on bone tissue regeneration, also studying cell viability for biomedical applications. They did not use plasticizer, and they did not take into account Mg(OH)$_2$ nanoparticles. On the other hand, Salaris et al. [24] reported a comparative study of both NPs but in another matrix, such as Poly(ϵ-caprolactone) PCL, electrospun-fiber mats, observing the presence of monocalcium phosphate, dicalcium phosphate, and tricalcium phosphate, obtaining very low Ca/P values.

In this work, we focused on the bioactivity and antibacterial behavior of PLA-based efibers with both MgO and Mg(OH)$_2$ NPs, comparing the results. The bioactivity will be discussed in terms of HA growth after 28 days of immersion in simulated body fluid, SBF, by analyzing the Ca/P ratio, SEM, and XRD results [25]. The antibacterial behavior of MgO and Mg(OH)$_2$ NPs in PLA-based efibers was tested against *Escherichia coli, E. coli*, as Gram-negative bacteria, and *Staphylococcus aureus, S. aureus*, as Gram-positive bacteria, considering potential biomedical applications.

2. Materials and Methods

Poly(lactic acid) (PLA3051D, 3% of D-lactic acid monomer, molecular weight of 14.2 × 10^4 g·mol^{-1}, density of 1.24 g·cm^{-3}) was supplied by NatureWorks®, Minneapolis, MN, USA. Lactic acid oligomer (Glyplast OLA8, ester content > 99%, density of 1.11 g·cm^{-3}, viscosity of 22.5 mPa·s, molecular weight of 1100 g·mol^{-1}) was kindly supplied by Condensia Quimica SA, Barcelona, Spain. Magnesium oxide nanoparticles (MgO NPs, average particle size of 20 nm, 99.9% purity, molecular weight of 40.30 g·mol^{-1}) and magnesium hydroxide nanoparticles (Mg(OH)$_2$ NPs, average particle size of 10 nm, 99.9% purity, density of 2.34 g·cm^{-3}, molecular weight of 58.32 g·mol^{-1}) were supplied by Nanoshel LLC, Willmington, IL, USA.

Each solution was prepared following the process described in our previous work [26]. In particular, PLA pellets were previously dried in an oven overnight at 60 °C. The polymer solutions were used at a final concentration of 10 wt% in $CHCl_3$:DMF (4:1). Both NPs were added at different concentrations with respect to the PLA matrix, such as 0.5, 1, 5, 10, 15, and 20 wt%. The different amounts of NPs were dispersed using a tip sonicator (Sonic Vibra-Cell VCX 750, Sonics & Materials, Newton, CT, USA), operating at 750 watts and an amplitude of 20% for 3 h. Once the different solutions were obtained, electrospun fiber mats were prepared in an Electrospinner Y-flow 2.2.D-XXX (Nanotechnology Solutions), following our previously described method [27].

The bioactivity of PLA-based electrospun fiber mats was studied by immersing a square of 1 cm^2 of each PLA-based electrospun fiber system in simulated body fluid (SBF) for 28 days at 37 ± 1 °C to study the bioactivity at different immersion times, named T0, indicating the sample prior to immersion; T14, indicating the samples after 14 days of immersion; and T28, indicating the samples after 28 days of immersion in SBF. The SBF solution was prepared by following the protocol described by Kokubo et al. [25]. The extraction days are named T_x, where x indicates the number of the corresponding day. The as-obtained electrospun mats are considered as time 0, T0, and are used as references. The samples were carefully washed with water and dried under vacuum for two weeks before characterization.

The antibacterial behavior of PLA-based electrospun fibers was studied against American Type Culture Collection (ATCC): *Escherichia coli* (*E. coli*, ATCC 25922), as Gram-negative bacteria, and *Staphylococcus aureus* (*S. aureus*, ATCC 29213), as Gram-positive bacteria, obtained from Oxoid^{TM}. Microorganisms were incubated for 24 h at 37 °C. The optical density of the microorganism suspensions was measured in McFarland units, proportional to microorganism concentration, using a DEN-1B densitometer (Biosan, Madrid, Spain). The bacteria suspensions of about 10^8 colony-forming units (CFU) were prepared by adjusting the concentration with saline solution to a McFarland turbidity standard of ca. 0.5. A suspension of ca. 5×10^3 CFU·mL^{-1} for each bacteria was finally obtained by further dilution with PBS. The antibacterial behavior of PLA-based electrospun fibers was determined following the E2149-20 standard method of the American Society for Testing and Materials [28]. Each sample was placed in a sterile falcon tube, and then the bacterial suspension was added. Falcon tubes containing only the inoculum were prepared as control experiments. The samples were shaken at 37 °C at 150 rpm for 24 h. Bacterial concentrations at time 0 and after 24 h were calculated using the counting method. Each sample was measured and counted at least twice.

Scanning electron microscopy, SEM, (PHILIPS XL30 Scanning Electron Microscope, Phillips, Eindhoven, The Netherlands) was used to study the morphology and the diameters of the efibers. The images are shown at ×8000 magnification, with a scale of 2 μm, 25.0 kV, and a spot size 3.0. All the samples were previously gold-coated (~5 nm thickness) in a Polaron SC7640 Auto/Manual Sputter Coater (Polaron, Newhaven, East Sussex, UK). The diameters were calculated as the average value of 30 random measurements for each sample using ImageJ 1.51k software. Energy-dispersive X-ray spectroscopy, EDX, analyses were carried out in a FE-SEM Hitachi SU 8000 (Hitachi, Tokyo, Japan) device with a Bruker XFlash Detector 5030 operating at 15 Kv.

Attenuated total reflectance-Fourier transform infrared spectroscopy (ATR-FTIR) measurements were conducted using a Spectrum One FTIR spectrometer (Perkin Elmer instruments, Shelton CT, USA). Spectra were obtained in the 4000–400 cm^{-1} region at room temperature in transmission mode, with a resolution of 4 cm^{-1} and an accumulation of 16 scans.

XRD measurements were performed using a Bruker D8 Advance instrument (Bruker, Billerica, MA, USA) with a CuK as the source (0.154 nm) and a Detector Vantec1 detector. The scanning range was 5° and 60°, and the step size and count time per step were 0.023851° and 0.5 s, respectively.

The results were statistically analyzed by one-way analysis of variance (ANOVA) and Tukey's test, with a 95% confidence level, using the statistical computer package Statgraphics Centurion XVII v16.01 (Statpoint Technologies, Inc., Warrenton, VA, USA) [9].

3. Results

First of all, neat PLA electrospun fiber mats, as well as plasticized PLA:OLA electrospun fiber mats, were successfully obtained. Moreover, different concentrations of MgO and Mg(OH)$_2$ NPs, such as 0,5, 1, 5, 10, 15, and 20 wt%, were added to the polymeric solution, obtaining reinforced electrospun systems based on plasticized PLA. In this work, we focus on their bioactivity and antimicrobial response, comparing both systems. Therefore, in order to study the bioactivity of PLA-based efibers in terms of hydroxyapatite growth, SEM images of neat PLA, PLA:OLA, and PLA-based efibers with MgO and Mg(OH)$_2$ NPs at T0, and after 14 and 28 days of immersion in SBF, are shown in Figure 1, Figure 2, and Figure 3, respectively.

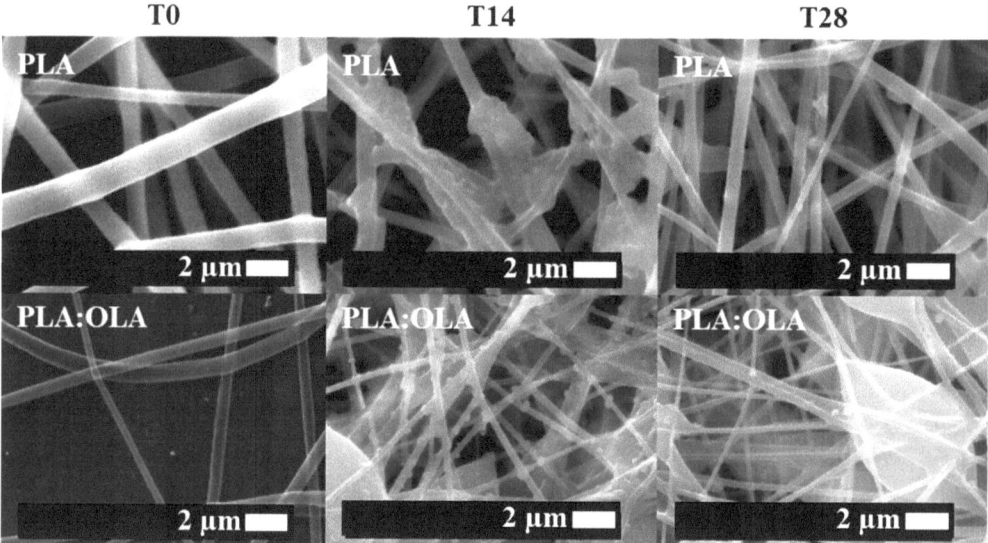

Figure 1. SEM images of PLA and PLA:OLA efibers in SBF at T0, T14, and T28.

As can be seen, the different efibers obtained by the electrospinning technique showed a 3D scaffold structure, with highly interconnected porosity. Both neat PLA and PLA:OLA efibers, our polymeric matrices at T0, showed smooth and regular surfaces, without the presence of beads. On the other hand, PLA:OLA-MgO and PLA:OLA-Mg(OH)$_2$ efibers showed some beads as the amount of NPs increased from 5 wt%, due to the presence of NP agglomeration.

In order to better visualize the fiber size distribution, in Figure 4, the diameter distribution of PLA and PLA:OLA efibers, as well as the different MgO and Mg(OH)$_2$ electrospun nanocomposites at T0, is reported, while the evolution of the average diameter at the different immersion times is reported in Figure 5.

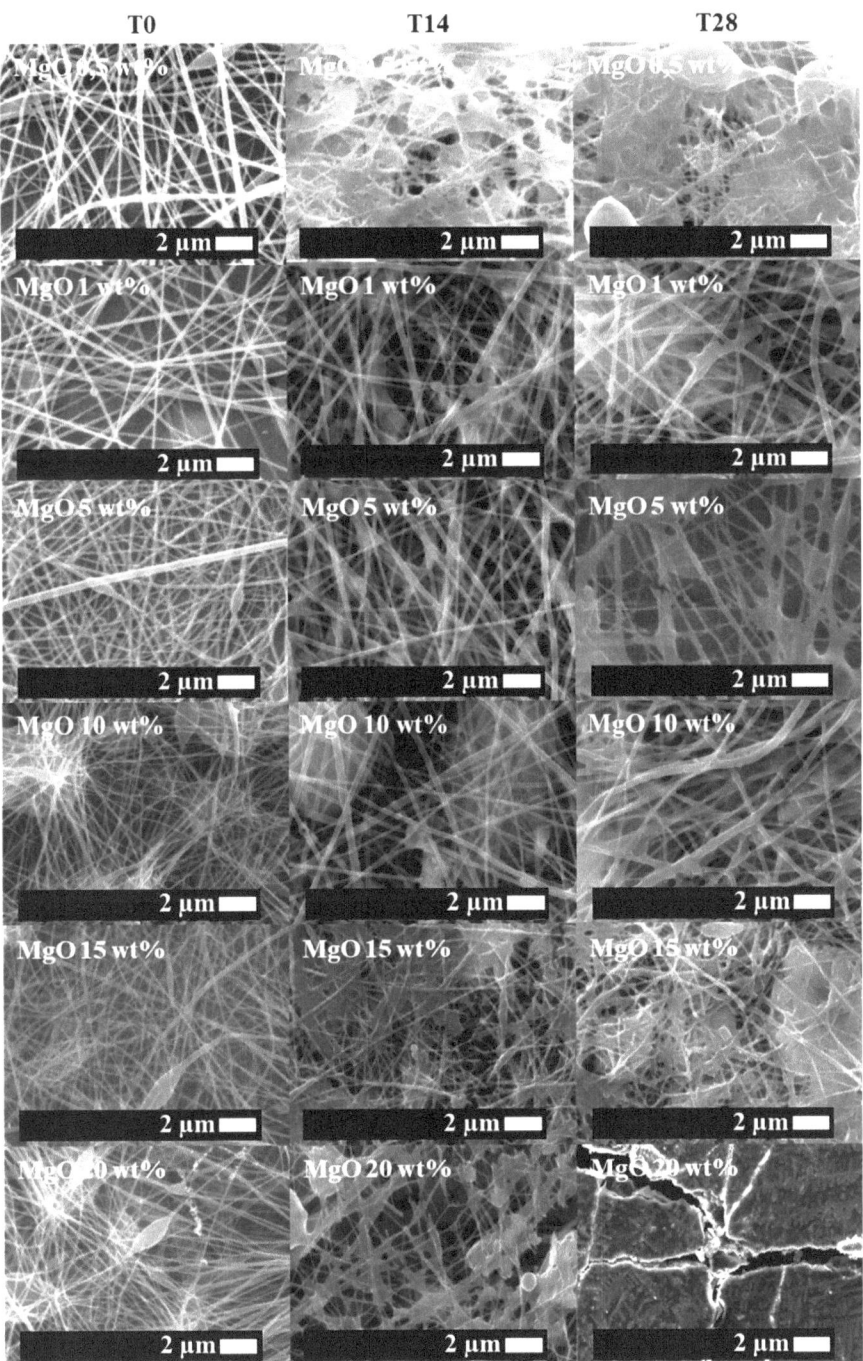

Figure 2. SEM images of PLA:OLA-MgO efibers in SBF at T0, T14, and T28.

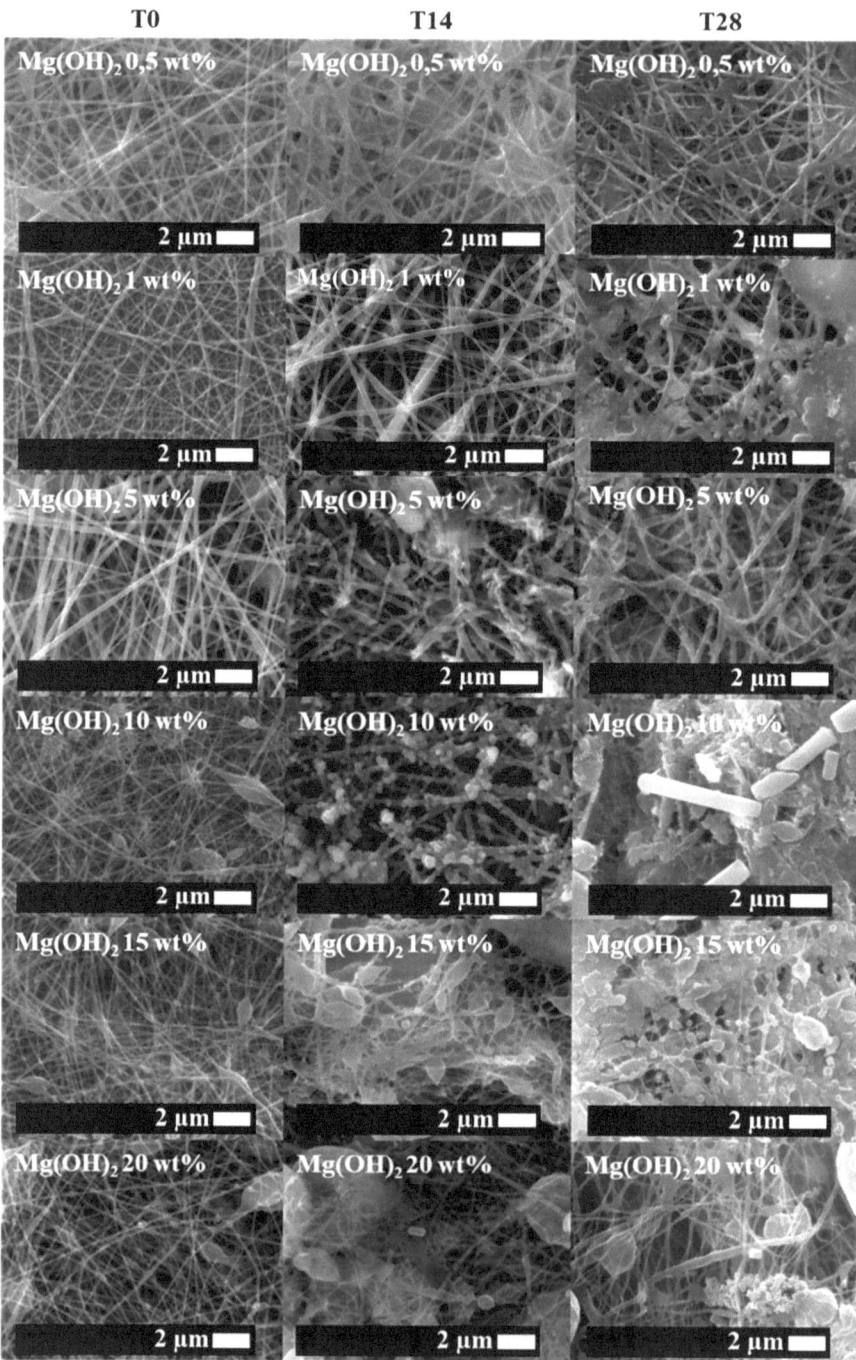

Figure 3. SEM images of PLA:OLA-Mg(OH)$_2$ efibers in SBF at T0, T14 and T28.

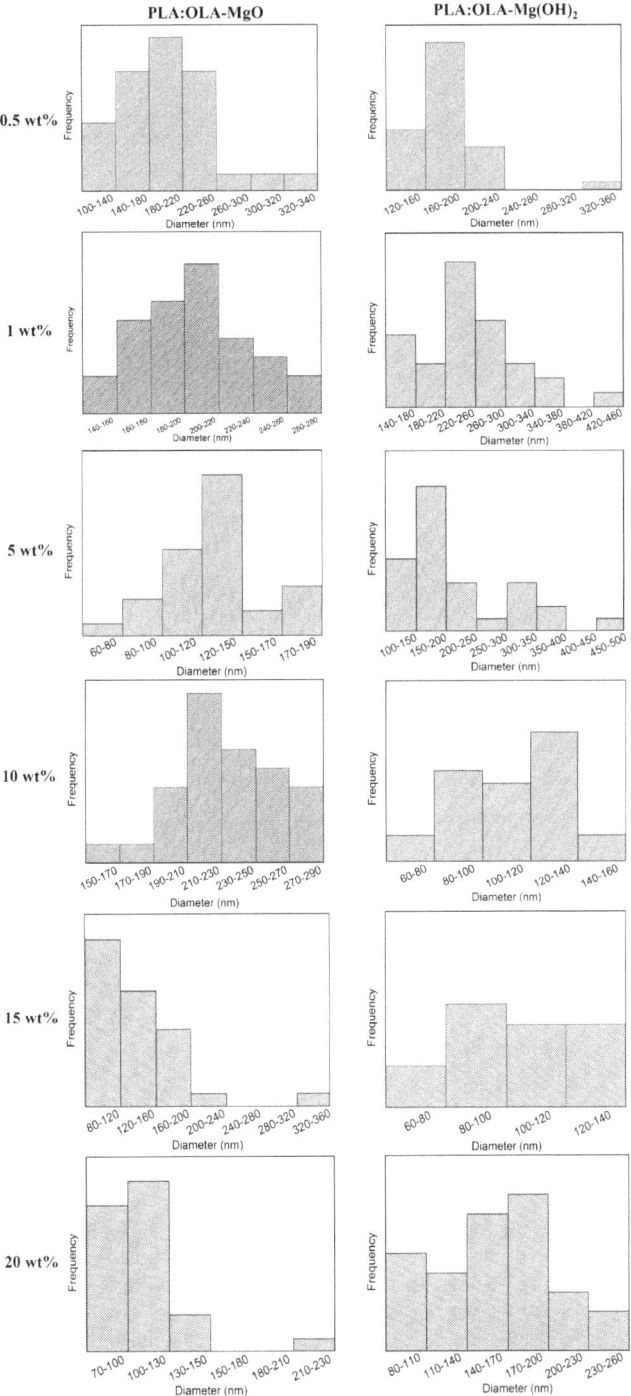

Figure 4. Diameter size distributions at T0 for the different electrospun fibers studied.

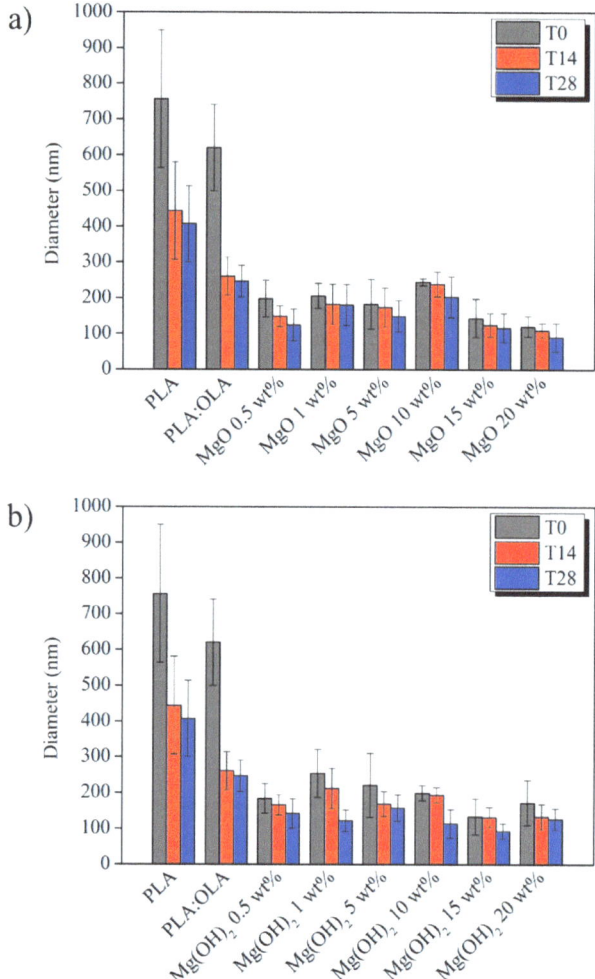

Figure 5. Average diameter evolution of PLA-based efibers in SBF at T0, T14, and T28 of (**a**) PLA:OLA-MgO and (**b**) PLA:OLA-Mg(OH)$_2$ efibers.

The biomineralization process, indicated by the presence of precipitated crystals on the surface of PLA-based efibers, started after 14 days of immersion in SBF in all PLA-based efibers, and continued until 28 days. In addition, it is worth noting that the immersion in SBF media affected the surface and morphologies of PLA-based efibers. Moreover, the amount of precipitated crystals increased by increasing the amount of both NPs, as can be observed in Figures 2 and 3. In particular, for the PLA:OLA-MgO efibers, the presence of precipitated crystals started after 14 days of immersion in SBF, being more abundant after 28 days for PLA:OLA-MgO efibers, with NPs in the range of 5–20 wt%. On the other hand, in the PLA:OLA-Mg(OH)$_2$ efibers, the precipitated crystals are clearly observed after 28 days in all the ranges of NP concentrations studied. The addition of both types of NPs, as well as the biomineralization process after immersion in SBF, affect the diameter evolution of PLA-based efibers, as can be observed in Figure 5 and as summarized in Table 1.

Table 1. Percentage of average diameter reduction (%D Reduction) for PLA, PLA-OLA, PLA:OLA-MgO, and PLA:OLA-Mg(OH)2 efibers after immersion in SBF.

Sample	%D Reduction	
	T14	T28
PLA	41	46
PLA:OLA	56	60
MgO 0.5 wt%	25	37
MgO 1 wt%	11	12
MgO 5 wt%	5	19
MgO 10 wt%	2	17
MgO 15 wt%	13	18
MgO 20 wt%	9	25
Mg(OH)$_2$ 0.5 wt%	10	23
Mg(OH)$_2$ 1 wt%	17	52
Mg(OH)$_2$ 5 wt%	24	29
Mg(OH)$_2$ 10 wt%	3	42
Mg(OH)$_2$ 15 wt%	1	30
Mg(OH)$_2$ 20 wt%	12	26

Simultaneously with the biomineralization process, a reduction in the average diameter of PLA-based efibers was observed, reported in Figure 5 and summarized in Table 1 in term of percentage of average diameter reduction (%D Reduction). In particular, after 28 days of immersion in SBF, a reduction of 46 and 60% in the average diameter of neat PLA and PLA:OLA efibers, respectively, was calculated. For PLA:OLA-MgO 0.5, 1, 5, 10, 15, and 20 wt% efibers, this reduction was noted as 37, 12, 19, 17, 18 and 25%, and for PLA:OLA-Mg(OH)$_2$ 0.5, 1, 5, 10, 15, and 20 wt% efibers, it was 23, 52, 29, 42, 30, and 26%, respectively. This average diameter reduction indicated the simultaneous degradation of efibers under SBF conditions. This degradation in aqueous media is widely reported in the literature, indicating the breaking of polymeric chains by hydrolytic degradation [8].

The precipitated crystals on the surface of PLA-based efibers were characterized by EDX analysis. The chemical composition was analyzed, and the Ca/P molar ratio was calculated after 28 days of immersion in SBF for all efibers, as summarized in Table 2. In addition, for a better comparison, the Ca/P ratios of HA from different natural sources and different calcium phosphates are summarized in Tables 3 and 4, respectively.

Table 2. Ca/P ratios of precipitated crystals in neat PLA, PLA-OLA, PLA:OLA-MgO, and PLA:OLA-Mg(OH)$_2$ efibers.

Sample	Ca/P Ratio
	T28
PLA	3.33
PLA:OLA	2.83
MgO 0.5 wt%	1.5
MgO 1 wt%	1.66
MgO 5 wt%	1.73
MgO 10 wt%	1.4
MgO 15 wt%	1.5
MgO 20 wt%	1.33
Mg(OH)$_2$ 0.5 wt%	1.26
Mg(OH)$_2$ 5 wt%	1.25
Mg(OH)$_2$ 10 wt%	1.6
Mg(OH)$_2$ 15 wt%	1.5
Mg(OH)$_2$ 20 wt%	1.5

The precipitated crystals on the surface of neat PLA and PLA:OLA showed a Ca/P ratio of 3.33 and 2.83, respectively, which is not consistent with any calcium phosphates summarized in Tables 2 and 3. However, this Ca/P ratio is described in the literature as carbonate apatite, if the hydroxyapatite is not carbonate-containing hydroxyapatite but carbonate apatite (according to the definition, carbonate ions occupy more than half the phosphoric acid positions), its Ca/P ratio should be much higher than the Ca/P ratio of 1.67 for the hydroxyapatite [29].

On the contrary, the presence of HA was confirmed by EDX on both PLA:OLA-MgO and PLA:OLA-Mg(OH)$_2$ efiber systems, showing Ca/P ratios in the range of 1.25–1.73 [30]. It is important to note that Ca/P ratios range from 1.20 for amorphous calcium phosphate to 1.67 for HA compound. In fact, compared to the synthetic HAs summarized in Table 2, natural HAs data, collected in Table 3, are non-stoichiometric, since they contain traces of elements such as Na^+, Zn^{2+}, Mg^{2+}, K^+, Si^{2+}, Ba^{2+}, F^-, and CO_3^{2-}, which make them similar to the chemical composition of human bone [31]. In our systems, the Ca/P ratios of HA extracted from different mammalian sources, presented in Table 3, are in the same range as those calculated for the PLA:OLA-MgO and PLA:OLA-Mg(OH)$_2$ efibers, suggesting the appearance of HA on the surface of the PLA-based nanocomposites.

Table 3. Main Ca/P ratios of different calcium phosphates.

Source	Ca/P Ratio	Ref.
β-Tricalcium phosphate	1.50	[30]
Amorphous calcium phosphate	1.20–2.20	[30]
Hydroxyapatite deficient in calcium	1.50–1.67	[30]
Hydroxyapatite	1.67	[30]

Table 4. Main Ca/P ratios of HA from different natural sources.

Source	Ca/P Ratio	Ref.
Bovine bone	2.23–1.95	[32]
Camelus bone	1.65	[33]
Turkey femur bone	1.66	[34]
Porcine bone	1.64	[35]
Fish bone	1.65–1.83	[36]
Cow, goat, and chicken bone	1.57–1.65	[37]

The precipitated crystals observed on the surface of the PLA-based efibers were characterized by XRD and FTIR analysis in order to determine their chemical structure. Specifically, XRD patterns of PLA-based efibers at T0 and T28 are reported in Figure 6.

At T0, previous to immersion in SBF, both neat PLA and PLA:OLA efibers show amorphous structures, whereas in the nanocomposites, the presence of the main peaks related to the crystallographic plane of MgO and Mg(OH)$_2$ NPs can be clearly observed.

For PLA:OLA-MgO efibers, the MgO XRD pattern shows peaks at 2θ = 36.9°, 42.9°, 62.3°, 74.6°, and 78.6°, which are attributed to the [111], [200], [220], [311], and [222] crystallographic planes; all diffraction peaks can be indexed to the cubic crystalline system for MgO [38]. In particular, the peaks located at 42.9° and 62.3° increase by increasing the amount of MgO NPs in the efibers, and these were used to corroborate the presence of MgO NP in the efibers.

The Mg(OH)$_2$ XRD pattern shows peaks at 2θ = 18.6°, 38.0°, 51.0°, and 58.6°, which are attributed to the [011], [101], [102], and [110] crystallographic planes and corresponded to the hexagonal structure of Mg(OH)$_2$ [39]. Moreover, it is important to note the effect of both MgO and Mg(OH)$_2$ NPs at 10 wt%. For MgO NPs, the crystallographic peak at 2θ = 18.7°, attributed to the [203] crystallographic plane of the α crystal of PLA [8], increased by increasing the amount of MgO NPs from 10 to 20 wt%, evidencing a higher crystallinity of the PLA:OLA-MgO efibers. However, this effect is not observed for Mg(OH)$_2$ NPs; in this case, the crystallographic peak at 2θ = 18.7° decreased for the PLA:OLA-Mg(OH)$_2$ efibers once it surpassed the 10 wt% of NPs.

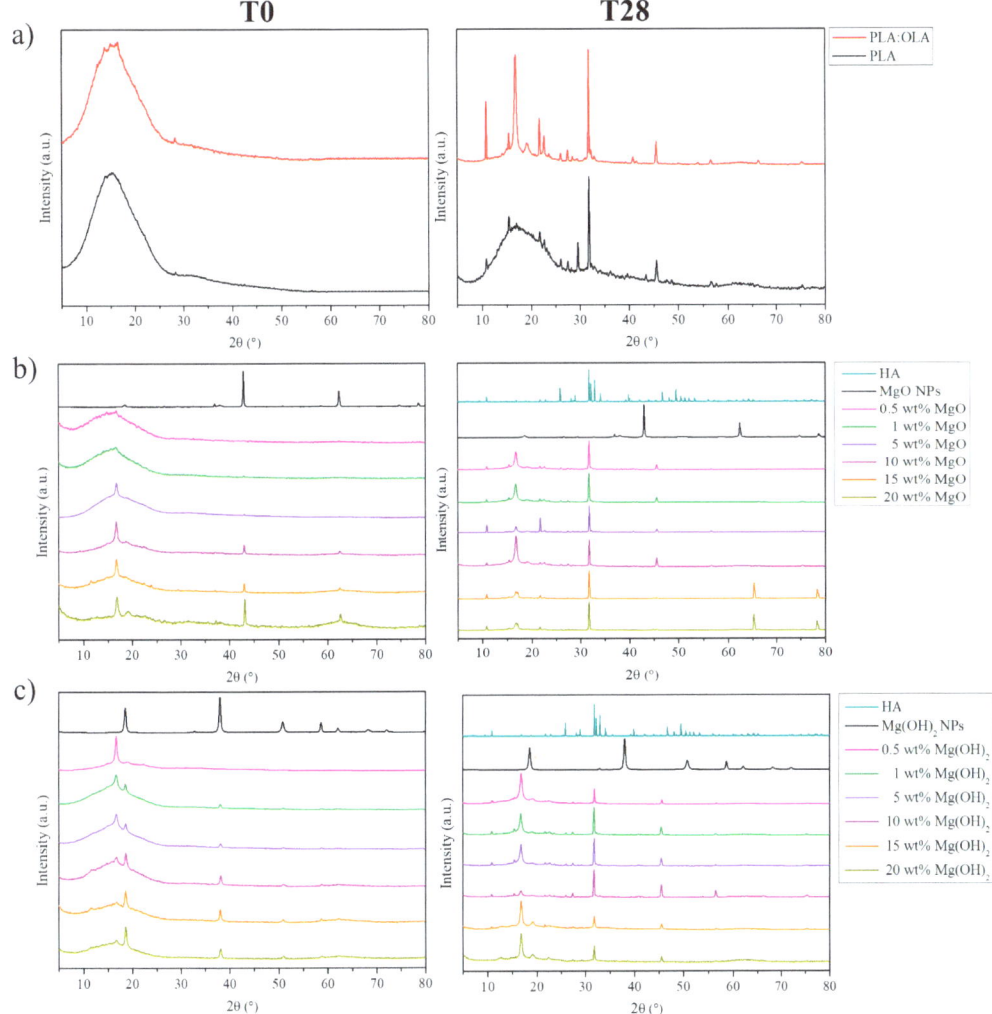

Figure 6. XRD patterns of (**a**) PLA and PLA:OLA, (**b**) PLA:OLA-MgO, and (**c**) PLA:OLA-Mg(OH)$_2$ efibers in SBF at T0 (**left**) and T28 (**right**).

The presence of peaks related to the crystallographic planes of the mineralization products is observed with both MgO and Mg(OH)$_2$ NPs. To properly study those peaks related to HA, the main XRD peaks of PLA, MgO NPs, Mg(OH)$_2$ NPs, and HA, as reported in the literature, are summarized in Table 5.

After 28 days of immersion in SBF, the PLA and PLA:OLA efibers showed crystallographic peaks at 2θ = 15.1, 16.6, 18.7, 22.1, 27.5, and 29.0°, which can be attributed to the [010], [200/110], [203], [015], [207], and [216] crystallographic planes of the α crystal conformation of PLA [8,40]. Comparing the XRD results before and after 28 days in SBF, we can conclude that immersion in SBF increases the degree of crystallinity in both the PLA and PLA:OLA efibers. An aqueous-based degradation media, such as SBF, provokes a hydrolytic reaction in the PLA chains, yielding a higher degree of crystallinity in PLA-based efibers [8]. This behavior was also observed in PLA:OLA-MgO and PLA:OLA-Mg(OH)$_2$ efibers, and in addition, new crystallographic peaks appear after 28 days in SBF, which can be related to the presence of mineralization products. Firstly, it is important to note that the mineralization of carbonate apatite in the PLA and PLA:OLA samples is in good agreement with the Ca/P ratio previously described. The XRD patterns of HA and carbonated apatite are very similar; however, in carbonate apatite, the crystallographic peaks at $2\theta = 26°$, attributed to the [002] crystallographic plane of the hexagonal structure of HA, shifts to a lower value of $2\theta = 25.5°$ [41]. In our PLA and PLA:OLA electrospun fibers, this crystallographic peak showed a higher intensity in comparison with those obtained for the electrospun fibers reinforced with both MgO and Mg(OH)$_2$.

In particular, crystallographic peaks at $2\theta = 22.0$, 23.0, and 26.0° can be observed in both the PLA:OLA-MgO and PLA:OLA-Mg(OH)$_2$ efibers, which are attributed to the [200], [111], and [002] crystallographic planes of the hexagonal structure of HA [42–44]. Moreover, a high peak at $2\theta = 32.3°$ is observed, attributed to the presence of phosphate crystals (PO$_4^{3-}$) from HA [44]. Additionally, crystallographic peaks related to other mineralization products can be observed, such as $2\theta = 45.0°$, attributed to the [220] crystallographic plane of NaCl [45].

Salaris et al. [24] studied both MgO and Mg(OH)$_2$ nanoparticles dispersed in PCL electrospun fiber mats. They observed the presence of monocalcium phosphate, dicalcium phosphate, and tricalcium phosphate, obtaining very low Ca/P values, varying from 0.18 at 0.5 wt% of MgO to 1.17 at 10 wt% of MgO and varying from 0.27 at 0.5 wt% of Mg(OH)$_2$ to 0.43 at 10 wt% of Mg(OH)$_2$. Only in the electrospun nanofiber mats reinforced with 20 wt% of NPs Ca/P obtained values of 1.76 and 1.33 for MgO and Mg(OH)$_2$, respectively. This marks a significant difference from the results of our work, where we demonstrated that in plasticized PLA electrospun fiber mats, low concentrations of both MgO and Mg(OH)$_2$ nanoparticles allow HA growth on the surface of the electrospun fiber mats. This fact is very important because it confirms that every system is different, and a deep study is required for each one in order to determine the correct response.

Table 5. Main XRD peaks of PLA, MgO NPs, Mg(OH)$_2$ NPs, and HA reported in the literature.

2θ (°)	Crystalline System	hkl	Ref. No.
15.1	α crystals	010	[8,40]
16.6	α crystals	200/110	[8,40]
18.7	α crystals	203	[8,40]
22.1	α crystals	015	[8,40]
27.5	α crystals	207	[8,40]
29.0	α crystals	216	[8,40]
22.0	Hexagonal	200	[42,44]
23.0	Hexagonal	111	[42,44]
26.0	Hexagonal	002	[42,44]
28.5	Hexagonal	102	[42,44]
29.2	Hexagonal	210	[42,44]
31.8	Hexagonal	211	[42,44]
32.3	Hexagonal	112	[42,44]
32.9	Hexagonal	300	[42,44]
34.6	Hexagonal	202	[42,44]

Table 5. Cont.

	2θ (°)	Crystalline System	hkl	Ref. No.
MgO NPs	36.9	Cubic	111	[38]
	42.9	Cubic	200	[38]
	62.3	Cubic	220	[38]
	74.7	Cubic	311	[38]
	78.6	Cubic	222	[38]
Mg(OH)$_2$ NPs	18.6	Hexagonal	001	[39]
	38.0	Hexagonal	101	[39]
	51.0	Hexagonal	102	[39]
	58.6	Hexagonal	110	[39]

Once the XRD patterns of the PLA-based efibers were characterized, the functional groups and chemical interactions were studied by FTIR analysis, Table 6, and their spectra are shown in Figure 7. All PLA-based efibers showed the main characteristic FTIR bands of PLA at T0, in particular, the C=O stretching band at 1749 cm^{-1} attributed to the carbonyl group, the CH$_3$ bond asymmetric vibration at 1452 cm^{-1}, and the CH$_3$ bond symmetric motion at 1182 cm^{-1}, as well as both the C-O asymmetric stretching band and the C-O symmetric stretching band at 1128 cm^{-1} and 1084 cm^{-1}, respectively [8,46]. Moreover, the presence of MgO and Mg(OH)$_2$ NPs at high concentrations, i.e., 10, 15, and 20 wt%, increased the 2882–3000 cm^{-1} bands related to the C-H stretching (asymmetric, symmetric vibrations of –CH$_3$ and CH modes) [47].

Table 6. Most significant infrared bands related to HA and PLA.

	Frequency (cm^{-1})	Assignment	Ref. No.
Hydroxyapatite	567–604	P-O asymmetric vibrations of PO$_4^{3-}$ groups	[42,48]
	632 and 1625	Hydroxyl liberation mode	[42,48]
	418 and 575	Antisymmetric bending of PO$_4^{3-}$ groups	[42,48]
	1089–1039	Stretching mode in PO$_4^{3-}$ groups	[42,48]
	650	Bending mode in PO$_4^{3-}$ groups	[42,48]
	1550	Stretching mode in CO$_3^{2-}$ groups	[42,48]
PLA	2882 and 3000	Asymmetric–Symmetric vibrations of –CH$_3$ and CH modes	[8,46]
	1452	CH$_3$ bond asymmetric vibration	[8]
	1749	C=O stretching band	[8,46]
	1182	CH$_3$ bond symmetric motion	[8,46]
	1128 and 1084	C-O asymmetric stretching band and C-O symmetric stretching band	[8,46]

After 28 days of immersion in SBF, a series of strong peaks can be seen in the spectra of both series, as reported in Figure 8, which could be related to HA, specifically, a new band at 575 cm^{-1} that can be associated with the antisymmetric bending of the PO$_4^{3-}$ groups [42,48]. In addition, the band at 650 cm^{-1} is well associated with the bending mode in the PO$_4^{3-}$ groups [42,48]. On the other hand, other mineralization products, such as carbonate, CO$_3^{2-}$, which showed a band at 1550 cm^{-1} associated with stretching mode in the CO$_3^{2-}$ groups [42,48], can be observed on the surface of the PLA-based efibers. These signals confirm the bioactivity of both the MgO and Mg(OH)$_2$ efibers after immersion in SBF. Moreover, considering that SBF is an aqueous media, the hydrolytic degradation of PLA-based electrospun fibers after 28 days can be observed. Based in our previous results [8], the hydrolytic degradation through the PLA matrix provokes the breakage of the ester group, and new bands at 1650 cm^{-1}, attributed to the carboxylate groups, can be observed for both the MgO and Mg(OH)$_2$ NPs.

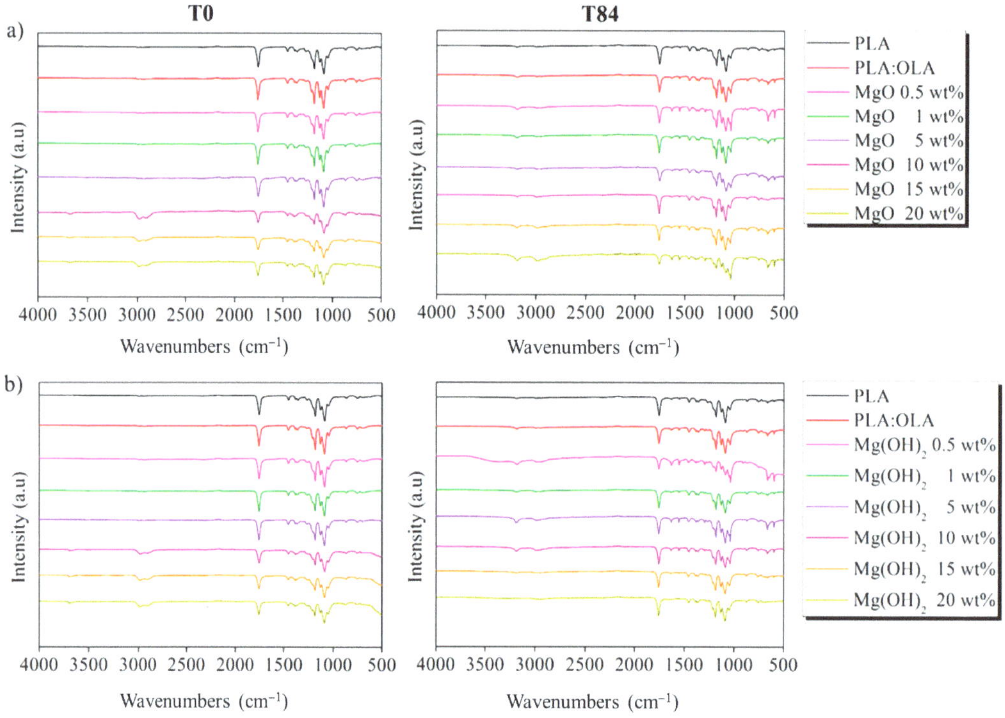

Figure 7. FTIR spectra of (**a**) PLA:OLA-MgO and (**b**) PLA:OLA-Mg(OH)$_2$ efibers after immersion in SBF at T0 and T28. Spectra of PLA and PLA:OLA are also included in the graphs.

The antibacterial activity of the PLA-based electrospun nanocomposites was tested against *E. coli* (Gram-negative) and *S. aureus* (Gram-positive), and the relative antibacterial activity was calculated, as summarized in Tables 7 and 8, respectively.

Table 7. Relative antibacterial activity for PLA:OLA-MgO efibers.

Sample	*E. coli*	*S. aureus*
PLA	0 [a]	0 [a]
PLA:OLA	0 [a]	0 [a]
MgO 0.5 wt%	2 ± 2 [a]	0 [a]
MgO 1 wt%	2 ± 1 [a]	0 [a]
MgO 5 wt%	2 ± 2 [a]	0 [a]
MgO 10 wt%	5 ± 1 [a]	6 ± 2 [b]
MgO 15 wt%	15 ± 2 [b]	5 ± 1 [b]
MgO 20 wt%	21 ± 2 [c]	5 ± 1 [b]

Different letters in the column indicate significant differences, according to Tukey's test ($p < 0.05$).

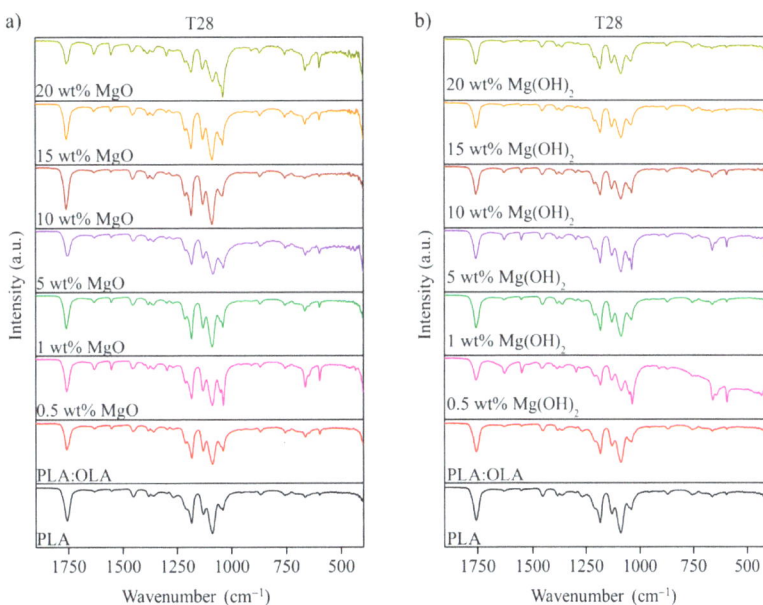

Figure 8. FTIR spectra 1800–400 cm^{-1} of (**a**) PLA:OLA-MgO and (**b**) PLA:OLA-Mg(OH)$_2$ efibers after immersion in SBF at T28. Spectra of PLA and PLA:OLA are also included in the graphs.

Table 8. Relative antibacterial activity for PLA:OLA-Mg(OH)$_2$ efibers.

Sample	E. coli	S. aureus
PLA	0 [a]	0 [a]
PLA:OLA	0 [a]	0 [a]
Mg(OH)$_2$ 0.5 wt%	1 ± 1 [a]	0 [a]
Mg(OH)$_2$ 1 wt%	2 ± 1 [a]	0 [a]
Mg(OH)$_2$ 5 wt%	2 ± 2 [a]	0 [a]
Mg(OH)$_2$ 10 wt%	16 ± 2 [b]	2 ± 1 [b]
Mg(OH)$_2$ 15 wt%	17 ± 3 [b]	2 ± 1 [b]
Mg(OH)$_2$ 20 wt%	34 ± 6 [c]	2 ± 1 [b]

Different letters in the column indicate significant differences, according to Tukey's test ($p < 0.05$).

As expected, the results indicated that the neat PLA and PLA:OLA efibers did not show antibacterial activity against either *E. coli* or *S. aureus*, in agreement with previous results reported in the literature [49]. From the MgO and Mg(OH)$_2$ NPs point of view, some considerations can be made. No statistically significant differences ($p < 0.05$) were observed in the relative antibacterial activity of fibers with MgO NPs at low concentrations such as 0.5, 1, 5, and 10 wt% against *E. coli* in comparison with neat PLA and PLA:OLA efibers. However, for the PLA:OLA-MgO 15 and 20 wt% efibers, a slight antibacterial activity percentage of 15 ± 2 and 21 ± 2 was obtained, respectively. The antibacterial activity of MgO NPs has been recently studied by some authors, with similar results. In particular, Swarrop et al. studied PLLA/MgO nanocomposite films, reporting 46% antibacterial activity against *E. coli* [50]. MgO NPs showed different antibacterial activity against *S. aureus* (Gram-positive). In general, no inhibitory effect was observed in the range of 0.5–5 wt%, but from this concentration on, the antibacterial activity starts to increase slightly. Thus, the PLA:OLA-MgO efibers were able to fight against Gram-negative bacteria more effectively than against Gram-positive bacteria.

On the other hand, Mg(OH)$_2$ NPs showed statistically significant differences ($p < 0.05$) in the antibacterial activity against *E. coli* at concentrations in the range of 10–20 wt% in comparison with neat PLA and PLA:OLA. In particular, the achieved relative antibacterial activity of 16 ± 2, 17 ± 3, and 34 ± 6% for PLA:OLA-Mg(OH)$_2$ 10, 15, and 20 wt% efibers, respectively, are slightly higher than those for PLA:OLA-MgO efibers in the same range of concentrations. In addition, poor antibacterial activity against *S. aureus* was observed for all PLA:OLA-Mg(OH)$_2$ efibers. In Figure 9, qualitative comparison of the antibacterial activity obtained for PLA-based electrospun fibers at the highest concentration of both nanoparticles, that is 20 wt%, is reported, compared with the control for better visualization of the qualitative antibacterial activity of our PLA-based electrospun fibers. As far as we know, almost no investigations regarding the antibacterial activity of Mg(OH)$_2$ NPs are found in the literature for comparison with our results. Only Meng et al. studied the antibacterial activity of Mg(OH)$_2$ NPs against oral bacteria (*Streptococcus mutans*) and concluded that Mg(OH)$_2$ NPs can be used to eradicate residual bacteria but with limited activity [15]. In our case, for a concentration at 20 wt% of both MgO and Mg(OH)$_2$ NPs, the plasticized PLA-based electrospun nanocomposites show antibacterial activity. Regarding its use in potential biomedical applications, this result suggests an optimal range of NPs concentration of about 20 wt% for improving the antibacterial activity of PLA-based systems.

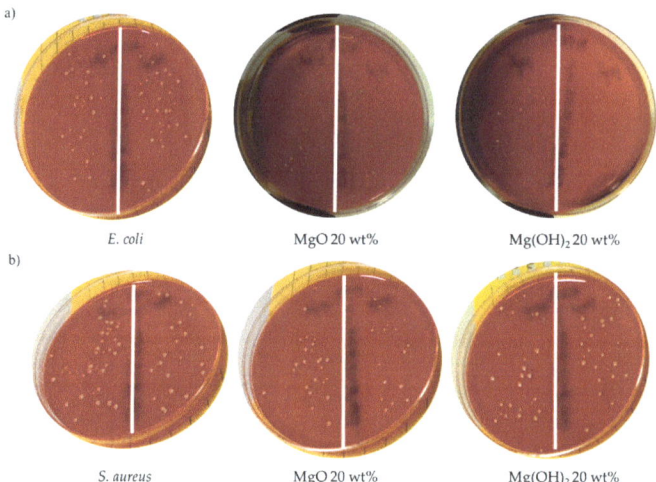

Figure 9. Images of the control plate and the plasticized PLA electrospun fiber at the highest concentrations of MgO and Mg(OH)$_2$ NPs against (**a**) *E. coli* and (**b**) *S. aureus*.

4. Conclusions

In this work, we studied the bioactivity and antibacterial behavior of PLA-based electrospun fibers, efibers, reinforced with both MgO and Mg(OH)$_2$ nanoparticles, with different concentrations such as 0.5, 1, 5, 10, 15, and 20 wt%. The bioactivity was discussed in terms of hydroxyapatite growth at T28 of immersion in simulated body fluid, SBF. In particular, at T14, the biomineralization process is evidenced in both reinforced systems, at concentrations higher than 5 wt% for MgO NPs, and at all the concentrations studied for Mg(OH)$_2$. However, the number of precipitated crystals increased by increasing the amount of both NPs. The chemical composition of the precipitated crystals was characterized in terms of Ca/P molar ratio after T28 of immersion in SBF, indicating the presence of hydroxyapatite on the surface of both reinforced efibers. Moreover, a reduction in the average diameter of PLA-based efibers was observed, reaching a maximum reduction of 46 and 60% in the average diameter of the neat PLA and PLA:OLA efibers, respectively, after 28 days of immersion in SBF, indicating their degradation process. The antibacterial behavior of MgO and Mg(OH)$_2$ NPs in PLA-based electrospun fibers was tested against Escherichia coli, *E. coli*, as the Gram-negative bacteria, and *Staphylococcus aureus*, *S. aureus*, as the Gram-positive bacteria, obtaining the best antibacterial activity against the Gram-negative bacteria *E. coli* of 21 ± 2% and 34 ± 6% for the highest concentration of MgO and Mg(OH)$_2$ NPs, respectively. This study suggests that these systems can potentially be used in biomedical applications.

Author Contributions: Conceptualization, L.P.; methodology, L.P., M.L. and A.M.-B.; investigation, A.L. and V.S.; resources, M.L., D.L. and L.P.; data curation, A.L. and V.S.; writing—original draft preparation, A.L. and V.S.; writing—review and editing, M.L., M.F-G., D.L. and L.P.; supervision, M.L. and L.P.; funding acquisition, M.L., D.L. and L.P. All authors have read and agreed to the published version of the manuscript.

Funding: This publication is part of the I+D+i project PID2021-123753NB-C31, funded by MCIN/AEI/ 10.13039/501100011033/, "ERDF—A Way of Making Europe".

Institutional Review Board Statement: Not applicable.

Data Availability Statement: The raw data supporting the conclusions of this article will be made available by the authors on request.

Acknowledgments: The authors thank MICIN for PID2021-123753NB-C31, Condensia Quimica S.A., for kindly supplying the plasticizers.

Conflicts of Interest: The authors declare no conflicts of interest.

References

1. Leonés, A.; Lieblich, M.; Benavente, R.; Gonzalez, J.L.; Peponi, L. Potential applications of magnesium-based polymeric nanocomposites obtained by electrospinning technique. *Nanomaterials* **2020**, *10*, 1524. [CrossRef] [PubMed]
2. Hickey, D.J.; Ercan, B.; Sun, L.; Webster, T.J. Adding MgO Nanoparticles to Hydroxyapatite-PLLA Nanocomposites for Improved Bone Tissue Engineering Applications. *Acta Biomater.* **2015**, *14*, 175–184. [CrossRef] [PubMed]
3. Habibovic, P. In Vitro and In Vivo Bioactivity Assessment of a Polylactic Acid/Hydroxyapatite Composite for Bone Regeneration. *Biomatter* **2014**, *4*, e27664. [CrossRef] [PubMed]
4. Peponi, L.; Puglia, D.; Torre, L.; Valentini, L.; Kenny, J.M. Processing of nanostructured polymers and advanced polymeric based nanocomposites. *Mater. Sci. Eng. R Rep.* **2014**, *85*, 1–46. [CrossRef]
5. Prasad, A.; Sankar, M.R.; Katiyar, V. State of Art on Solvent Casting Particulate Leaching Method for Orthopedic Scaffolds Fabrication. *Mater. Today Proc.* **2017**, *4*, 898–907. [CrossRef]
6. Kyle, S.; Jessop, Z.M.; Al-Sabah, A.; Whitaker, I.S. 'Printability' of Candidate Biomaterials for Extrusion Based 3D Printing: State-of-the-Art. *Adv. Healthc. Mater.* **2017**, *6*, 1700264. [CrossRef]
7. Salaris, V.; Leonés, A.; Lopez, D.; Kenny, J.M.; Peponi, L. Shape-Memory Materials via Electrospinning: A Review. *Polymers* **2022**, *14*, 995. [CrossRef] [PubMed]
8. Leonés, A.; Peponi, L.; Lieblich, M.; Benavente, R.; Fiori, S. In vitro degradation of plasticized PLA electrospun fiber mats: Morphological, thermal and crystalline evolution. *Polymers* **2020**, *12*, 2975. [CrossRef]
9. Leonés, A.; Salaris, V.; Mujica-Garcia, A.; Arrieta, M.P.; Lopez, D.; Lieblich, M.; Kenny, J.M.; Peponi, L. Pla electrospun fibers reinforced with organic and inorganic nanoparticles: A comparative study. *Molecules* **2021**, *26*, 4925. [CrossRef]
10. Kareem, M.M.; Tanner, K.E. Optimising Micro-Hydroxyapatite Reinforced Poly(Lactide Acid) Electrospun Scaffolds for Bone Tissue Engineering. *J. Mater. Sci. Mater. Med.* **2020**, *31*, 38. [CrossRef]
11. Maguire, M.E.; Cowan, J.A. Magnesium Chemistry and Biochemistry. *BioMetals* **2002**, *15*, 203–210. [CrossRef] [PubMed]
12. Esteban-Cubillo, A.; Pina-Zapardiel, R.; Moya, J.S.; Barba, M.F.; Pecharromán, C. The Role of Magnesium on the Stability of Crystalline Sepiolite Structure. *J. Eur. Ceram. Soc.* **2008**, *28*, 1763–1768. [CrossRef]
13. Wetteland, C.L.; Jesus Sanchez, J.; Silken, C.A.; Nguyen, N.Y.T.; Mahmood, O.; Liu, H. Dissociation of Magnesium Oxide and Magnesium Hydroxide Nanoparticles in Physiologically Relevant Fluids. *J. Nanoparticle Res.* **2018**, *20*, 215. [CrossRef]
14. Gonzalez, J.; Hou, R.Q.; Nidadavolu, E.P.S.; Willumeit-Römer, R.; Feyerabend, F. Magnesium Degradation under Physiological Conditions—Best Practice. *Bioact. Mater.* **2018**, *3*, 174–185. [CrossRef] [PubMed]
15. Meng, Y.; Zhang, D.; Jia, X.; Xiao, K.; Lin, X.; Yang, Y.; Xu, D.; Wang, Q. Antimicrobial Activity of Nano-Magnesium Hydroxide against Oral Bacteria and Application in Root Canal Sealer. *Med. Sci. Monit.* **2020**, *26*, e922920-1–e922920-10. [CrossRef] [PubMed]
16. Nguyen, N.Y.T.; Grelling, N.; Wetteland, C.L.; Rosario, R.; Liu, H. Antimicrobial Activities and Mechanisms of Magnesium Oxide Nanoparticles (NMgO) against Pathogenic Bacteria, Yeasts, and Biofilms. *Sci. Rep.* **2018**, *8*, 16360. [CrossRef] [PubMed]
17. Ghozali, M.; Fahmiati, S.; Triwulandari, E.; Restu, W.K.; Farhan, D.; Wulansari, M.; Fatriasari, W. PLA/Metal Oxide Biocomposites for Antimicrobial Packaging Application. *Polym. Technol. Mater.* **2020**, *59*, 1332–1342. [CrossRef]
18. Witkowska, D.; Słowik, J.; Chilicka, K. Review Heavy Metals and Human Health: Possible Exposure Pathways and the Competition for Protein Binding Sites. *Molecules* **2021**, *26*, 6060. [CrossRef] [PubMed]

19. Leonés, A.; Peponi, L.; García-Martínez, J.M.; Collar, E.P. Study on the Tensile Behavior of Woven Non-Woven PLA/OLA/MgO Electrospun Fibers. *Polymers* **2023**, *15*, 3973. [CrossRef]
20. Dehghan, S.F.; Golbabaei, F.; Maddah, B.; Latifi, M.; Pezeshk, H.; Hasanzadeh, M.; Akbar-Khanzadeh, F. Optimization of Electrospinning Parameters for Polyacrylonitrile-MgO Nanofibers Applied in Air Filtration. *J. Air Waste Manag. Assoc.* **2016**, *66*, 912–921. [CrossRef]
21. Du, P.; Song, L.; Xiong, J.; Wang, L.; Li, N. A Photovoltaic Smart Textile and a Photocatalytic Functional Textile Based on Co-Electrospun TiO_2/MgO Core–Sheath Nanorods: Novel Textiles of Integrating Energy and Environmental Science with Textile Research. *Text. Res. J.* **2013**, *83*, 1690–1702. [CrossRef]
22. Rijal, N.P.; Adhikari, U.; Khanal, S.; Pai, D.; Sankar, J.; Bhattarai, N. Magnesium oxide-poly(ε-caprolactone)-chitosan-based composite nanofiber for tissue engineering applications. *Mater. Sci. Eng. B Solid-State Mater. Adv. Technol.* **2018**, *228*, 18–27. [CrossRef]
23. Canales, D.A.; Reyes, F.; Saavedra, M.; Peponi, L.; Leonés, A.; Palza, H.; Boccaccini, A.R.; Grünewald, A.; Zapata, P.A. Electrospun fibers of poly (lactic acid) containing bioactive glass and magnesium oxide nanoparticles for bone tissue regeneration. *Int. J. Biol. Macromol.* **2022**, *210*, 324–336. [CrossRef] [PubMed]
24. Salaris, V.; Leonés, A.; López, D.; Kenny, J.M.; Peponi, L. A Comparative Study on the Addition of MgO and $Mg(OH)_2$ Nanoparticles into PCL Electrospun Fibers. *Macromol. Chem. Phys.* **2022**, *224*, 2200215. [CrossRef]
25. Kokubo, T.; Takadama, H. How useful is SBF in predicting in vivo bone bioactivity? *Biomaterials* **2006**, *27*, 2907–2915. [CrossRef]
26. Leonés, A.; Peponi, L.; Fiori, S.; Lieblich, M. Effect of the Addition of MgO Nanoparticles on the Thermally-Activated Shape Memory Behavior of Plasticized PLA Electrospun Fibers. *Polymers* **2022**, *14*, 2657. [CrossRef] [PubMed]
27. Leonés, A.; Sonseca, A.; López, D.; Fiori, S.; Peponi, L. Shape memory effect on electrospun PLA-based fibers tailoring their thermal response. *Eur. Polym. J.* **2019**, *117*, 217–226. [CrossRef]
28. ASTM E2149-01; Standard Test Method for Determining the Antimicrobial Activity of Immobilized Antimicrobial Agents under Dynamic Contact Conditions (Withdrawn 2010). ASTM International: West Conshohocken, PA, USA, 2010.
29. Kono, T.; Sakae, T.; Nakada, H.; Kaneda, T.; Okada, H. Confusion between Carbonate Apatite and Biological Apatite (Carbonated Hydroxyapatite) in Bone and Teeth. *Minerals* **2022**, *12*, 170. [CrossRef]
30. Akram, M.; Ahmed, R.; Shakir, I.; Ibrahim, W.A.W.; Hussain, R. Extracting hydroxyapatite and its precursors from natural resources. *J. Mater. Sci.* **2014**, *49*, 1461–1475. [CrossRef]
31. Mohd Pu'ad, N.A.S.; Koshy, P.; Abdullah, H.Z.; Idris, M.I.; Lee, T.C. Syntheses of Hydroxyapatite from Natural Sources. *Heliyon* **2019**, *5*, e01588. [CrossRef]
32. Ayatollahi, M.R.; Yahya, M.Y.; Asgharzadeh Shirazi, H.; Hassan, S.A. Mechanical and Tribological Properties of Hydroxyapatite Nanoparticles Extracted from Natural Bovine Bone and the Bone Cement Developed by Nano-Sized Bovine Hydroxyapatite Filler. *Ceram. Int.* **2015**, *41*, 10818–10827. [CrossRef]
33. Jaber, H.L.; Hammood, A.S.; Parvin, N. Synthesis and Characterization of Hydroxyapatite Powder from Natural Camelus Bone. *J. Aust. Ceram. Soc.* **2017**, *54*, 1–10. [CrossRef]
34. Esmaeilkhanian, A.; Sharifianjazi, F.; Abouchenari, A.; Rouhani, A.; Parvin, N.; Irani, M. Synthesis and Characterization of Natural Nano-Hydroxyapatite Derived from Turkey Femur-Bone Waste. *Appl. Biochem. Biotechnol.* **2019**, *189*, 919–932. [CrossRef] [PubMed]
35. Vuong, B.X.; Linh, T.H. Extraction of Pure Hydroxyapatite from Porcine Bone by Thermal Process. *Metall. Mater. Eng.* **2019**, *25*, 47–58. [CrossRef] [PubMed]
36. Pon-On, W.; Suntornsaratoon, P.; Charoenphandhu, N.; Thongbunchoo, J.; Krishnamra, N.; Tang, I.M. Hydroxyapatite from Fish Scale for Potential Use as Bone Scaffold or Regenerative Material. *Mater. Sci. Eng. C* **2016**, *62*, 183–189. [CrossRef] [PubMed]
37. Ramesh, S.; Loo, Z.Z.; Tan, C.Y.; Chew, W.J.K.; Ching, Y.C.; Tarlochan, F.; Chandran, H.; Krishnasamy, S.; Bang, L.T.; Sarhan, A.A.D. Characterization of Biogenic Hydroxyapatite Derived from Animal Bones for Biomedical Applications. *Ceram. Int.* **2018**, *44*, 10525–10530. [CrossRef]
38. Singh, A.K.; Pramanik, K.; Biswas, A. MgO Enables Enhanced Bioactivity and Antimicrobial Activity of Nano Bioglass for Bone Tissue Engineering Application. *Mater. Technol.* **2019**, *34*, 818–826. [CrossRef]
39. Kumari, L.; Li, W.Z.; Vannoy, C.H.; Leblanc, R.M.; Wang, D.Z. Synthesis, Characterization and Optical Properties of $Mg(OH)_2$ Micro-/Nanostructure and Its Conversion to MgO. *Ceram. Int.* **2009**, *35*, 3355–3364. [CrossRef]
40. Tashiro, K.; Kouno, N.; Wang, H.; Tsuji, H. Crystal Structure of Poly(Lactic Acid) Stereocomplex: Random Packing Model of PDLA and PLLA Chains As Studied by X-ray Diffraction Analysis. *Macromolecules* **2017**, *50*, 8048–8065. [CrossRef]
41. Midorikawa, K.; Hiromoto, S.; Yamamoto, T. Carbonate content control in carbonate apatite coatings of biodegradable magnesium. *Ceram. Int.* **2024**, *50*, 6784–6792. [CrossRef]
42. Chavan, P.N.; Bahir, M.M.; Mene, R.U.; Mahabole, M.P.; Khairnar, R.S. Study of nanobiomaterial hydroxyapatite in simulated body fluid: Formation and growth of apatite. *Mater. Sci. Eng. B Solid-State Mater. Adv. Technol.* **2010**, *168*, 224–230. [CrossRef]
43. Adam, M.; Ganz, C.; Xu, W.; Sarajian, H.R.; Götz, W.; Gerber, T. In Vivo and In Vitro Investigations of a Nanostructured Coating Material—A Preclinical Study. *Int. J. Nanomed.* **2014**, *9*, 975–984. [CrossRef] [PubMed]
44. Spanos, N.; Misirlis, D.Y.; Kanellopoulou, D.G.; Koutsoukos, P.G. Seeded Growth of Hydroxyapatite in Simulated Body Fluid. *J. Mater. Sci.* **2006**, *41*, 1805–1812. [CrossRef]

45. Rodriguez-Navarro, C.; Linares-Fernandez, L.; Doehne, E.; Sebastian, E. Effects of Ferrocyanide Ions on NaCl Crystallization in Porous Stone. *J. Cryst. Growth* **2002**, *243*, 503–516. [CrossRef]
46. Chen, X.; Kalish, J.; Hsu, S.L. Structure evolution of α′-phase poly(lactic acid). *J. Polym. Sci. Part B Polym. Phys.* **2011**, *49*, 1446–1454. [CrossRef]
47. Swaroop, C.; Shukla, M. Development of Blown Polylactic Acid-MgO Nanocomposite Films for Food Packaging. *Compos. Part A Appl. Sci. Manuf.* **2019**, *124*, 10548. [CrossRef]
48. Nahanmoghadam, A.; Asemani, M.; Goodarzi, V.; Ebrahimi-Barough, S. Design and Fabrication of Bone Tissue Scaffolds Based on PCL/PHBV Containing Hydroxyapatite Nanoparticles: Dual-Leaching Technique. *J. Biomed. Mater. Res. Part A* **2020**, *109*, 981–993. [CrossRef]
49. Tawakkal, I.S.M.A.; Cran, M.J.; Miltz, J.; Bigger, S.W. A Review of Poly(Lactic Acid)-Based Materials for Antimicrobial Packaging. *J. Food Sci.* **2014**, *79*, R1477–R1490. [CrossRef]
50. Swaroop, C.; Shukla, M. Nano-Magnesium Oxide Reinforced Polylactic Acid Biofilms for Food Packaging Applications. *Int. J. Biol. Macromol.* **2018**, *113*, 729–736. [CrossRef]

Disclaimer/Publisher's Note: The statements, opinions and data contained in all publications are solely those of the individual author(s) and contributor(s) and not of MDPI and/or the editor(s). MDPI and/or the editor(s) disclaim responsibility for any injury to people or property resulting from any ideas, methods, instructions or products referred to in the content.

MDPI AG
Grosspeteranlage 5
4052 Basel
Switzerland
Tel.: +41 61 683 77 34

Polymers Editorial Office
E-mail: polymers@mdpi.com
www.mdpi.com/journal/polymers

Disclaimer/Publisher's Note: The statements, opinions and data contained in all publications are solely those of the individual author(s) and contributor(s) and not of MDPI and/or the editor(s). MDPI and/or the editor(s) disclaim responsibility for any injury to people or property resulting from any ideas, methods, instructions or products referred to in the content.

www.ingramcontent.com/pod-product-compliance
Lightning Source LLC
LaVergne TN
LVHW072332090526
838202LV00019B/2403